A Book of
Home Plans

A Book of Home Plans

SECOND EDITION

D. N. GHOSE
MCE, FIE, FIWWA,
FIPHE, MIAWPC

CBS

CBS Publishers & Distributors Pvt. Ltd.

New Delhi • Bengaluru • Chennai • Kochi • Kolkata • Mumbai • Pune

ISBN: 978-81-239-1657-6

First Edition: 1989
Reprint: 1997, 2000, 2002, 2004, 2005
Second Edition: 2008
Reprint: 2010, 2011, 2013, 2015

Published by:
Satish Kumar Jain for CBS Publishers & Distributors Pvt. Ltd.,
4819/XI Prahlad Street, 24 Ansari Road, Daryaganj, New Delhi - 110002
delhi@cbspd.com, cbspubs@airtelmail.in • www.cbspd.com
Ph.: 23289259, 23266861, 23266867 • Fax: 011-23243014

Corporate Office: 204 FIE, Industrial Area, Patparganj, Delhi - 110 092
Ph: 49344934 • Fax: 011-49344935
E-mail: publishing@cbspd.com • publicity@cbspd.com

Branches:
• *Bengaluru:* 2975, 17th Cross, K.R. Road, Bansankari 2nd Stage,
  Bengaluru - 70 • Ph: +91-80-26771678/79 • Fax: +91-80-26771680
  E-mail: cbsbng@gmail.com, bangalore@cbspd.com
• *Chennai:* No. 7, Subbaraya Street, Shenoy Nagar, Chennai - 600030
  Ph: +91-44-26681266, 26680620 • Fax: +91-44-42032115
  E-mail: chennai@cbspd.com
• *Kochi:* 36/14, Kalluvilakam, Lissie Hospital Road, Kochi - 682018
  Ph: +91-484-4059061-65 • Fax: +91-484-4059065
  E-mail: cochin@cbspd.com
• *Mumbai:* 83-C, Dr. E. Moses Road, Worli, Mumbai - 400018
  Ph: +91-9833017933, 022-24902340/41 • E-mail: mumbai@cbspd.com
• *Pune:* Bhuruk Prestige, Sr. No. 52/12/2+1+3/2,
  Narhe, Haveli (Near Katraj-Dehu Road Bypass), Pune - 411041
  Ph: +91-20-64704058/59, 32342277 • E-mail: pune@cbspd.com

Representatives:
• Hyderabad: 0-9885175004   • Kolkata: 0-9831437309, 0-9051152362
• Nagpur: 0-9021734563   • Patna: 0-9334159340
• Vijayawada: 0-9000660880

Printed at:
J.S. Offset Printers, Delhi

*"Knowledge leads to unity ;
Ignorance to diversity."*

—Shri Ramakrishna

**To
Smt. Sarbani Ghose**
Who inspired me
in writing this book

Preface to the Second Edition

"A Book of Home Plans" was prepared and published based on the concept of the facilities required by the people to make a home like home. It is fact that architecture is mother all arts. A properly planned and designed home with architectural composition (geometrical pattern, shape, appendages, colour, texture, etc.) may be compared with frozen music.

My most fervent thanks are due to the publisher in bringing out the second edition of the book.

D.N. Ghose

Preface to the First Edition

One of the basic needs of man is a home. Everyone desires to have a comfortable, good-looking house of his own taste suitable to the environment he lives in. While planning a residential building, all the desired facilities should be provided as far as practicable depending on the land availability, climatic condition, availability and cost of building materials and skilled labours. The designer should keep in mind the maximum utilisation of space, provision of natural air and light, comfort with beautiful architectural composition and cost economics. It may be mentioned here that a simple building can be made good-looking by providing offsets, finges, appendages and by bringing contrast in colour and surface texture.

In this volume of "A Book of Home Plans" every attempt has been made to prepare cost-effective, beautiful home plans for different categories of people having land areas of various shapes and sizes. My most fervent thanks are due to my architect friends, whose architectural compositions on Indian soil in the corridors of India inspired me to produce this book.

D.N. Ghose

CONTENTS

INTRODUCTION

Prior to making a 'home plan', it is essential for a planner to consider the following salient points:

- Size,shape and location of the plot.
- Specific requirements of the house owner.
- Fund resource available.
- Locally available materials for construction.
- Meteorological status of the area.

The units which are must for a home plan are Bed room, Kitchen, Dining space, W.C. & Bath and a stair if more than one storey is needed. In addition to these, a number of units of the following are provided as per desire of the house owner: Drawing room, Study room, Home Library, Recreation room, Guest room, Dressing room, Box room, Store room, Parlour, Verandah, Prayer room, etc.

Normally, for a family of economically weaker section the units to be provided are two small Bed rooms, a Kitchenette, a Dinette, W.C. & Bath and a front Verandah. The verandah serves the Purpose of a drawing room.

For a family of middle income group, the units commonly provided are two Bed rooms, one drawing cum dining room, a Verandah, Kitchen, W.C. and Bath room.

The units mostly desired by a family of high income group are: two to three nos of Bed rooms with attached toilet (W.C. & Bath), a Drawing cum Dining hall, Kitchen with store, a Box room, Guest room/Study room, one or two Verandahs preferably attached to the bed rooms and a Garage.

The requirements of a rich family are of varied nature, The choice of a rich family is to go for quite a number of spacious Bed rooms with attached Toilets, Dressing room and Verandahs, a large Drawing room, a dining hall, Guest room, Kitchen, store, Box room, Servant room, Garage, Parlour and a spacious staircase.

Usually a corner space in a Bed room or passage or Verandah or any other suitable place or the attic (a small room on the top of the staircase) is used as a prayer room. A rich family however desires a separate prayer room.

Medical, Engineering and Law practitioners desire to have a consultation room and a home Library, while Artists ask for a Studio and a Study room. The writers convert the Drawing room into a home Library. A family having good taste converts the roof into a Garden. Childrens search a place in the house to build their world. Youths desire to have a recreation room in the house.

Everyone has a specific desire of his/her interest. Planning may suitably be done to meet all the requirements, provided the fund resource is available. If adequate land is not available in the area of desire, split-level planning may be done to provide the required facilities.

In the perspective of present day cost of land, labour and materials, most of the people run for low-cost housing. When fund and land are prime factors, there must be some compromise between the requirements and the constraints, in making a home plan. It is needless to mention that attention should be paid for maximum utilisation of space with dual function, where it is possible. The building plan should be compact with common walls as far as possible. The doors and windows share an appreciable cost of a building. But, for entry of natural air and light and proper ventilation, provision of more nos. of window is recommended in a compact plan.

BED ROOM : About one-third of one's life is spent in a bed room. It is the place for taking rest and sleeping. The room should therefore be made confortable with provision of natural air and light. The room should be spacious to accommodate one's requirements. It must have privacy and security. The space for placement of beds should be kept such that cross air currents pass over the beds and when the door is open, the beds do not come in the view of the outsider. A bed room should accommodate two beds one wardrobe, an almirah, a rack,or hanger, one bed side table, a radio, a T.V. Set and other handy articles of use.

The minimum floor area of a bed room should be 100 sft. However, for a single bed room, a dimension of 8'x 10' is adequate. The dimensions normally used for a bed room are 10'x 12', 10'x14', 10' x 16', 12' x 14', 12' x 16', 12' x 18'. The room height should be minimum 10'-6".

A bed room should have an attached toilet (W.C. & Bath) and a dressing room if possible. Alternatively, the toilet should be placed at such a location that it is independently and conveniently approachable from each bed room.

Windows should be placed in such a manner that they admit air and light and ensure privacy. When a verandah is accessible from two rooms, no window on the Verandah side should be provided. Also, no window should be placed on the corridor side.

The bed rooms should be placed on the side of the direction of the prevailing wind.

DRAWING ROOM : A Drawing room in essence is a multipurpose room. This is a room for reception, discussions, relaxation, dining on special occasions, holding social functions, family meeting, get together, religious discourse and for accommodating guests in the night, if no separate guest room is provided. This room according to the utility should be spacious and comfortable. The desired size of a drawing room is 12' x 18'. Sometimes, a larger room is provided as a drawing cum dining room. Owing to space restriction and planning difficulty, one may make a small drawing room of 8'x 10' with a separate dinette of 10' x 10'. The drawing room should be located on one side of the building with an entrance through a front verandah. Due to, too much of space restriction, sometimes the staircase is housed within the drawing room for maximum utilisation of space. In certain cases it may be required to combine drawing room, dining room and staircase in one enclosure. But care should be taken as regards free movement, easy accessibility, privacy, safety and security and natural air and light. The drawing room should be independently approachable from all the bed rooms. It should have toilet facility. The toilet may be provided under the staircase landing. If it is a drawing cum dining room, the kitchen should be located close to it but without creating any smoke or fly nuissance. The doors should not be placed too close or too far.

DINING ROOM : A dining room which is used only for one or two hours a day exclusively for daily family meals, seems to be a luxury in the present hard days in developing countries. The economically weaker section and middle income group of people can not afford a separate dining room. Normally a small space called 'Dinette' is provided very close to the kitchen.

Some people combine the drawing room and dining room in one. It is expected that no stranger will come during lunch or dinner time. The probability of coming of a stranger during these two hours is remote. Thus, the same room can be used as a drawing room and a dining room. Some designers with this idea provide a hall of comparatively large size to serve as a drawing cum dining room.

A dinette is usually 7'x9' or 8'x10'. A small dining room measures 10'x12'. The other dimensions of a dining room are : 12'x12', 12'x14', 12'x15', 12'x16' and 14'x16'. The dimensions of a drawing cum dining room may be : 12'x16', 12'x18', 14'x16', 14'x18', and 14'x20'.

The dining space in a drawing cum dining room may be separated from the drawing room by providing a folding or movable decorated wooden partition, venetian blind, curtain, hard wood partition, go-down shelf or by arranging plants and creepers. Sometimes, a dining room is used as a parlour.

The dining room should have large windows with facilities of air, light and ventilation. The outdoor view through the windows of a dining or drawing room should be pleasing to the eye.

Some people enjoy the charms of open air dining during fair weather. Screens of hedge or creeper having a shade of evergreen trees are made on the lawn or on the roof. For this purpose, trellies or pargola is built. Some persons build a shed of ribbed shell structure which imparts beauty.

GARAGE : The provision of a garage for keeping a car is desired by the people of high income group and by a rich family. It should be located at one side of the building front. The size of a garage should be 8'x 18' or 10'x20'. The height should be restricted to a maximum of 6'-6", so that adequate headroom of the mezzanine room is available. A rolling or folding shutter is preferred for the garage door. An inside door may be provided for entering into the building directly from the garage. This door is of much use during rain, dead night and emergency.

MEZZANINE ROOM : This is a room between the garage and the first floor. Its size is that of the garage. Normally, its entry is given from staircase landing. This room is quite useful as a study room or recreation room or guest room. Its height is usually restricted to 7'-0".

GUEST ROOM : In a home plan a room should be kept to accommodate guests in occasions. In these hard days when space is precious, the guest room may be used as a study room or recreation room or for other purposes when there is no guest. A guest room should preferably be isolated from other bed rooms and it should have an independent access to common bath & W.C., if no attached toilet is provided. It should have a separate access to drawing room If possible, a verandah attached to the guest room should be provided. The room should be well ventilated with lighting arrangement.

VERANDAH : In an Indian house, a verandah - narrow or wide, small or large, is an essential feature and is desired by every house owner or tenant. A verandah facilitates one to come out from the enclosure of four walls and to enjoy the outdoor climate or environment. It also helps to keep the rooms out of the scorching rays of sun in the summer afternoon and wind-blown rains during stormy weather. It serves the purpose of a waiting room in ground floor, prior to entering into the drawing room or living room. Sometimes, a verandah or balcony serves as a passage to give independent access to rooms and maintains privacy. Light discussion with friends or other members of a family in a flow of cool breeze in summer evening or after-dinner sitting is another good use of a verandah. In special situations, a verandah becomes a place of watching an outside event. It is a place of making see-off to friends and relatives. Childrens also like to play in a verandah in the exposure of nature.

A suitably-placed verandah improves the appearance of a building. From the utilities mentioned above, it may be opined that a verandah is a must in a building. However, when rooms are quite spacious with a number of large windows, it may not be required to provide a verandah. Again, when the space is restricted, it may not be feasible to provide a verandah. Sometimes, a projected verandah of 3 to 4 feet width is provided.

Usually, the width of verandah varies from 3 feet to 6 feet. Too long a verandah does not look nice. Moreover, verandah-side windows cannot be provided for the sake of privacy, when the verandah is common to a number of rooms. The usual length of a verandah is 6 feet to 12 feet . The height of a verandah if needed, may be reduced to 8 feet maximum. In case of a verandah having sloped roof, a height of 7 feet near the eaves may be permitted.

In India normally, verandahs are placed on East or South of the building so that flood of morning sunshine is admitted. When road-facing verandahs are provided on the West of a building, sunbreakers or shutters are used. Sometimes, dense creepers are planted.

KITCHEN : This is a room for preparation of food stuff and cooking food for family members and occasional guests. Adequate light, proper ventilation and necessary arrangements must be made for the house wife or cook. Cleanliness is a must. Safety against fire hazards must be kept. The supply of water in the kitchen sink and to a tap within the kitchen must be ensured. A platform of suitable height should be provided in the kitchen for installation or placement of a cooking range or chullah or stove so that the cook can prepare the food in standing position with comfort. A cabinet or a rack for storing raw food stuff, grains and spices should be provied within the reach of the cook so that his/her movement during cooking is kept as minimum as possible.

The size of a kitchen and its shape should be planned by keeping in view the allocation of spaces as mentioned above. The kitchen shold be located close to the dinning space and it should have individual access from bed rooms. It is preferred not to locate the kitchen very close to the bed rooms. It should never be placed on the frontside of the building. It is normally built on North-East or North-West of a building. The minimum width of a kitchen is 5 to 6 feet. A small kitchen (kitchenette) size is 5 feet x 8 feet wih a separate storage space. The normal size of a kitchen is 8 feet X 10 feet or 8 feet X 12 feet. A kitchen must have a smoke outlet and if possible, should be provided with an exhaust fan.

STORE ROOM: A room space should be provided in a building for storing articles. Normally, a small store room is kept close to the kitchen for storing food grains, vegetables and fuel. If the kitchen is large, no such additional space is provided.

A store room in form of a box room attached to the bed room is provided to keep the winter clothes, blankets, unused garments, etc.'

For storing or keeping ceremonial articles, equipments of daily use and various other materials usually no separate room is provided. For this purpose, the extreme corner of a Verandah or corridor and the space below the flight of stairs may be used. Articles of frequent use may also be stored in these spaces.

Sometimes, lofts are provided in a building for dead storage of articles for their disposal afterwards.

Built-in wall cupboards with sliding shutters are also provided below window sills upto the floor level. Similarly, built-in small cupboards with shutters may be made above the door and upto the ceiling.

In some buildings, underground cupboards are kept for storage of valuables and secret articles.

In general, some storage space of any form must be kept in a building such that it does not get exposure to outsiders/visitors. On the other hand, it should be provided with facilities of adequate air and light.

TOILET : In modern-day bed rooms, attached toilet is a must. The toilet room should have adequate space to accommodate bathing place and a water closet (W.C.). A toilet room of 5' x 8' size seems to be adequate, if it is attached to a bed room. Sometimes, it is made smaller or larger than the size mentioned depending upon the availability of space. Normally, it accommodates a bath tub with shower, a wash hand basin and a W.C. pan or bidet. The toilet floor should be at least 3" lower than the outer floor. The toilet room should be well-ventilated by keeping ventilators or windows. Attempts should be made to place the toilet or latrine or bath room in a location opposite to the direction of the prevailing wind. No ventilator should be provided over the door, opening into a room.

Sometimes, only a W.C. is made attached to a bed room with a common bath room for a number of bed rooms. In general, there should be an additional toilet (common) for guests/visitors/servant/maid- servant. A guest room should have an attached toilet.

A common bath room serving a number of bed rooms should be placed centrally or equidistant from each room as far as practicable and it should have only one door opening from a common lobby. A common bath room must not have two doors connecting two rooms.

DRESSING ROOM : Normally, a corner space in a bed room is used as a dressing room. But, where space is available, a dressing room may be attached to a bed room preferably close to the toilet. It must have privacy with provision of diffused light falling on one's body in front of a full size mirror. For ventilation, high level windows should be provided. The room size may be as small as 6' x 8'.

BOX ROOM : It is a box like small space of 3' x 4' or 4' x 4' having same floor height or lower than the floor height. This room is used for storing winter garments, quilts, blankets, additional bedding, etc.

STUDY ROOM : In some domestic buildings, a small room is provided in a calm and quiet place within the building, which is preserved for study purpose. The room size normally kept is 6' x 8' or 6' x 10'.

HOME LIBRARY : The members of an educated and cultured family may desire to have a home library for study purpose. A good home library may be accommodated in the drawing room. But, a drawing room can not be used as a study room. In case a space is kept in the drawing room for home library, there should be a separate room for study purpose. A space of 8' x 10' or 8' x 12' may be kept for home library cum study room.

STAIRCASE : It is meant or intercommunication between floors in a building. A staircase should be conveniently located so that it has easy access from all rooms. Normally, a dog-legged staircase occupying a space of 6' x 12' or 6' x 14' is provided in a residential house. A geometrical staircase having three or more flights occupies less space compared to a dog-legged staircase. With a view to having maximum utilisation of space, various forms of staircase are designed to suit the different layout patterns of rooms in buildings. Sometimes, a staircase is housed within the drawing room in a building. Also, when the land area is very much restricted, an open staircase may be provided with a single flight and a landing attached to the outer wall of a building.

A staircase in general should be well-lighted and ventilated. There should not be more than 10 numbers of step in each flight and any flight in a geometrical staircase must have at least 3 steps. The number of steps in each flight should be kept same as far as practicable. The height of the riser should be uniform. Each riser should preferably be 6" high and not more than 8" in anycase. The width of each tread should preferably be 10" and not less than 9" in any case.

The width of steps in a winding flight should be same as in the straight flight. The width of each flight should be 2'-6" minimum and 4'-0" maximum. Normally, it is kept as 3'-0". There should be at least 6'-6" headroom above a step. The width of landing varies from 2'-6" to 4'-0. Normally it is 3'-0". The stair railing should be 3'-0" high. Normally, a stair should not have any triangular or winding steps.

- Staircase Layouts
- Concrete Jafriwork
- Chujjas and Hoods
- Sunbreakers
- Designs of Porch
- Verandahs
- Parapet Walls
- Wall Texture
- Water Reservoirs
- Forms of Sheds
- Forms of Grills
- Tiling Pattern
- Typical Design of Tiles

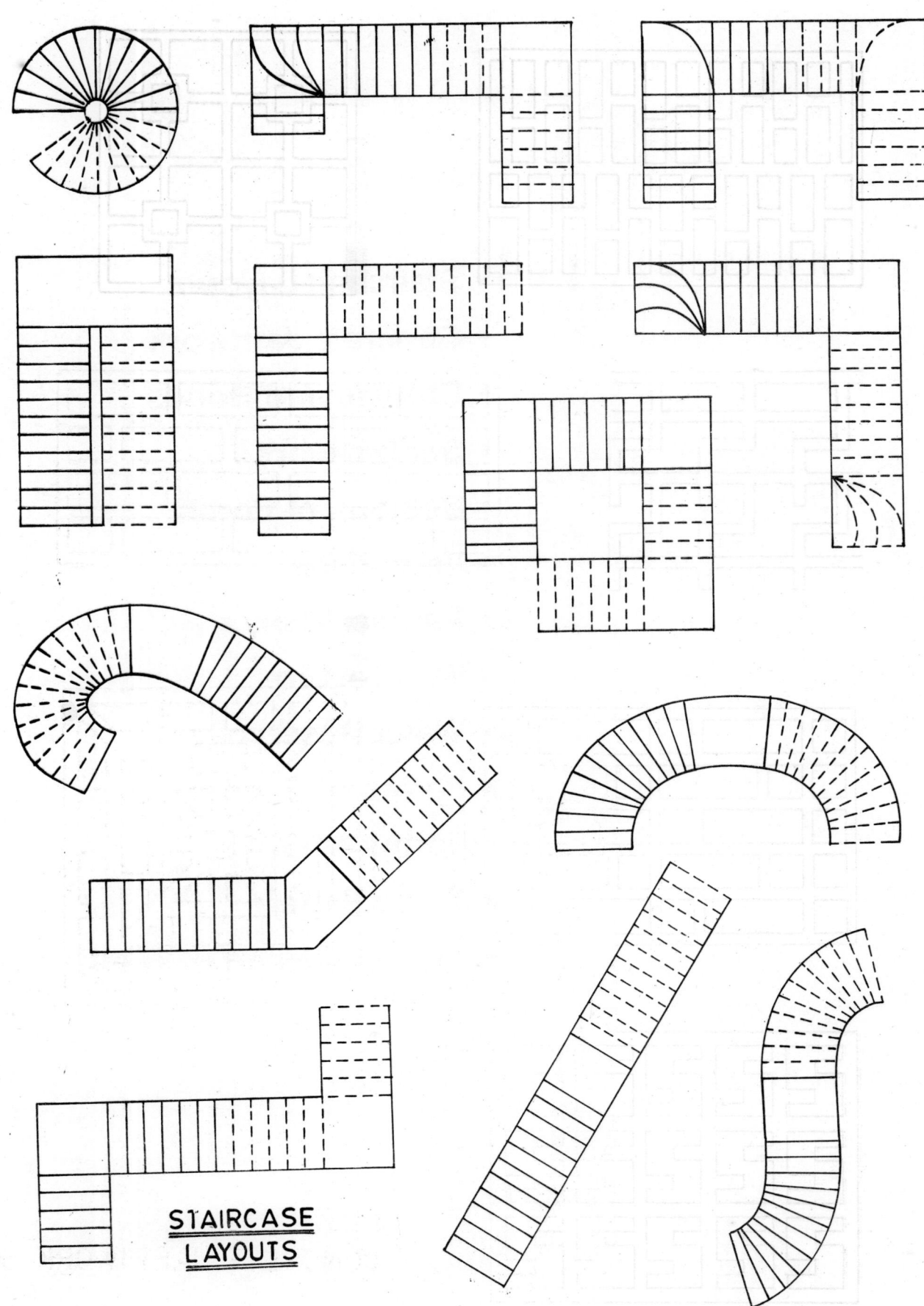

STAIRCASE
LAYOUTS

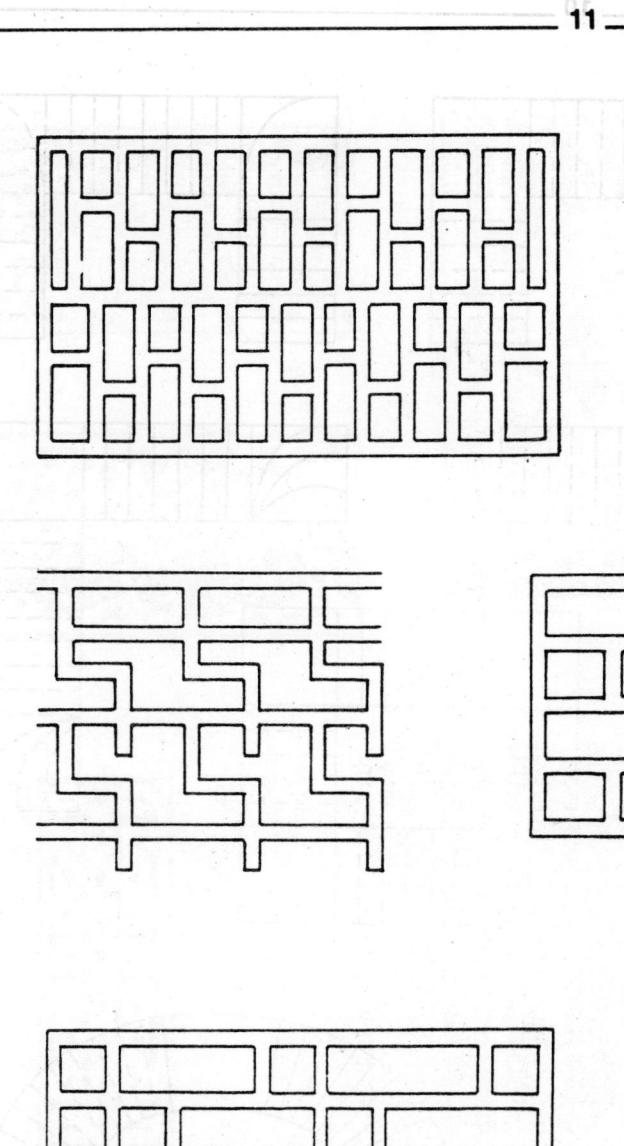

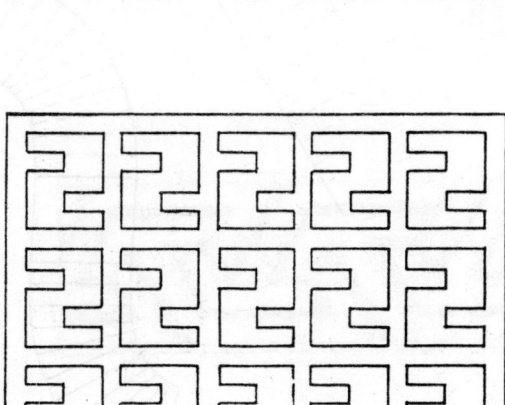

CONCRETE JAFRIWORK

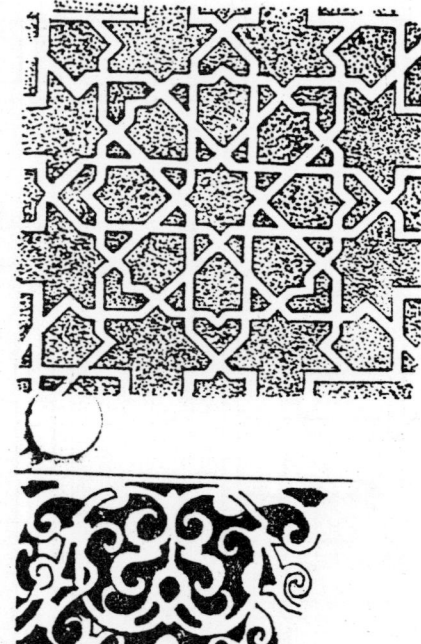

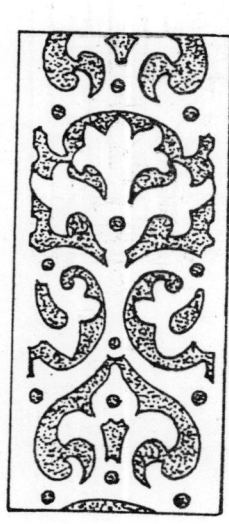

CONCRETE JAFRIWORK

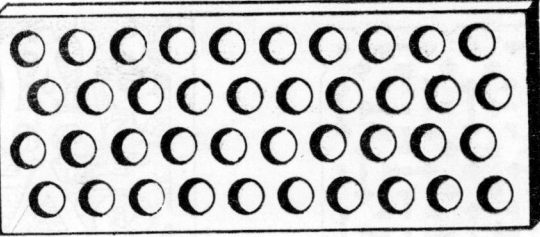

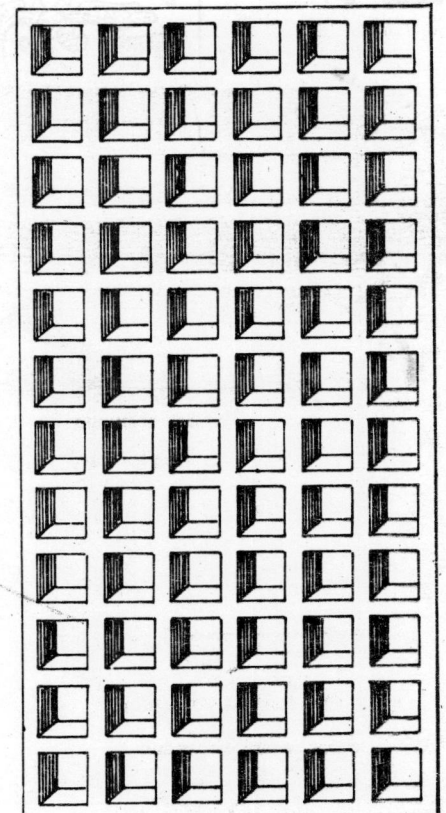

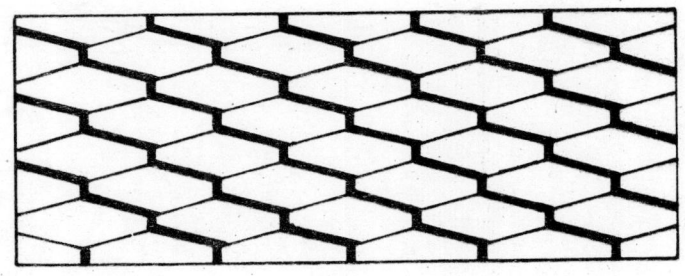

CONCRETE JAFRI WORK

CHUJJAS AND HOODS

CHUJJAS AND HOODS

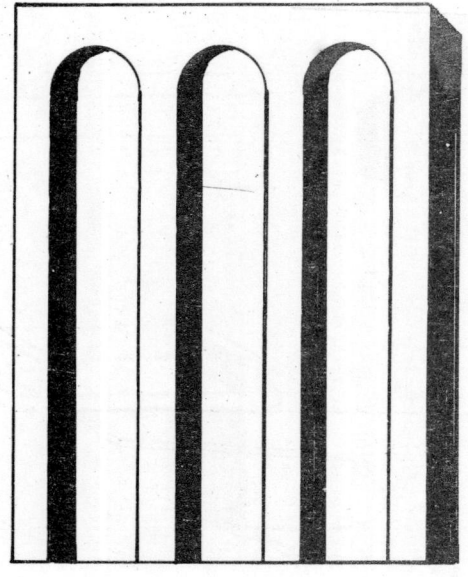

SUNBREAKERS

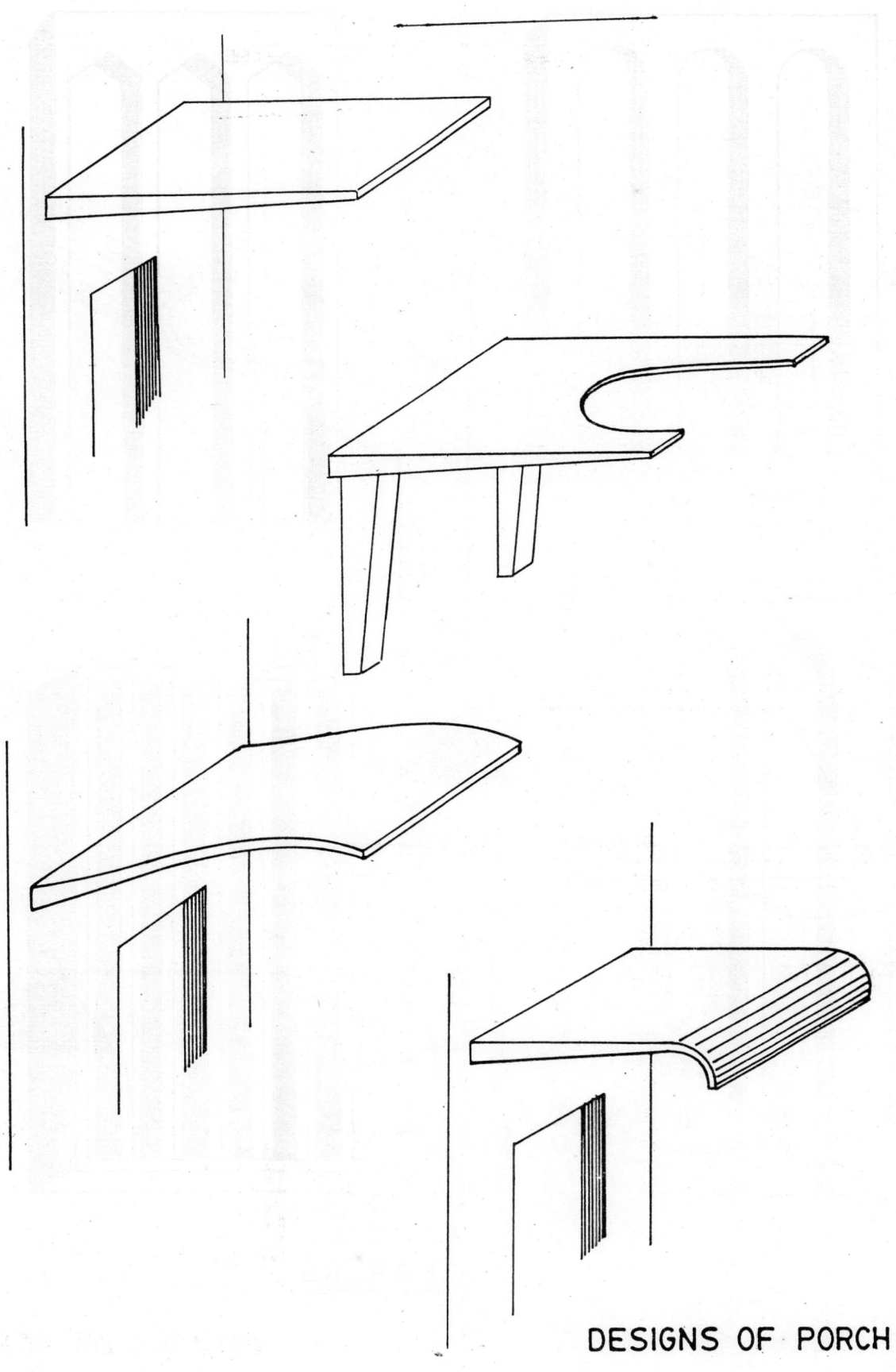

DESIGNS OF PORCH

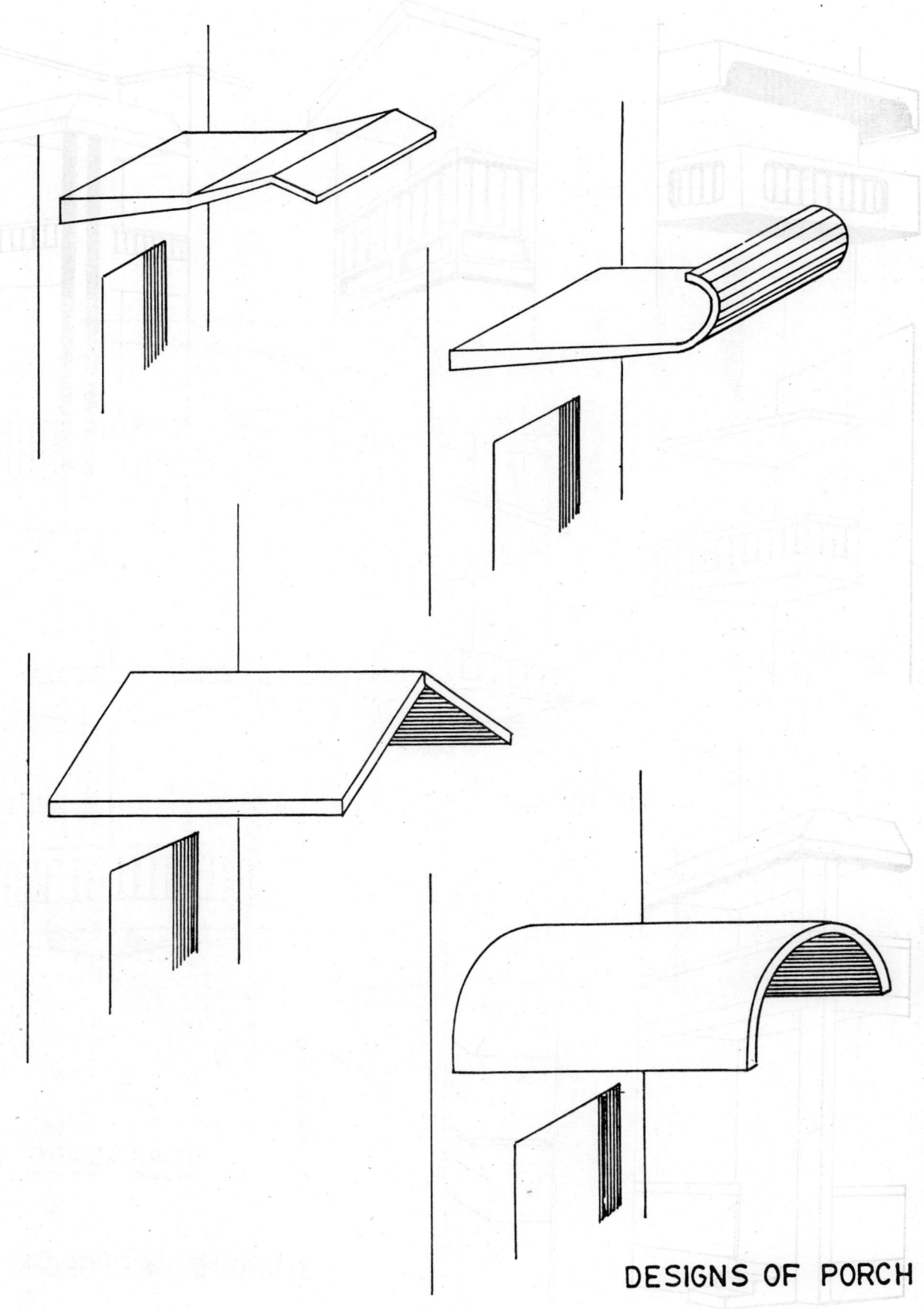

DESIGNS OF PORCH

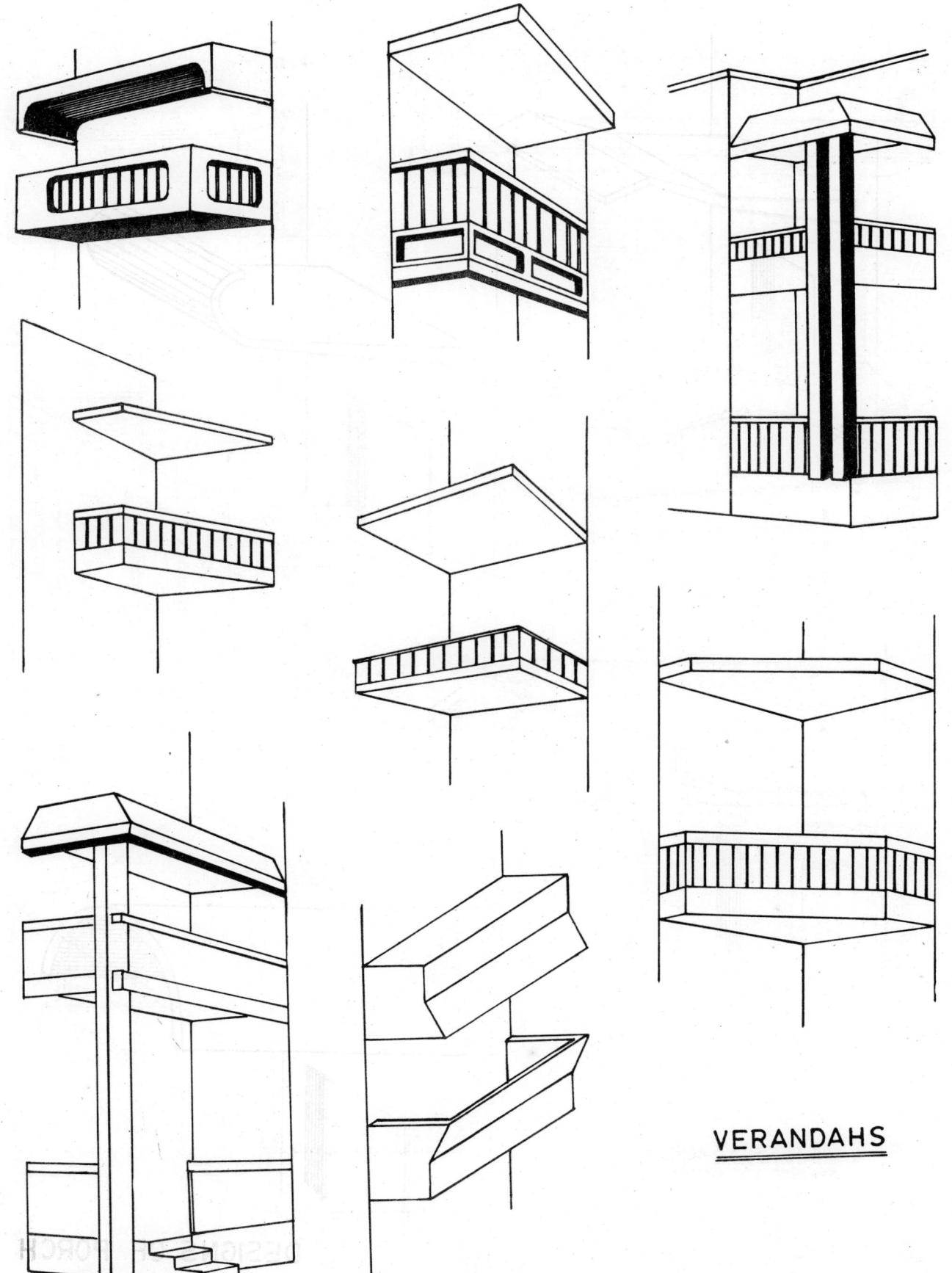

VERANDAHS

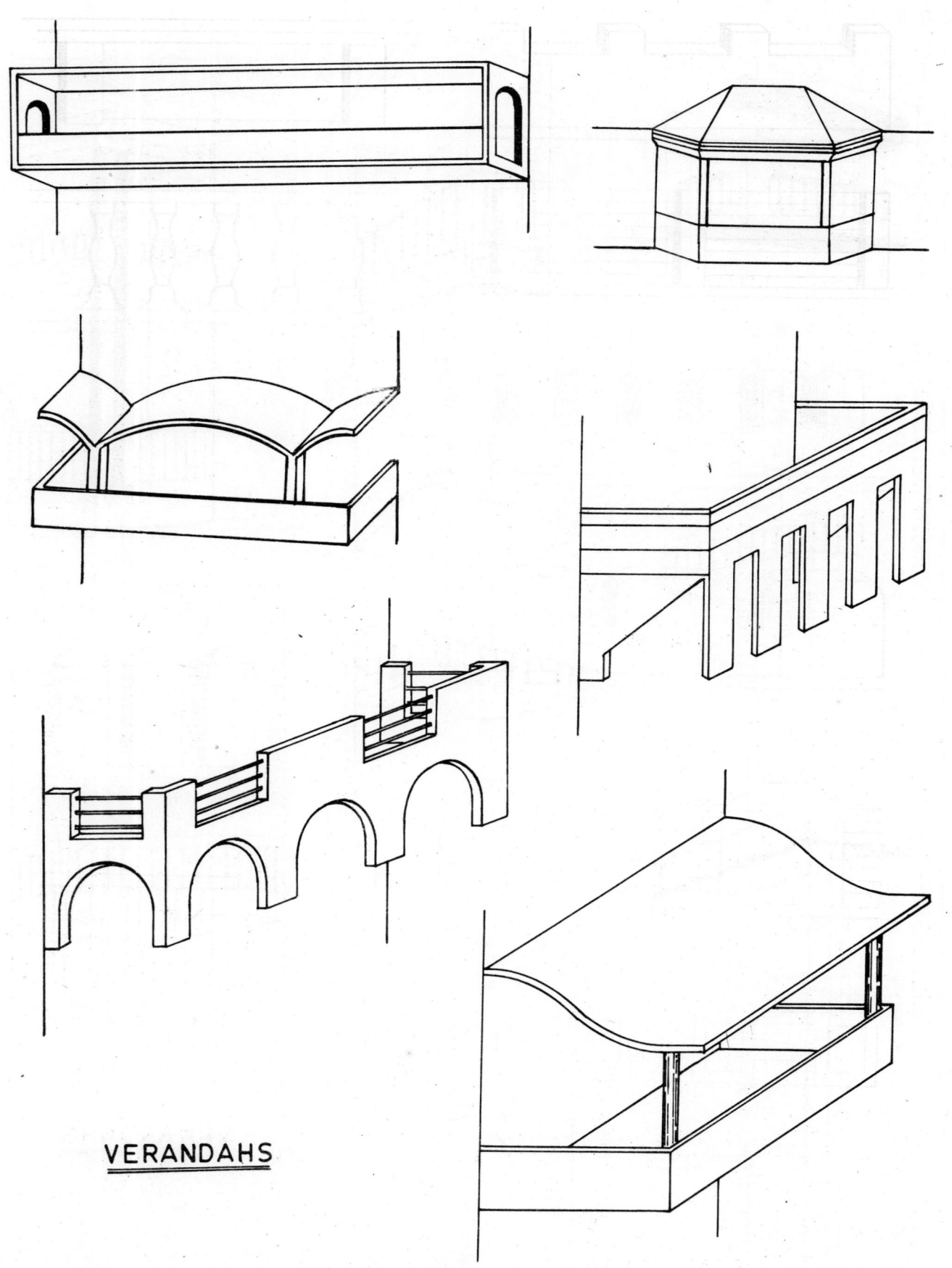

VERANDAHS

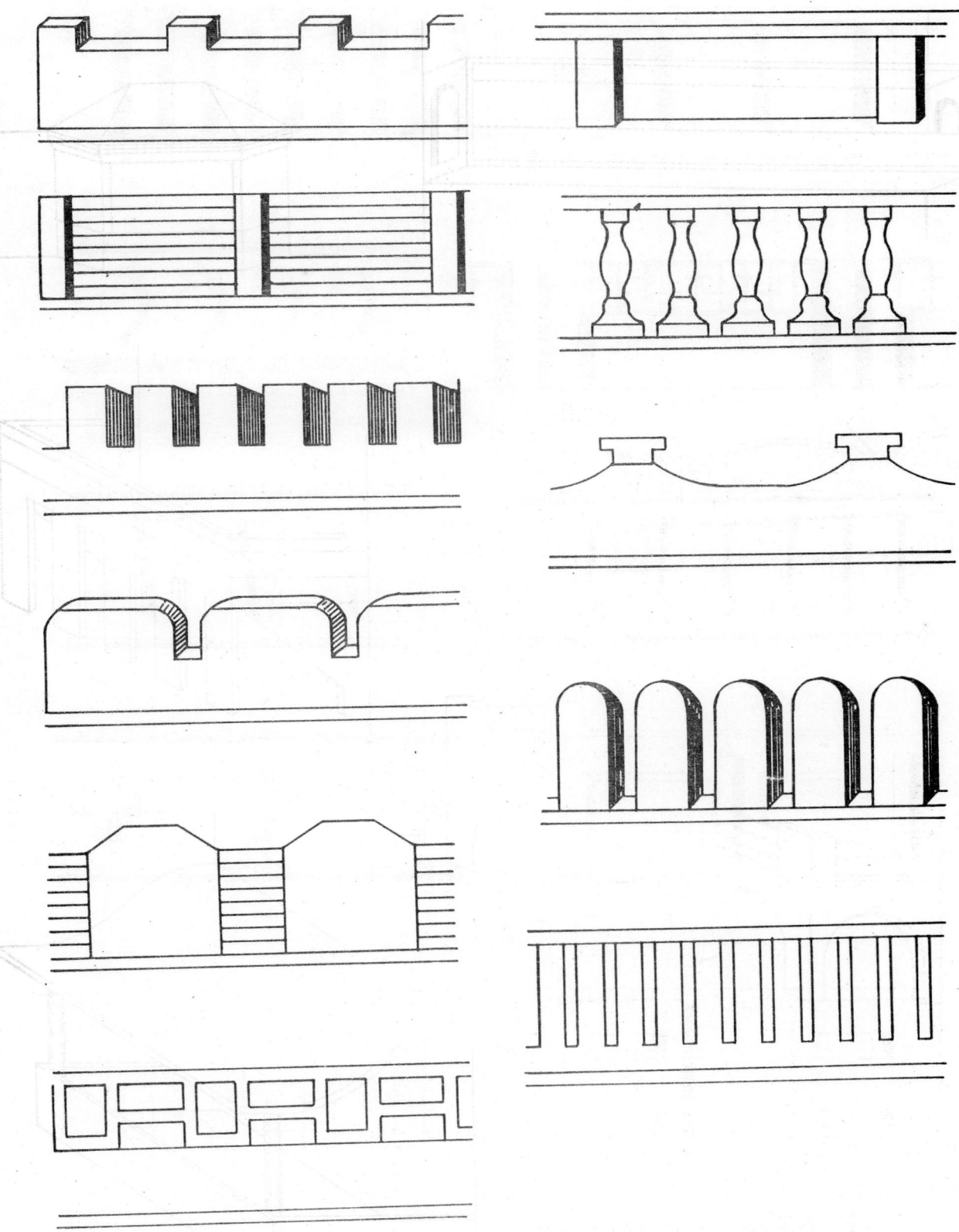

PARAPET WALLS

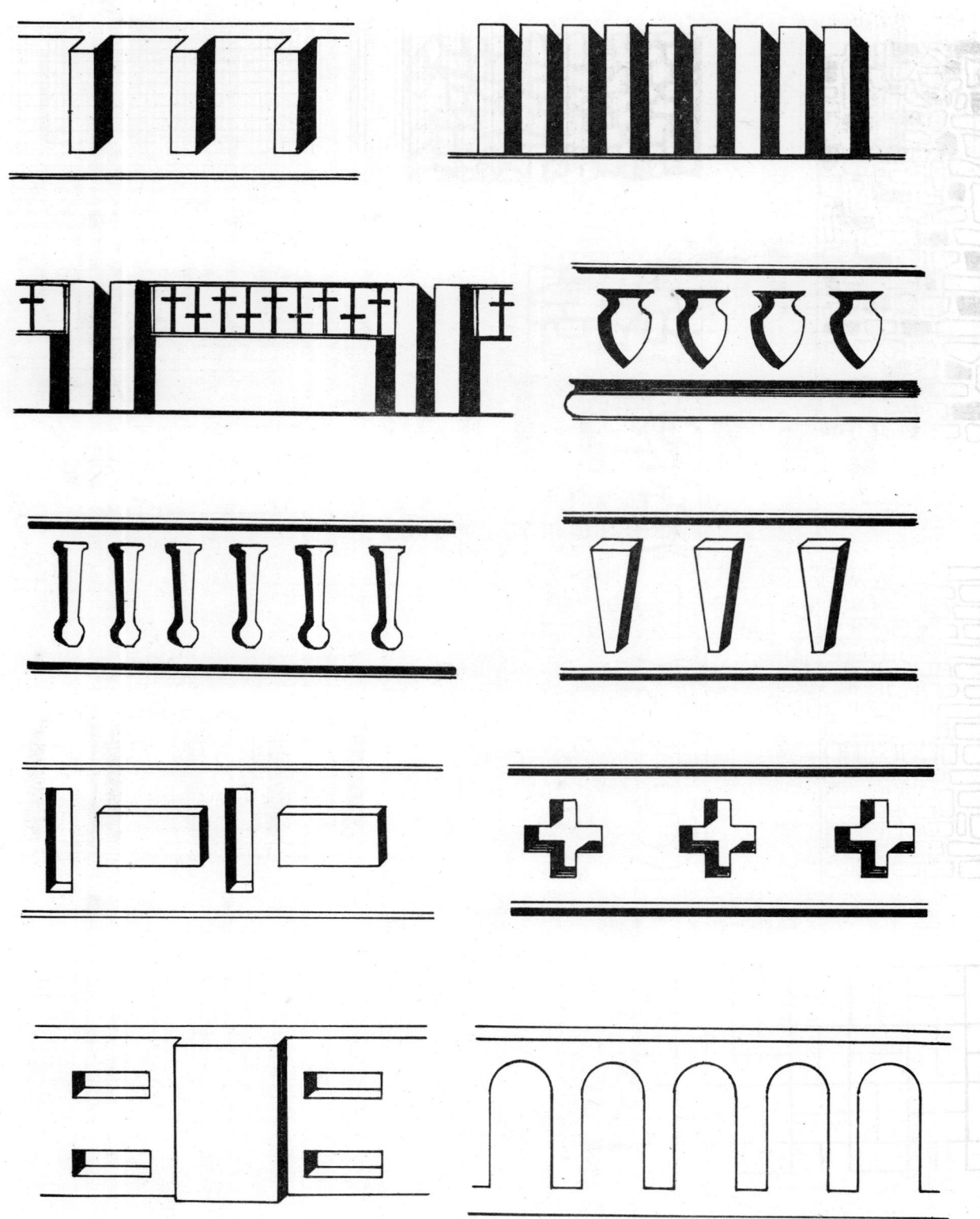

PARAPET WALLS

WALL TEXTURES

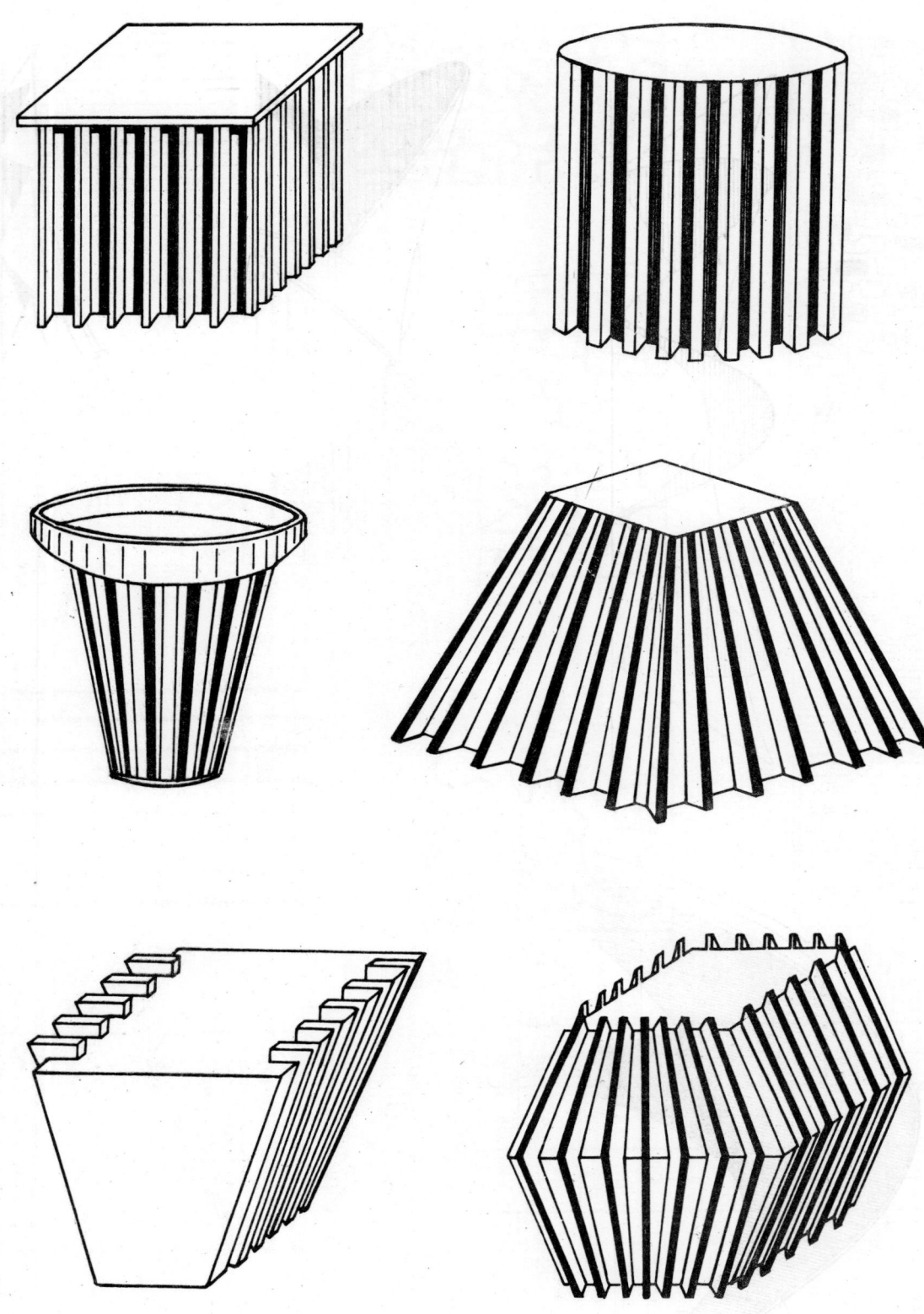

WATER RESERVOIRS

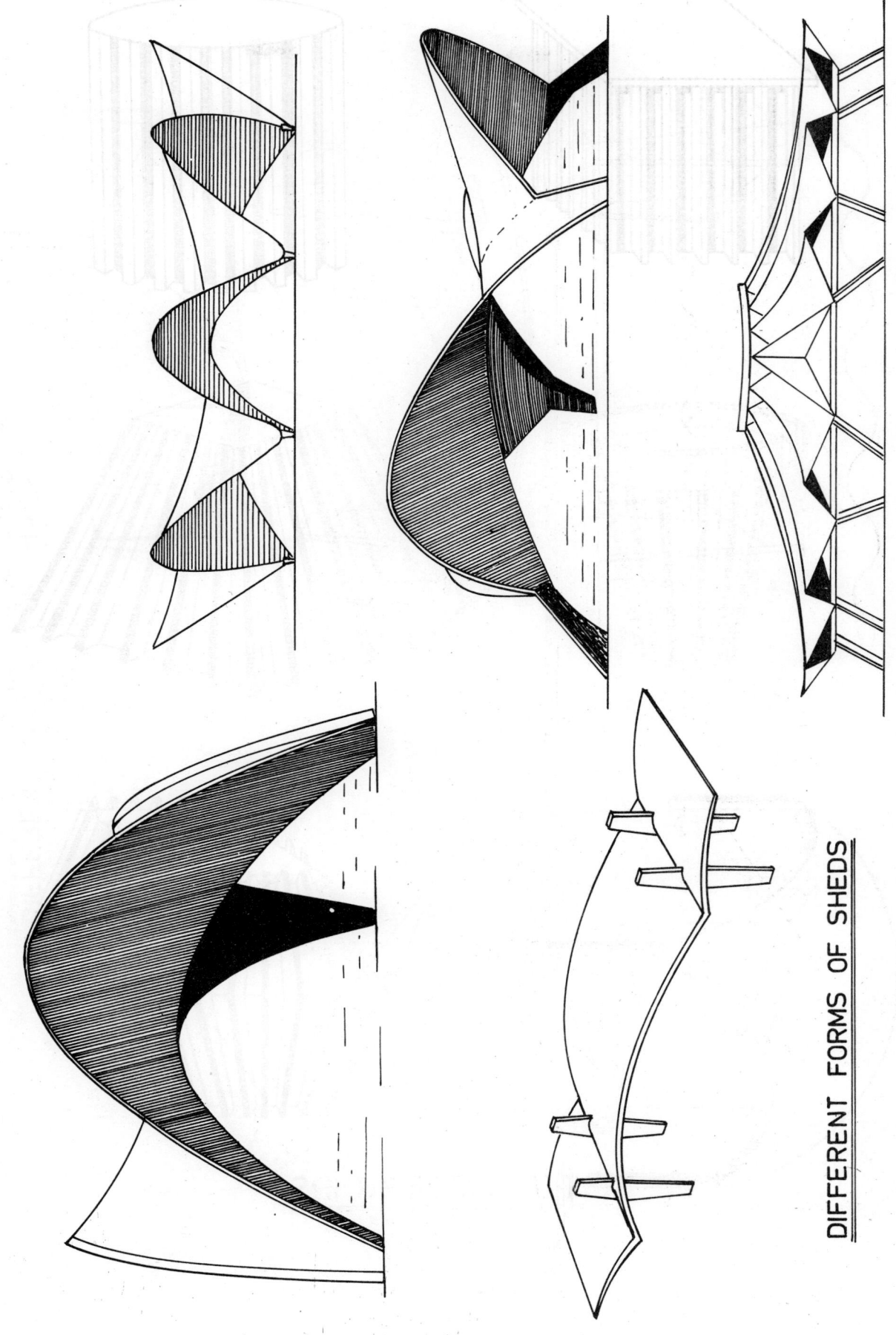

DIFFERENT FORMS OF SHEDS

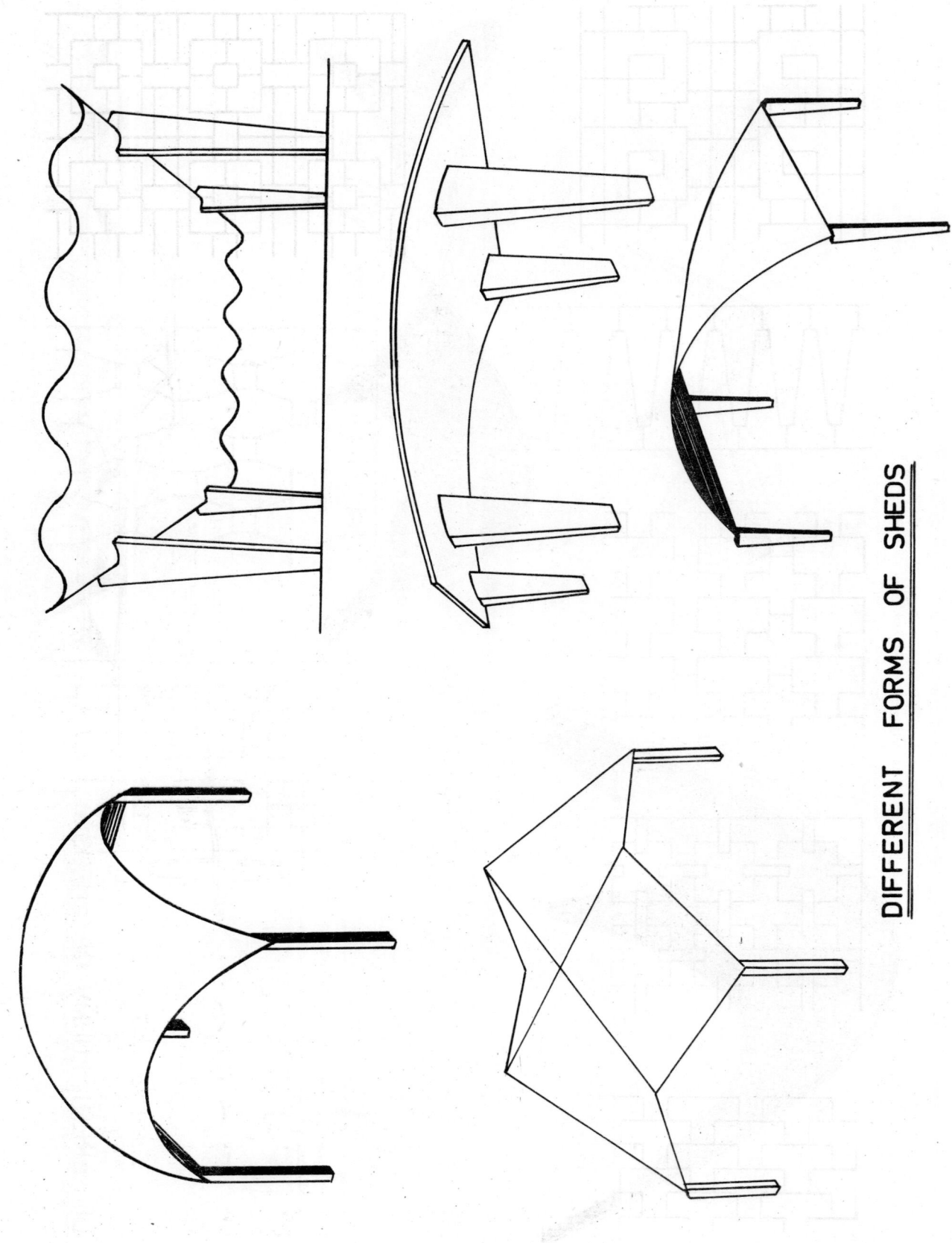

DIFFERENT FORMS OF SHEDS

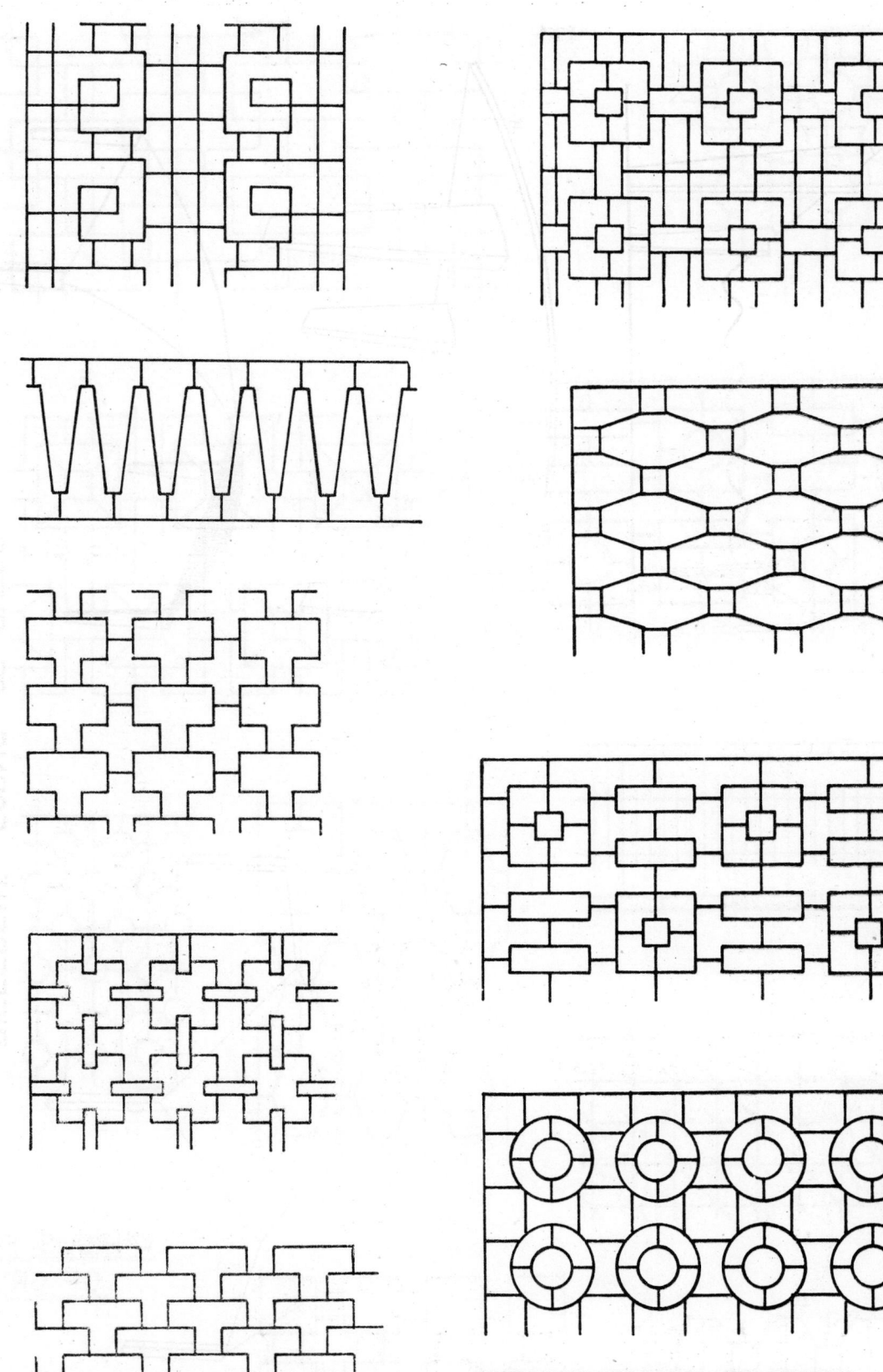

VARIOUS FORMS OF GRILLS

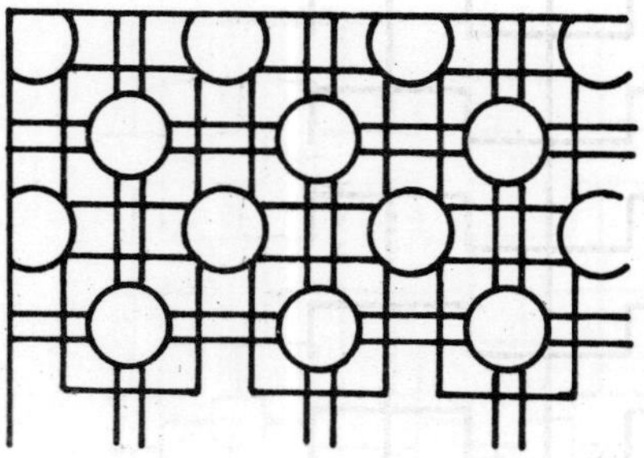

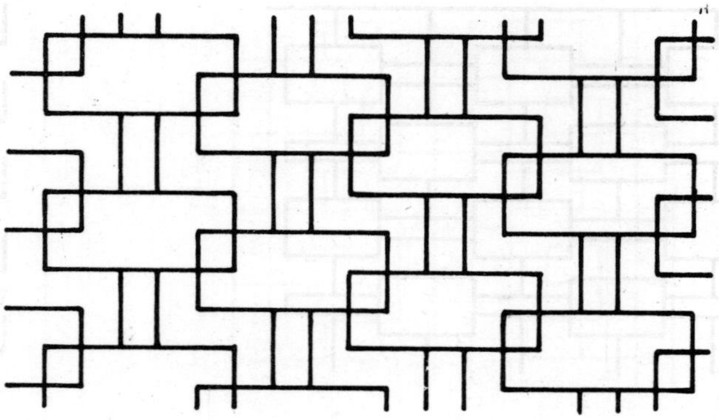

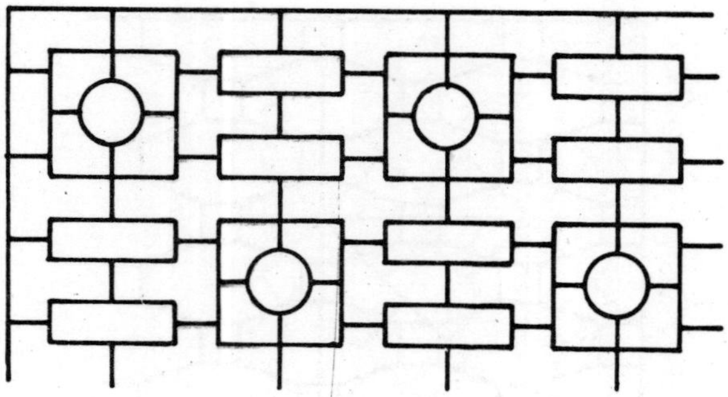

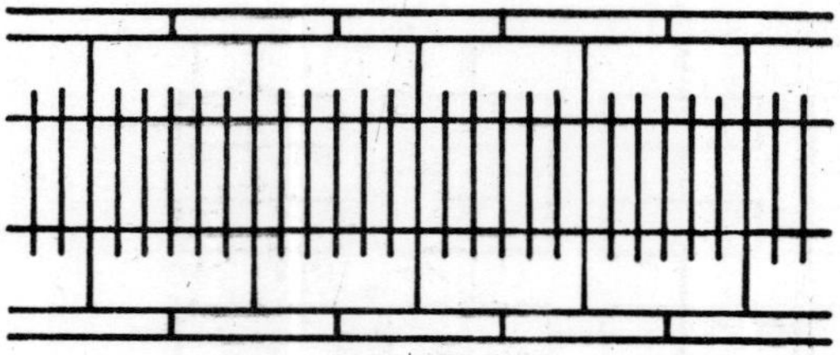

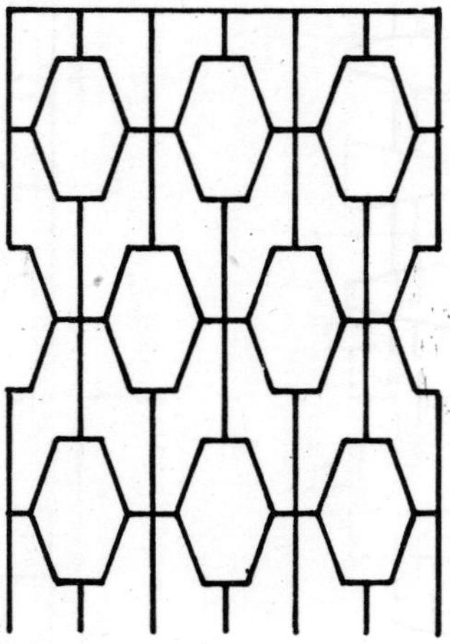

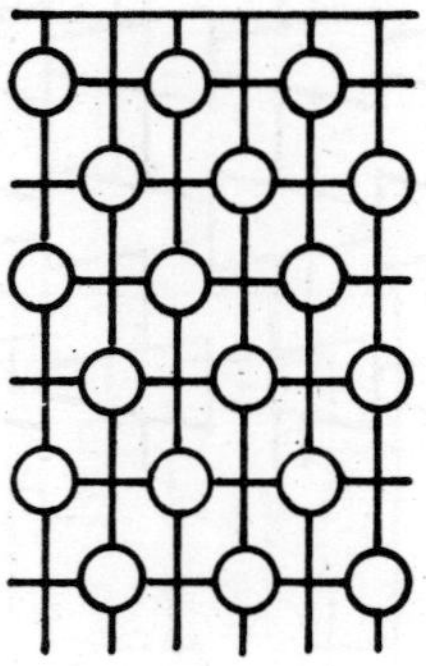

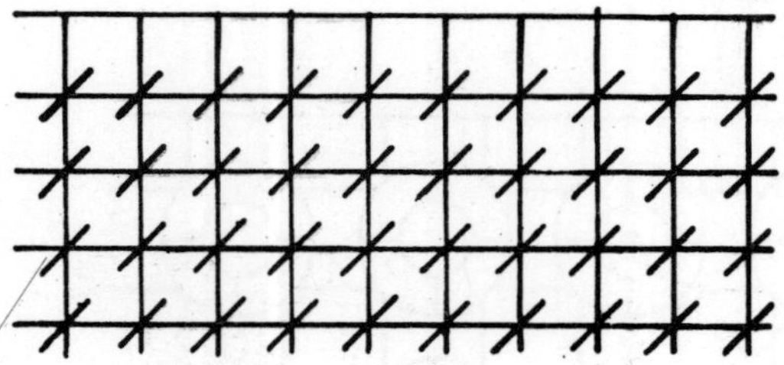

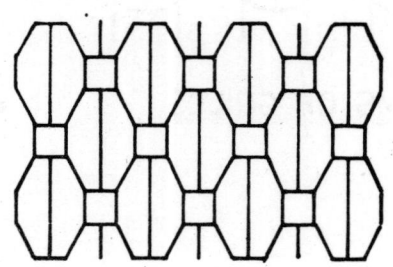

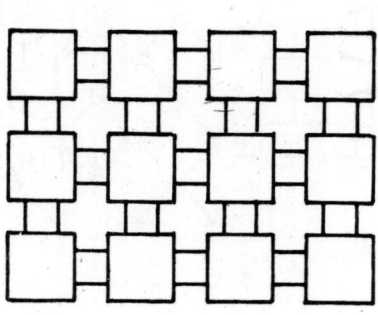

VARIOUS FORMS
OF GRILLS

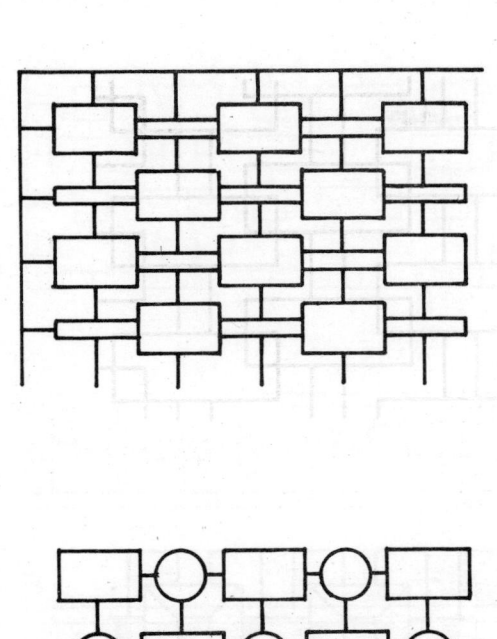
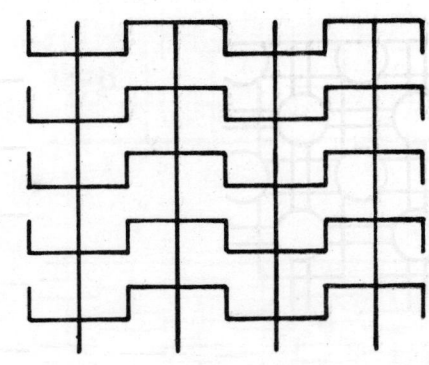
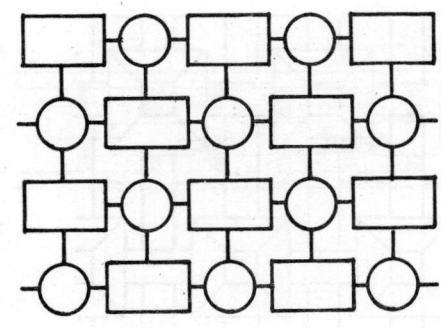

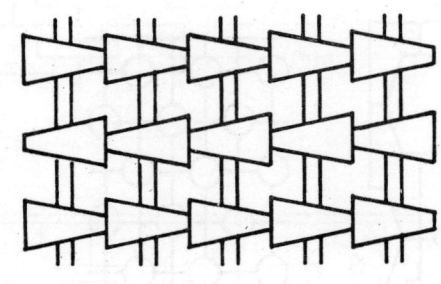

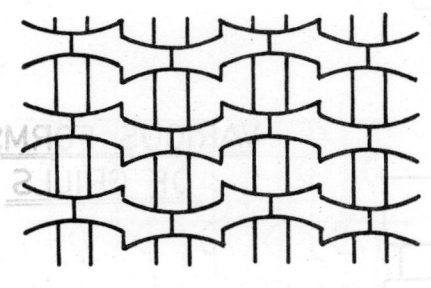

VARIOUS FORMS
OF GRILLS

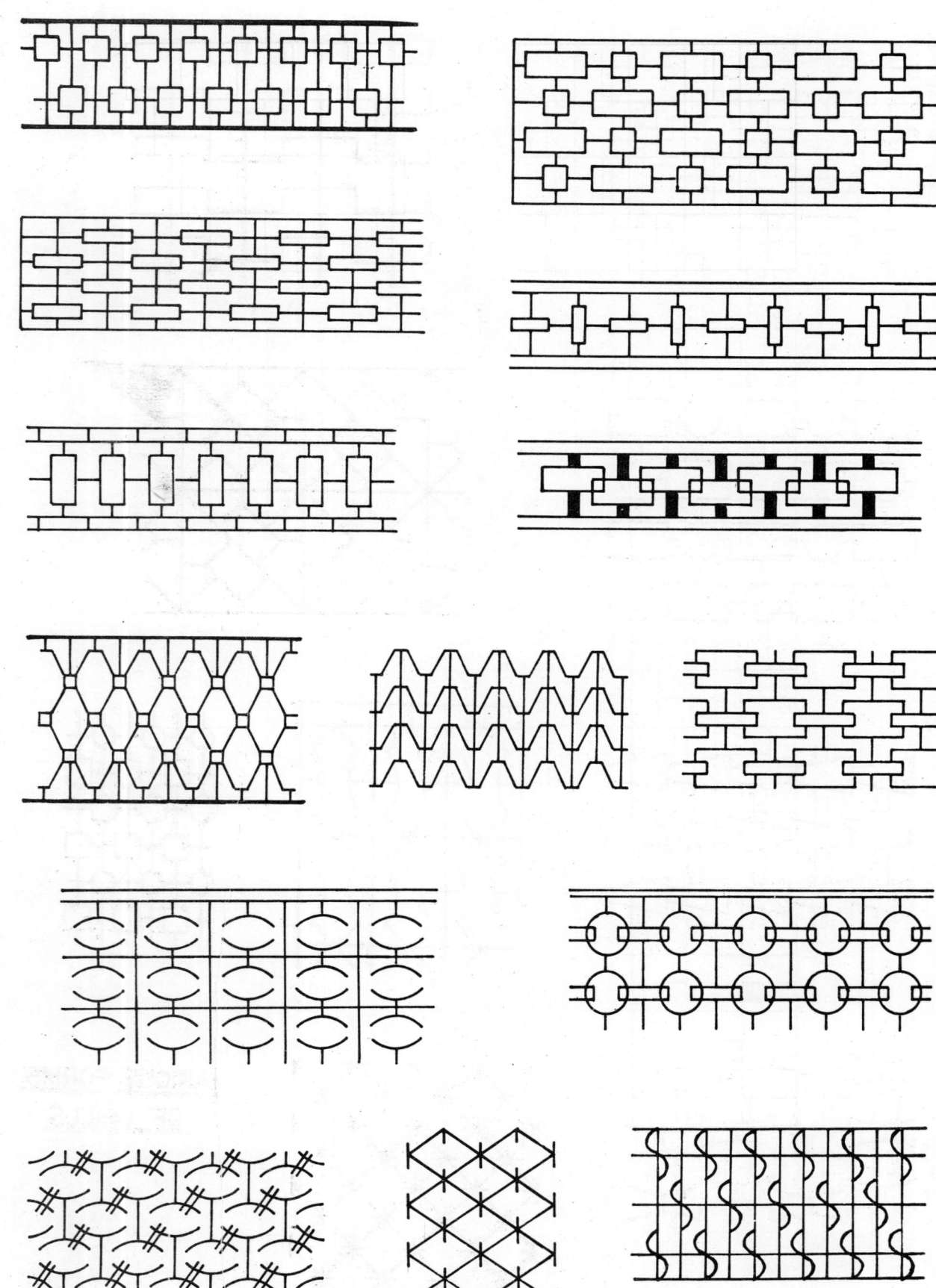

FORMS OF GRILLS

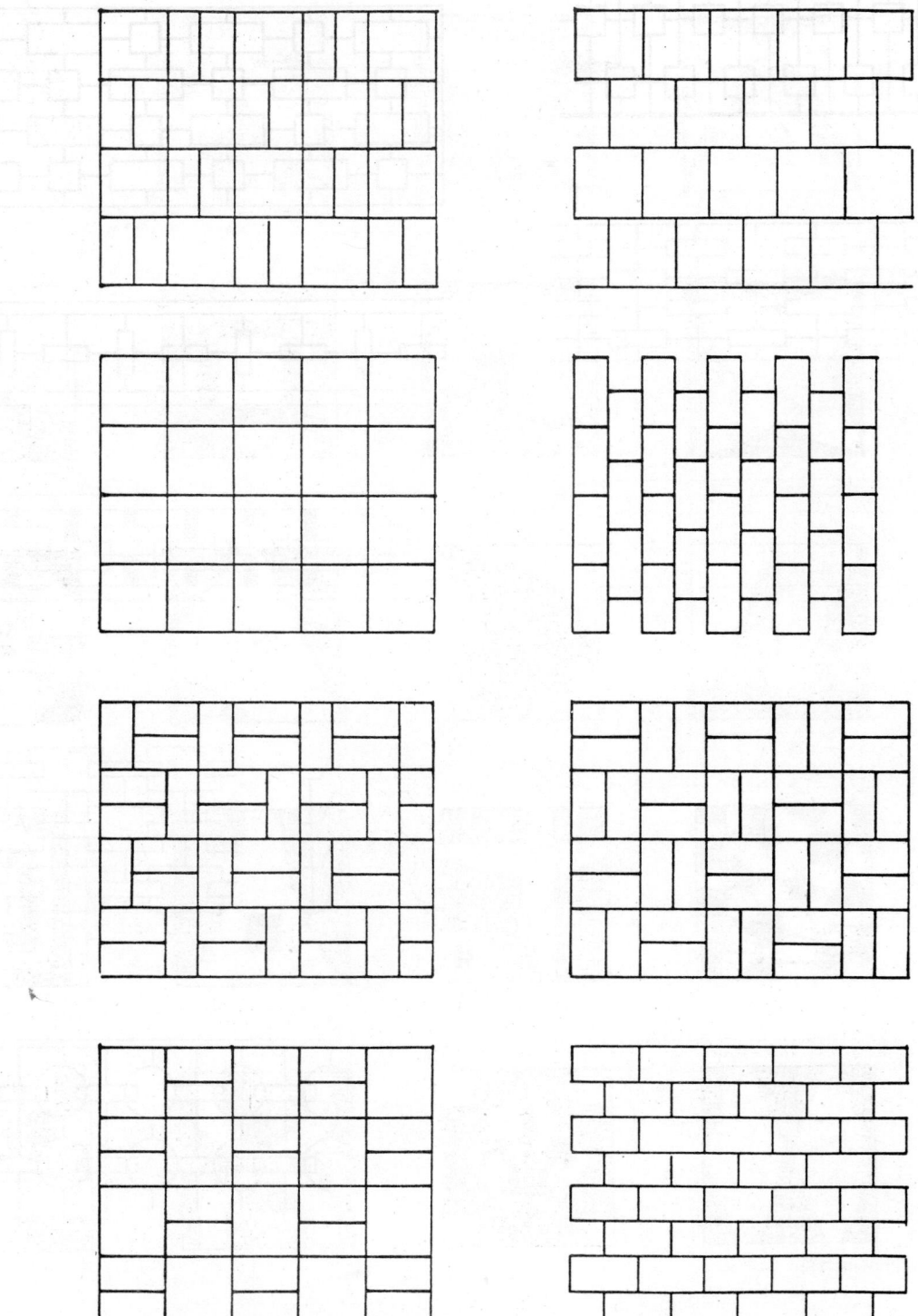

TILING PATTERN

TYPICAL DESIGNS OF TILES

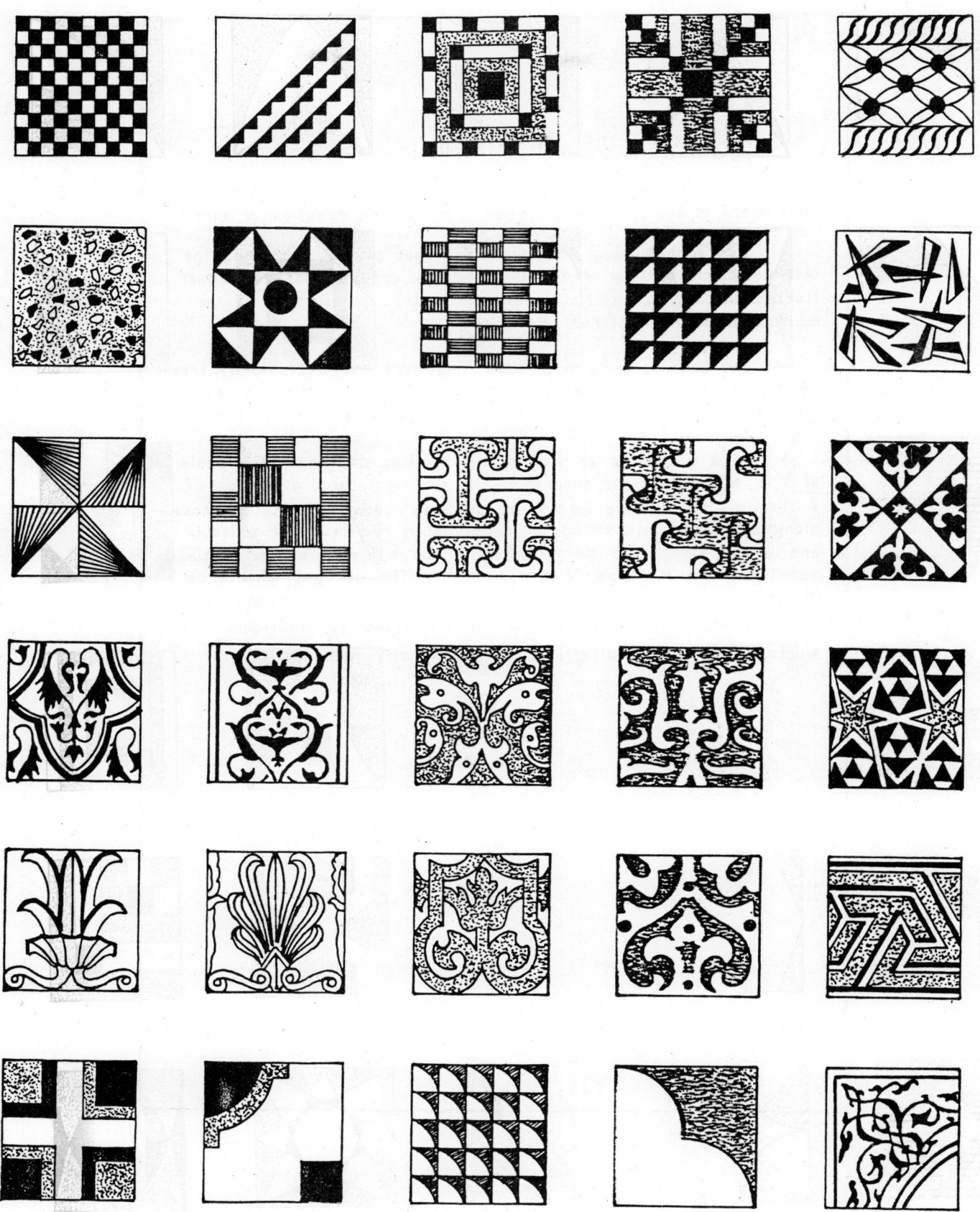

TYPICAL DESIGNS OF TILES

Building Plans:

Planning has been made according to the land area configuration, dimensions of the plot, available road frontage and direction of wind.

For each building, drawings have been prepared showing the building layout/built-up area within the plot, ground floor and upper floor plans and a pictorial view (perspective) or elevation of the building with architectural effects.

The building units with their individual use, accessibility, free circulation of air with entry of natural light and other aspects have been stated in short form, within the space in the building.

It is suggested to prepare the building drawings to a scale of 8 ft. to an inch for submission to Municipal Corporation/Board/ Authority. The building details viz. sectional veiw showing staircase, plumbing details, and structural details with reinforcement schedule - and a site plan must be furnished in the building plan for obtaining sanction from Municipal /Board/Authority. The site plan should be drawn to a scale of 50 ft. to an inch. The rules and regulations for sanctioning the building plan as prescribed by individual Municipality/Board/Authority must be strictly adhered to.

PLAN No. 1

Land Area : 726 sft ≈ 1 cottah ≈ 67.5 sq. m.

Plot Size : 17'-6" X 41'-6"

Plinth Area : 408 sft. ≈ 38 sq.m.

Floor Area (in each floor, including staircase) : 296 sft. ≈ 27.5 sq.m.

It is a narrow rectangular strip of land at the junction of two streets. The plan has to be made very compact for maximum utilisation of space. Although compact, provision has been made for cross ventilation and natural light. The house can well accomodate a small family.

In ground floor, drawing cum dining room, kitchen, toilet and staircase have been provided. The space for staircase is 6'x 12'. In first floor, provision has been made to accomodate two bed rooms and a toilet. The attic room may be used as prayer room.

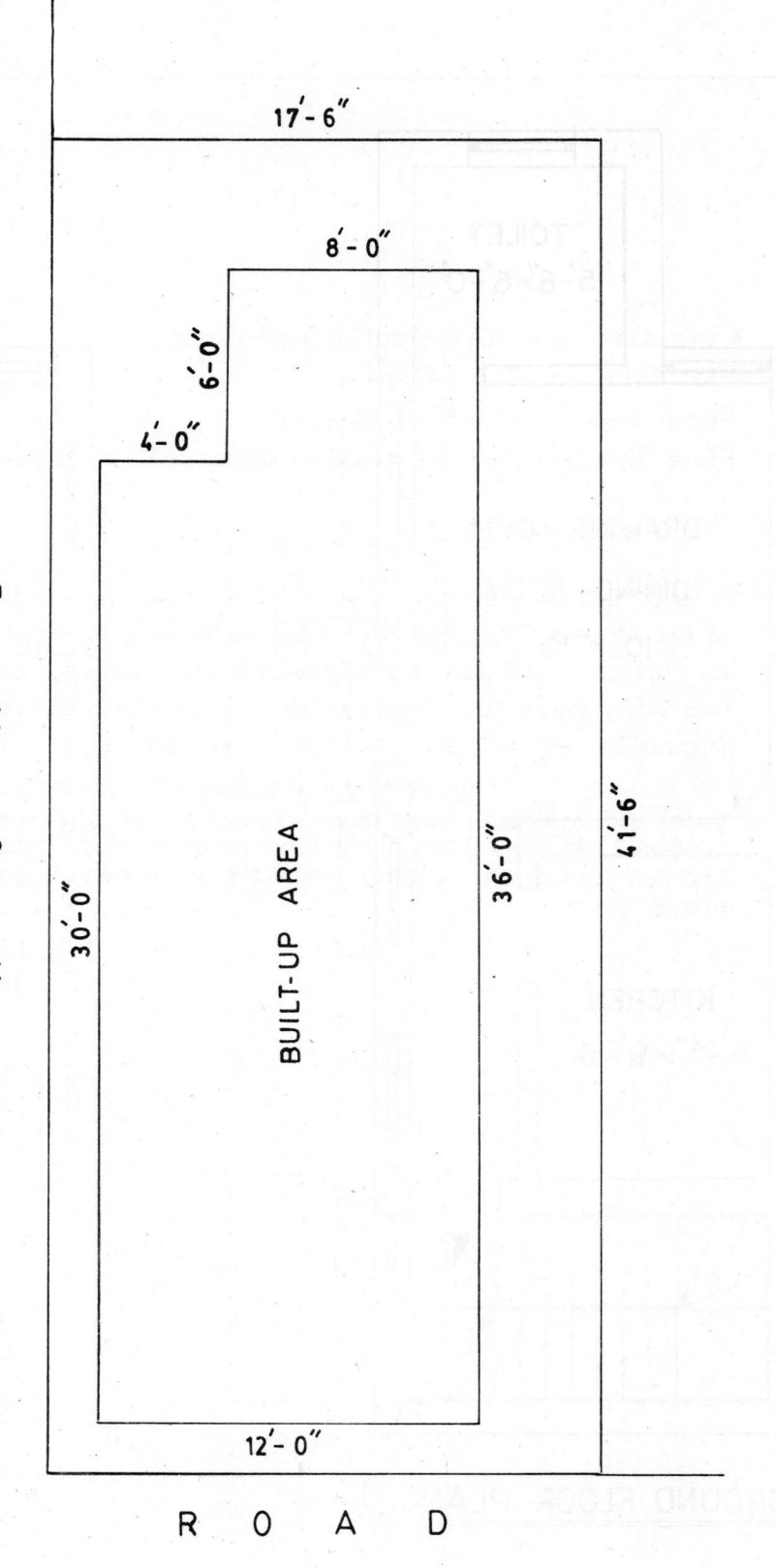

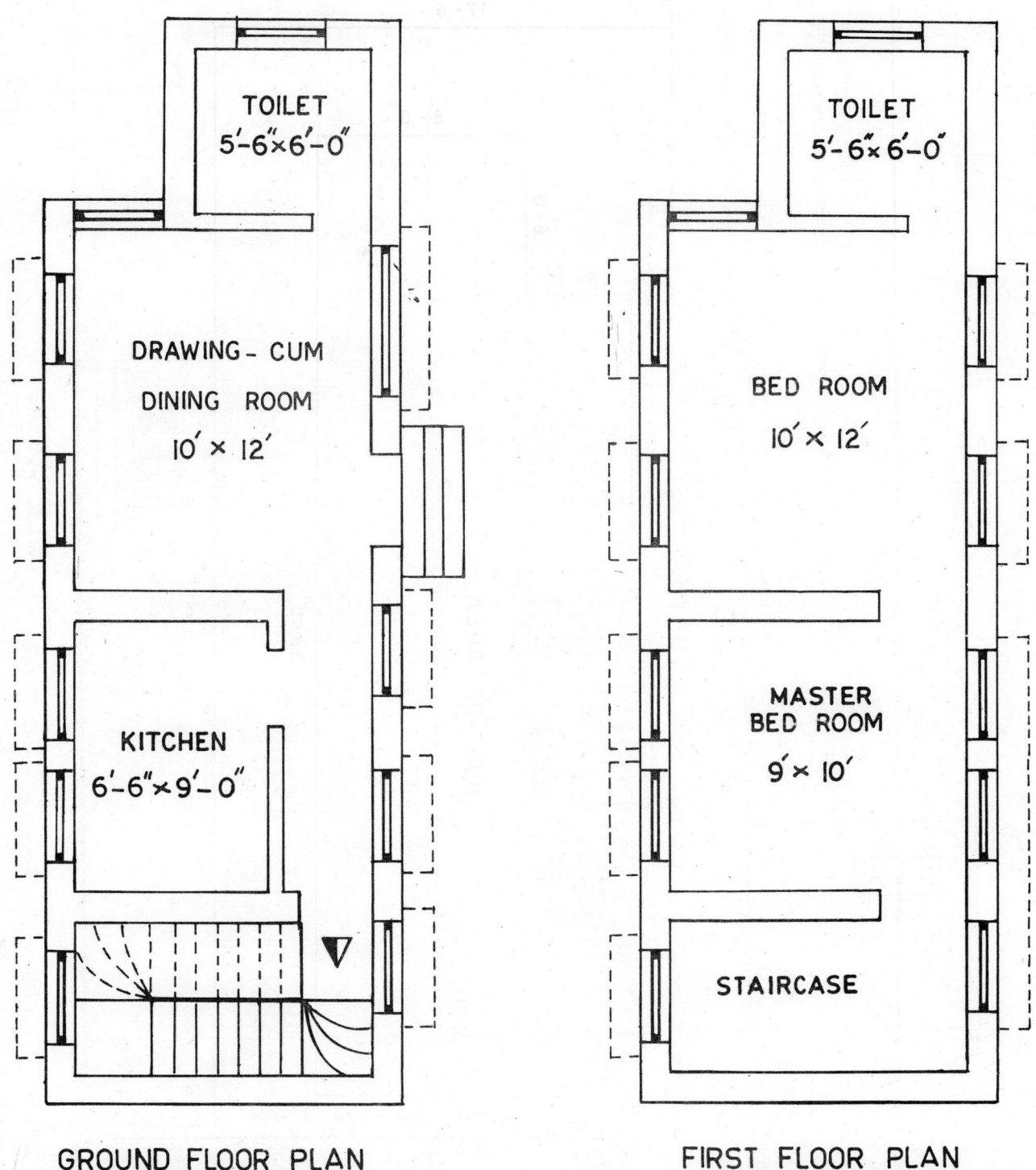

TOILET
5'-6"×6'-0"

DRAWING - CUM
DINING ROOM
10' × 12'

KITCHEN
6'-6"×9'-0"

GROUND FLOOR PLAN

TOILET
5'-6"×6'-0"

BED ROOM
10' × 12'

MASTER
BED ROOM
9' × 10'

STAIRCASE

FIRST FLOOR PLAN

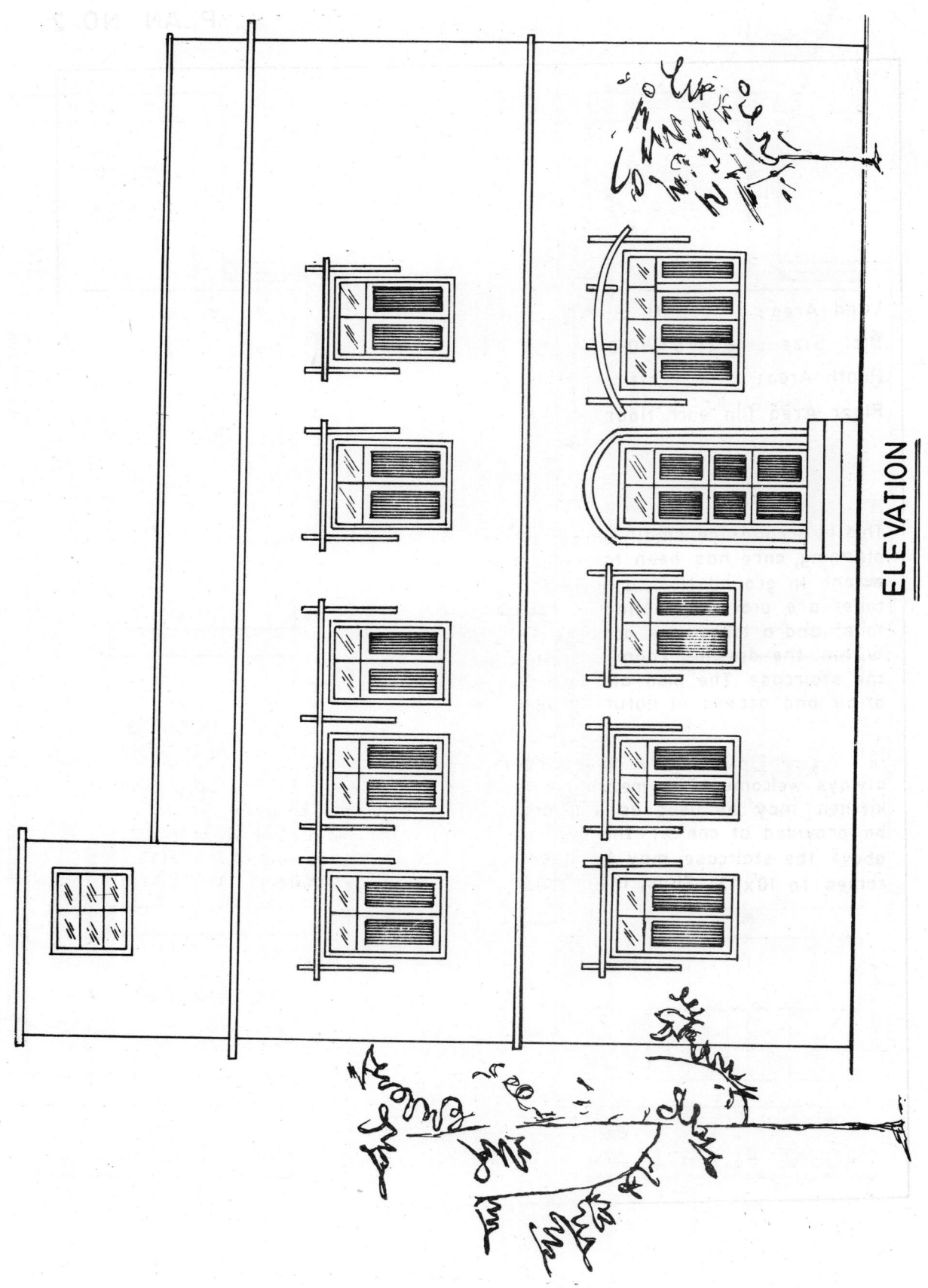

ELEVATION

PLAN NO. 2

Land Area :   1075 sft ≈ 1·50 cottahs ≈ 100 sq.m.

Plot Size :   21'-6" x 50'-0"

Plinth Area: 562·5 sft ≈ 52·3 sq.m.

Floor Area (in each floor): 377 sft ≈ 35 sq.m.

This is a roadside small corner plot. The land is a rectangular strip. In planning, care has been taken to utilise space to the maximum possible extent. In ground floor, a shop, drawing room, dining room, kitchen and a toilet are provided, while in first floor two bed rooms with attached toilet and a lobby are accomodated. The staircase with a quarter landing within the drawing room encasing is shown. A toilet is to be built under the staircase. The plan although very compact, facilitates free circulation of air and access of natural light.

A spare room even in such a small building is always welcome. The mezzanine room (11'x13') just above the shop and kitchen may be used as a recreation / guest room. An attached toilet can be provided at one of the rare corners of the room. The attic room above the staircase may be used as an ideal prayer room. Its size comes to 10'x13'. Thus, the total floor area in the building is 1027 sft.

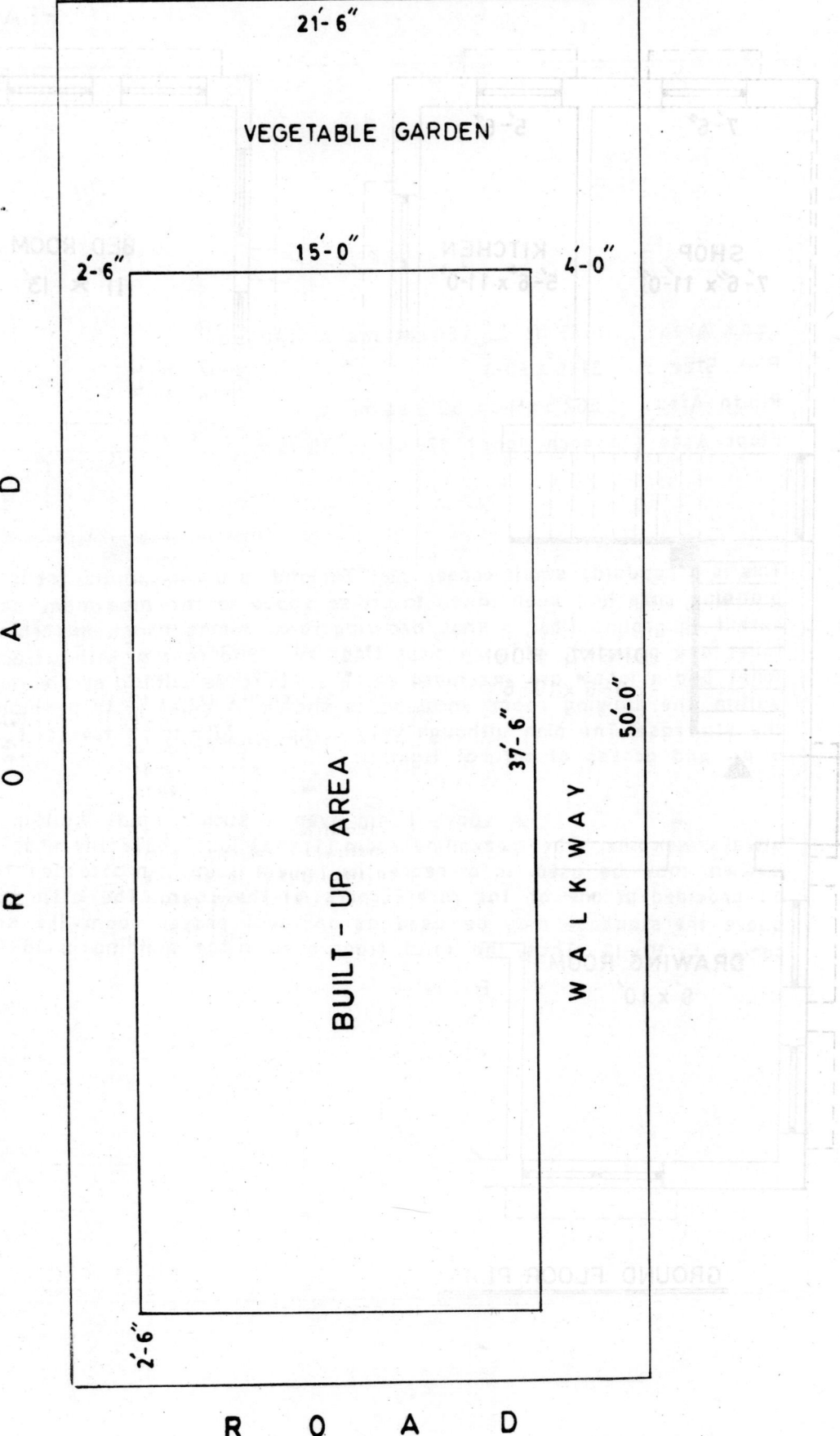

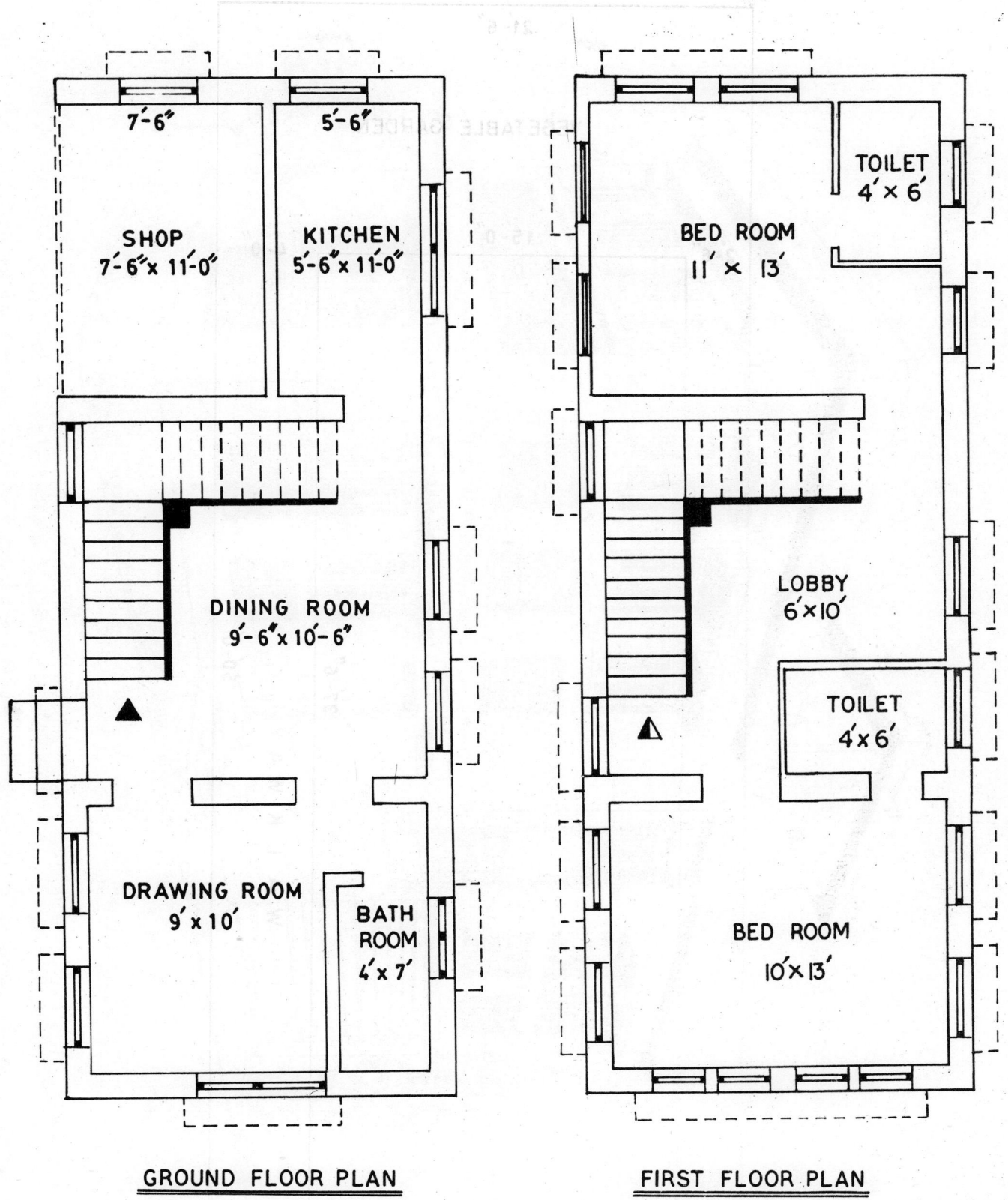

GROUND FLOOR PLAN

SHOP
7'-6" x 11'-0"

KITCHEN
5'-6" x 11'-0"

7'-6"

5'-6"

DINING ROOM
9'-6" x 10'-6"

DRAWING ROOM
9' x 10'

BATH ROOM
4' x 7'

FIRST FLOOR PLAN

BED ROOM
11' x 13'

TOILET
4' x 6'

LOBBY
6' x 10'

TOILET
4' x 6'

BED ROOM
10' x 13'

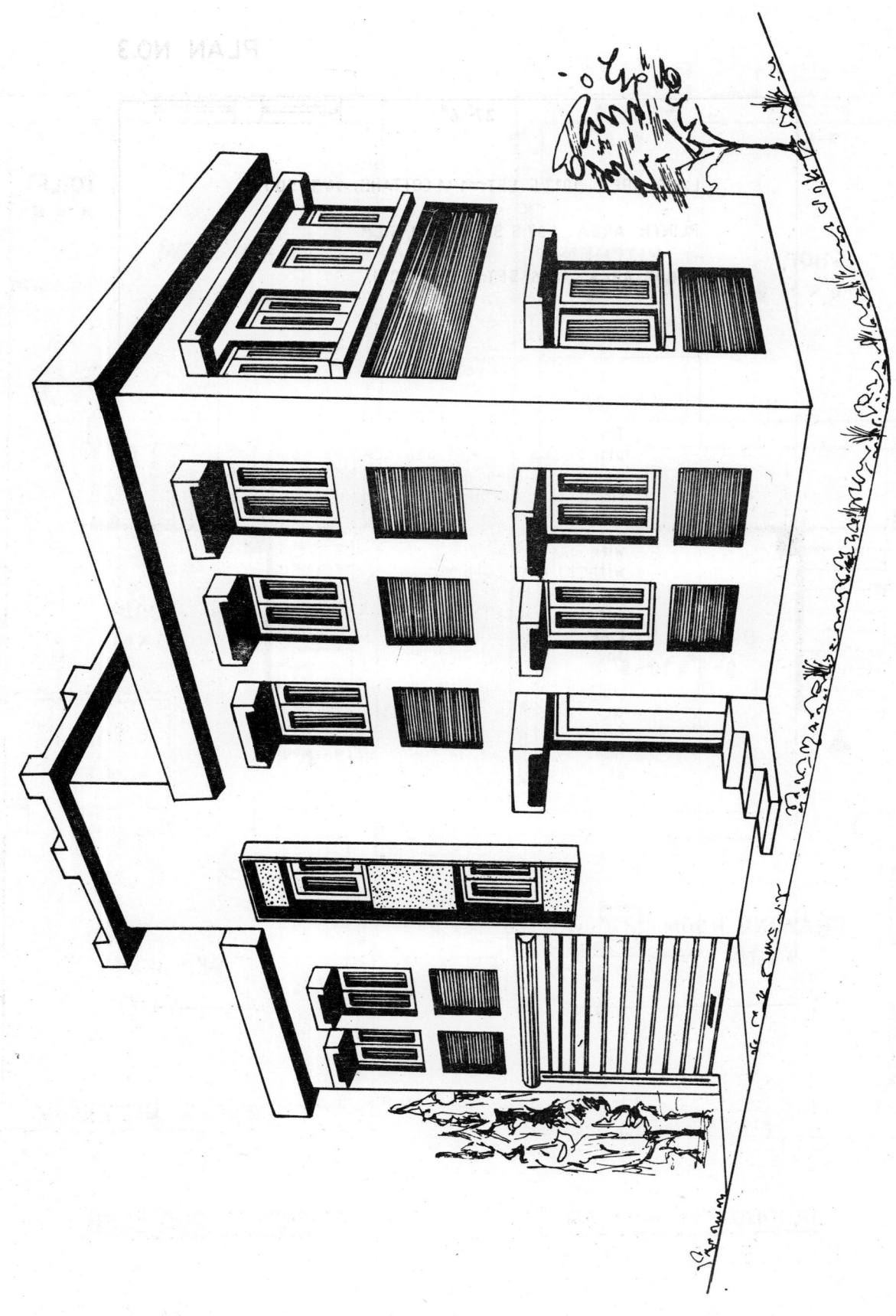

PLAN NO.3

27'- 6"

LAND AREA: 1017·5 SFT≈1·41 COTTAHS≈94·56 SQ.M.

PLINTH AREA: 487·5 SFT≈45·31 SQ.M.

FLOOR AREA: 416 SFT≈38·66 SQ.M.

19'-6"

The plan has been made on a small back plot having a 4' wide passage and no accessible road. construction of a single-storied building is permitted.

The plan is very compact with one Living room and one Bed room with Kitchen, Dining space, a front verandah, bath and w.c. All walls are 5"thick with 10"x 10" brick pillars, the roof is not accessible. this will make the construction economical.

This small house can well accomodate a low-income-group family of four members. simple efforts have been made for a beautiful elevation of the building.

37'-0"

25'-0"

2'

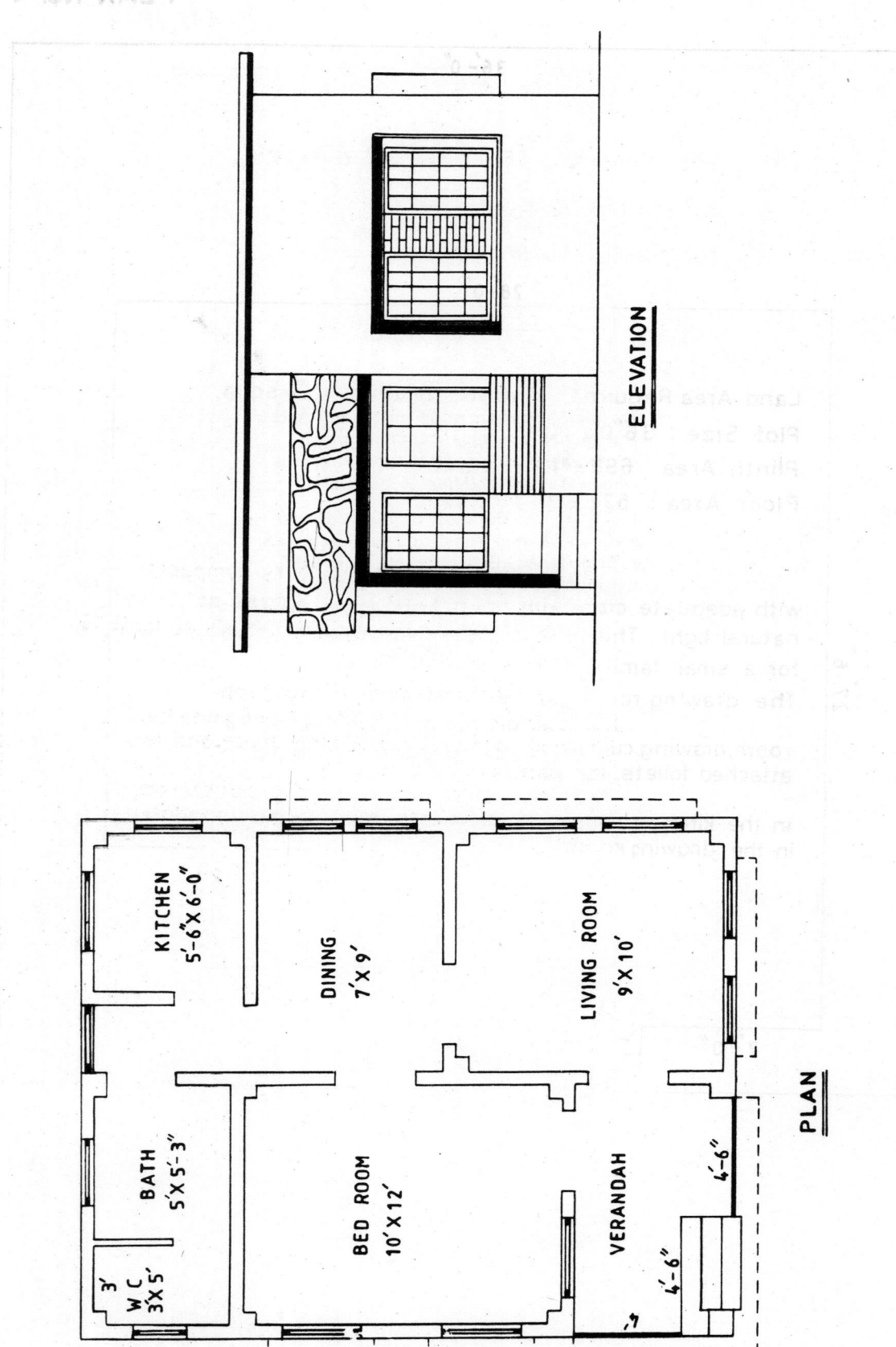

ELEVATION

KITCHEN
5'-6" X 6'-0"

DINING
7' X 9'

LIVING ROOM
9' X 10'

BATH
5' X 5'-3"

WC
3' X 5'

3'

BED ROOM
10' X 12'

VERANDAH

4'-6"

4'-6"

PLAN

PLAN NO. 4

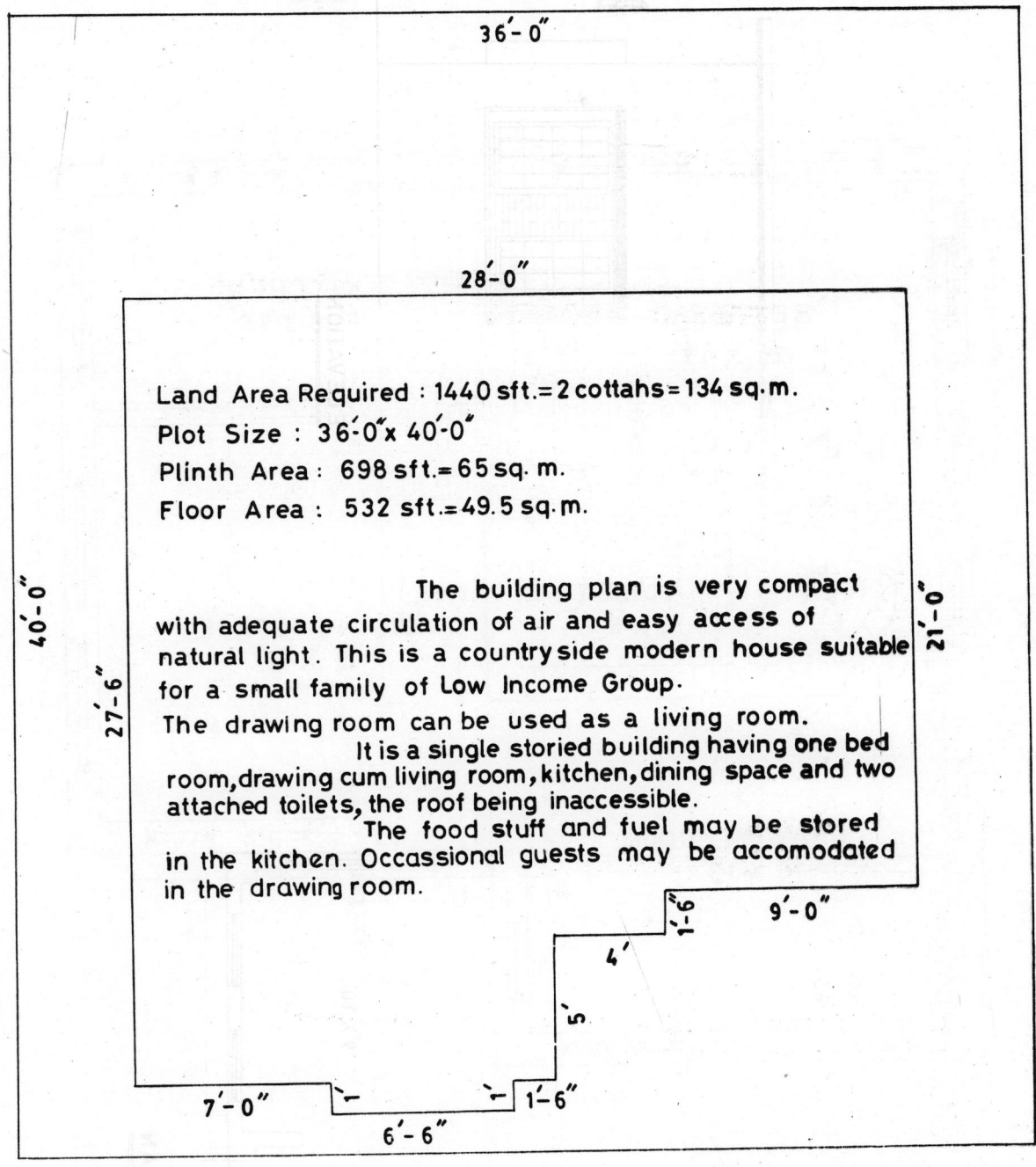

36'-0"

28'-0"

40'-0"

27'-6"

21'-0"

Land Area Required : 1440 sft.= 2 cottahs = 134 sq. m.

Plot Size : 36'-0" x 40'-0"

Plinth Area : 698 sft. = 65 sq. m.

Floor Area : 532 sft. = 49.5 sq. m.

        The building plan is very compact with adequate circulation of air and easy access of natural light. This is a country side modern house suitable for a small family of Low Income Group.

The drawing room can be used as a living room.
        It is a single storied building having one bed room, drawing cum living room, kitchen, dining space and two attached toilets, the roof being inaccessible.
        The food stuff and fuel may be stored in the kitchen. Occassional guests may be accomodated in the drawing room.

1'-6"

9'-0"

4'

5'

7'-0"

1'-6"

6'-6"

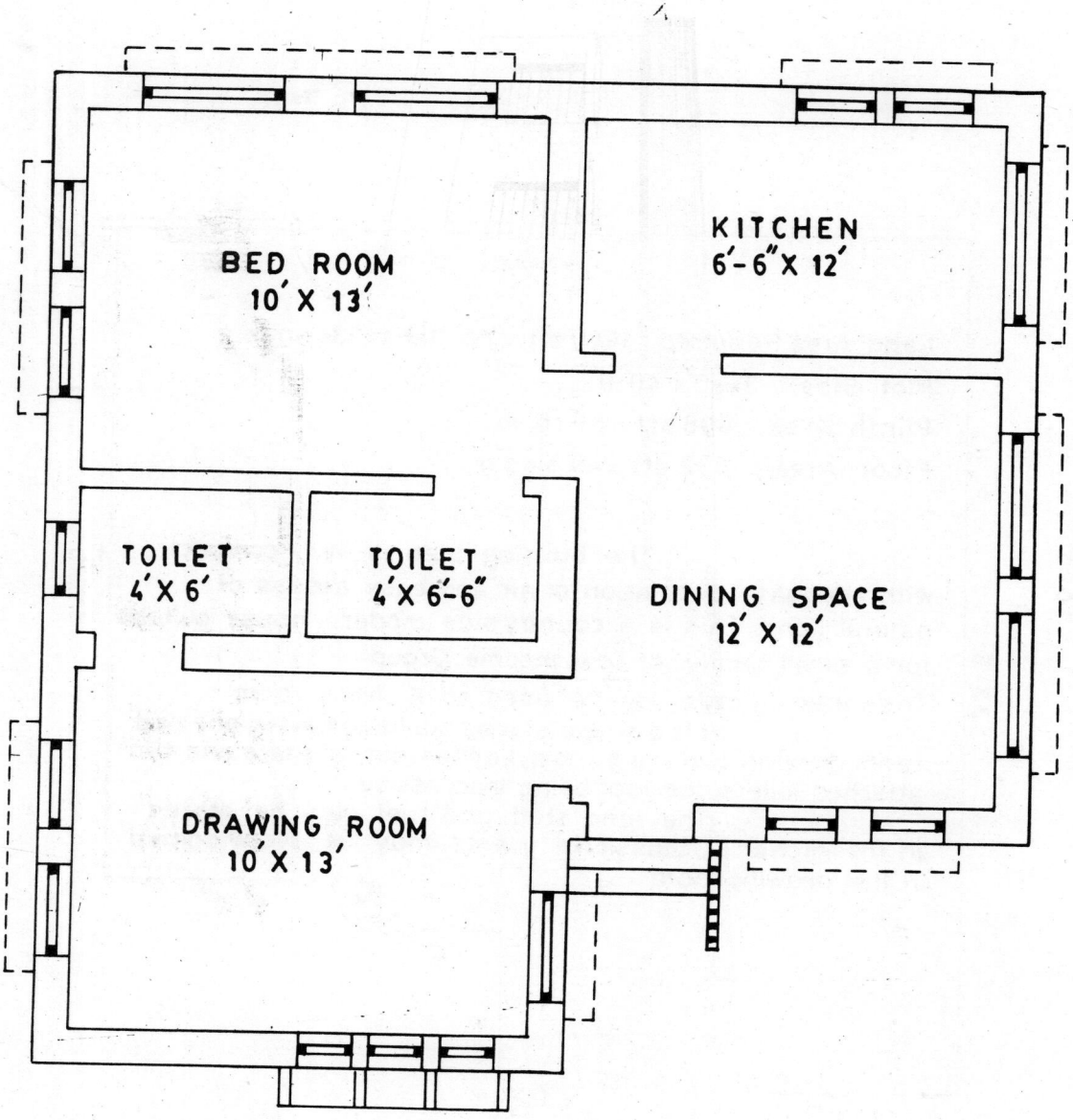

BED ROOM
10' X 13'

KITCHEN
6'-6" X 12'

TOILET
4' X 6'

TOILET
4' X 6-6"

DINING SPACE
12' X 12'

DRAWING ROOM
10' X 13'

GROUND FLOOR PLAN

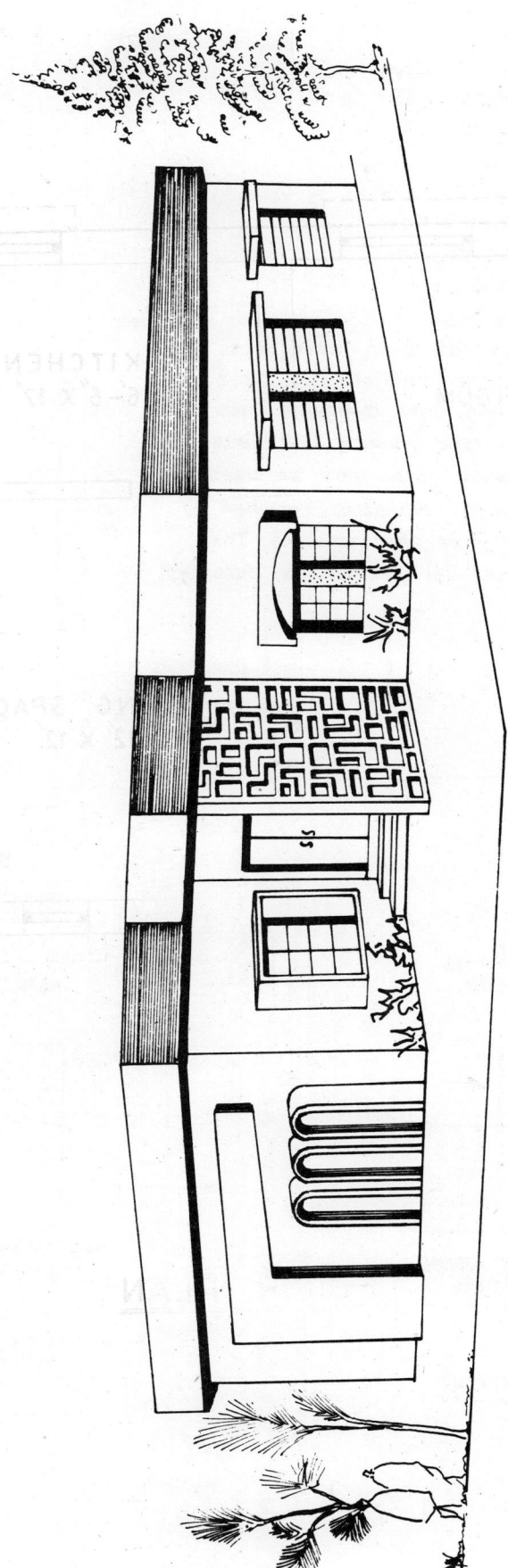

Land Area: 1008 Sft ≈ 1·4 Cottahs ≈ 93·7 Sq.m.

Plot Size: 18´x 56´

Plinth Area: 522 Sft ≈ 48·51 Sq.m.

Floor Area (Including Staircase):

    Ground Floor 395 Sft ≈ 36·71 Sq.m.

    First Floor 439 Sft ≈ 40·85 Sq.m.

Total Floor Area (Including Mezzanine)

    959 Sft ≈ 89 Sq.m.

        The building plan has been made on a narrow long strip of land at the junction of a main road and a side road. In ground floor, provision has been made for a shop, staircase, drawing cum dining kitchen and a toilet. The drawing room may be used to accomodate occassional guest. In first floor, two bed rooms with a common toilet have been provided. The mezzanine room may be used for recreational purposes.

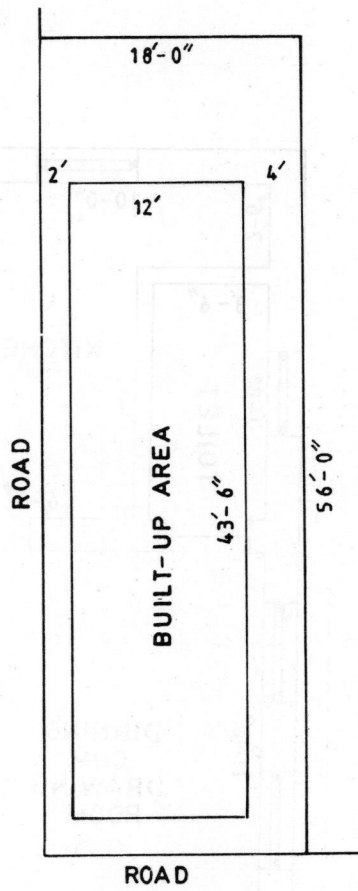

PLAN NO. 5

SITE PLAN

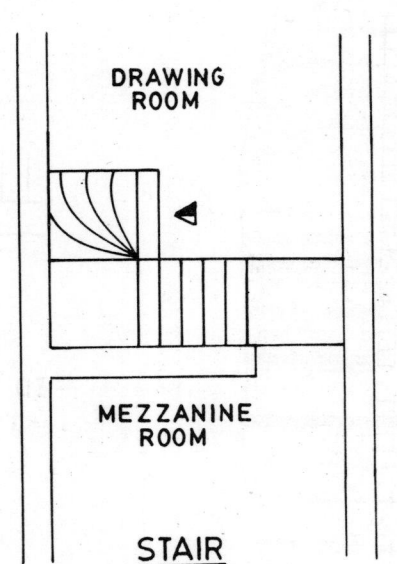

STAIR

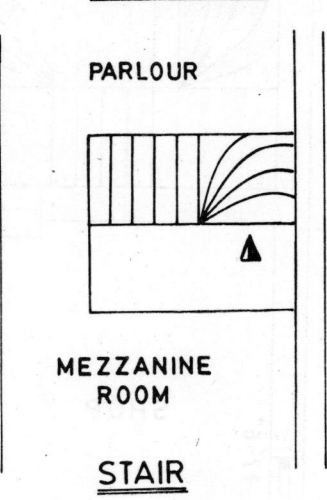

STAIR

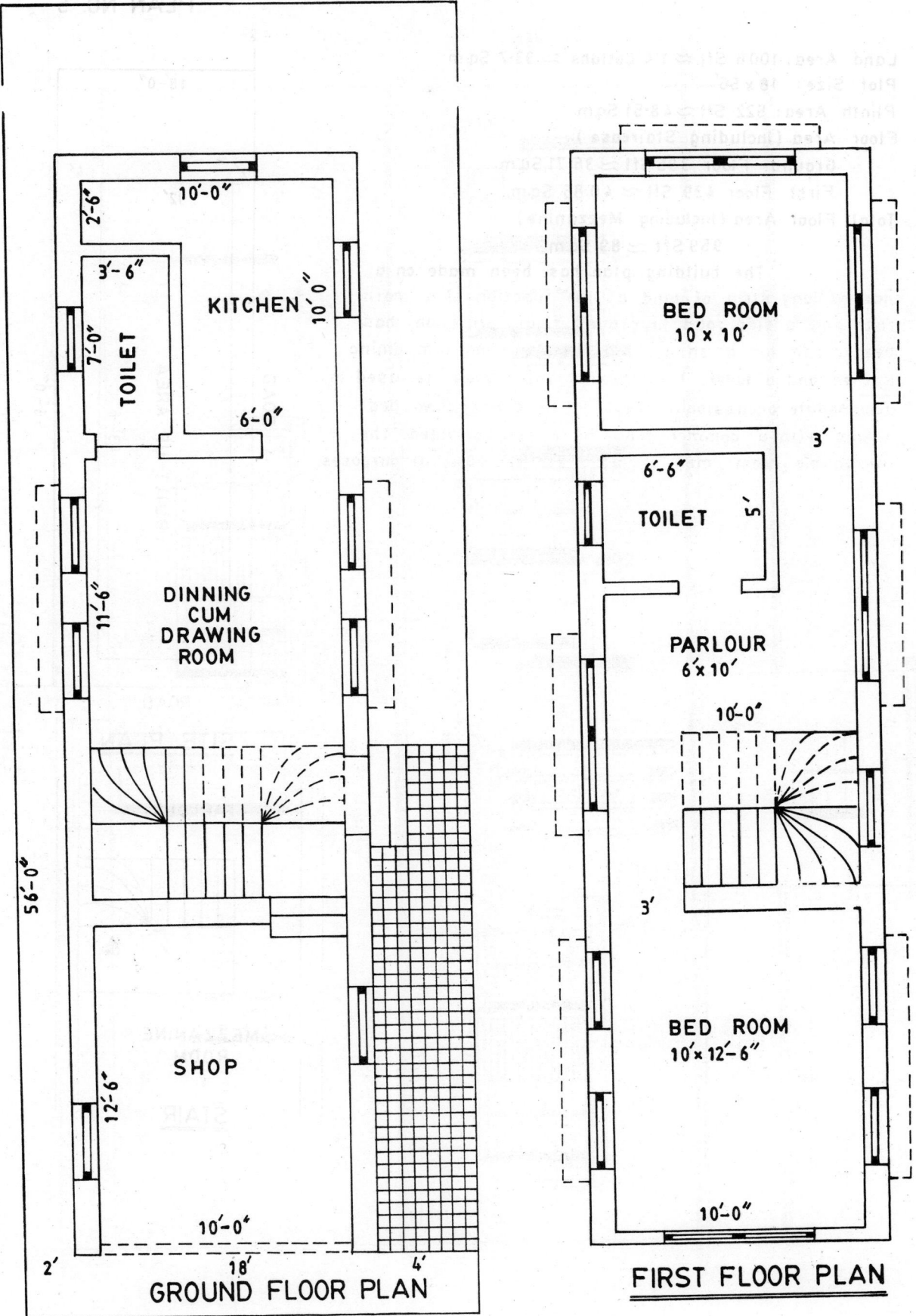

GROUND FLOOR PLAN

KITCHEN
10'-0"
2'-6"
3'-6"
TOILET
7'-0"
6'-0"
DINNING CUM DRAWING ROOM
11'-6"
56'-0"
SHOP
12'-6"
10'-0"
2'
18'
4'

FIRST FLOOR PLAN

BED ROOM
10' x 10'
3'
6'-6"
TOILET
5'
PARLOUR
6' x 10'
10'-0"
3'
BED ROOM
10' x 12'-6"
10'-0"

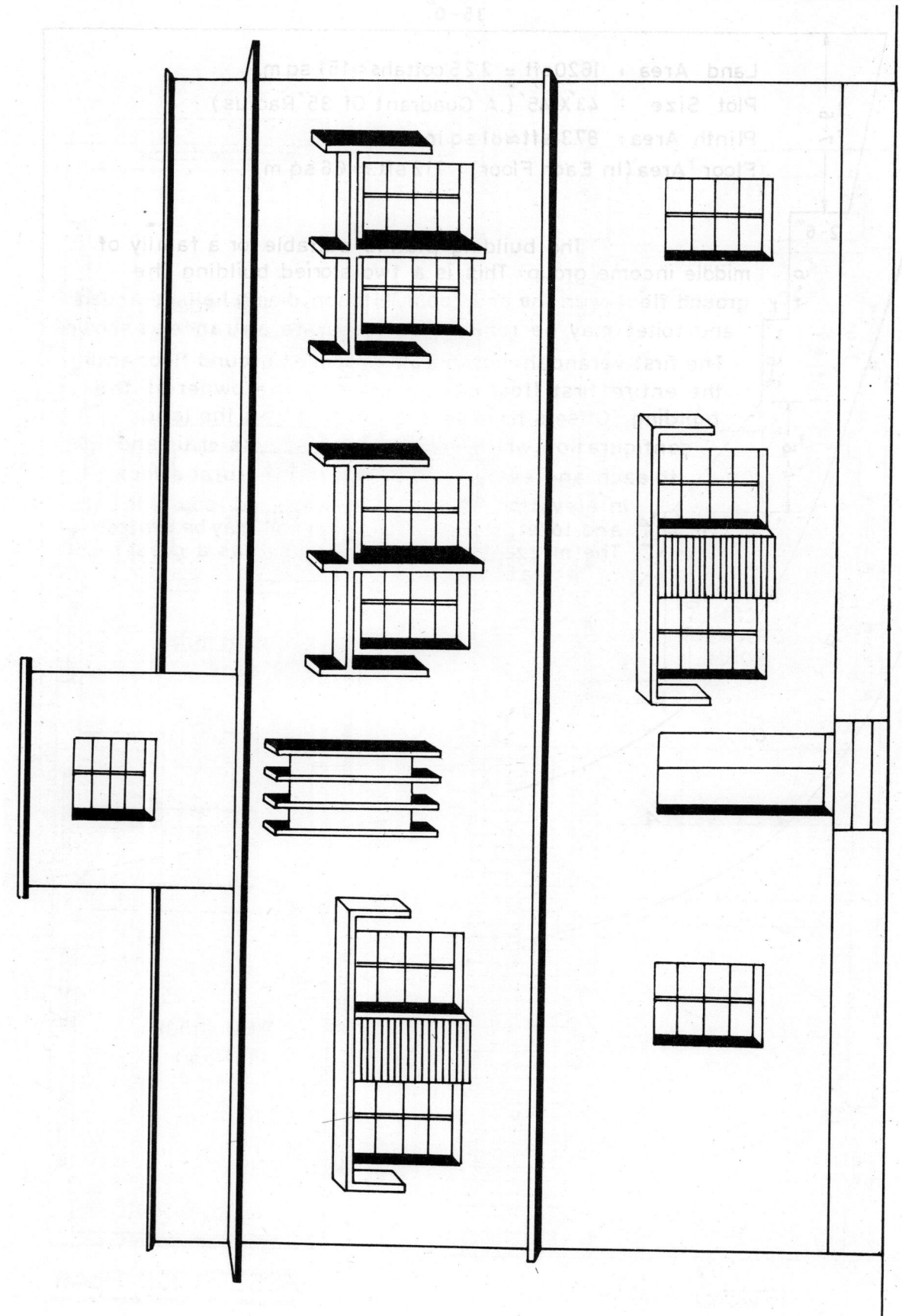

ELEVATION

PLAN NO. 6

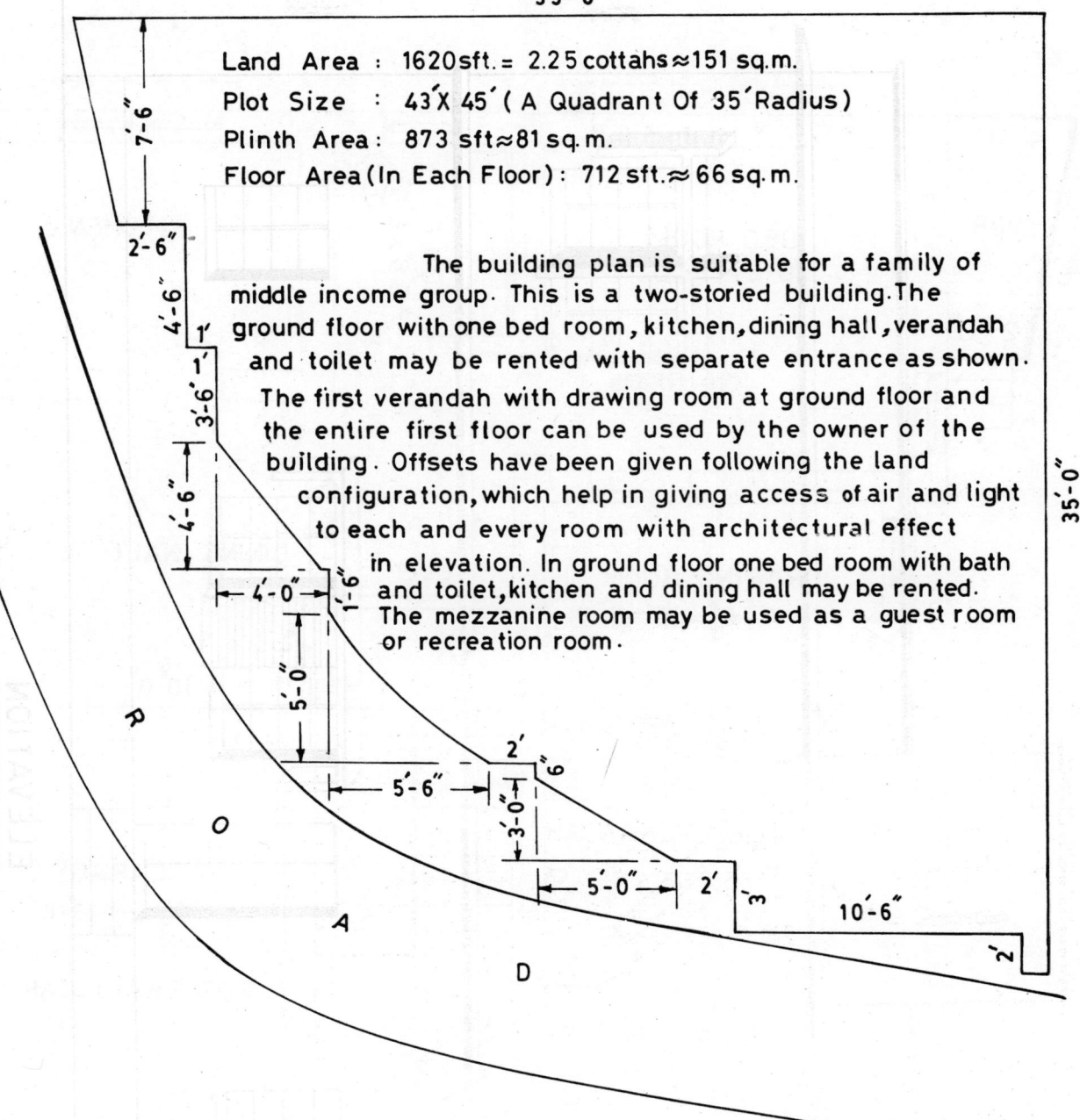

Land Area : 1620 sft. = 2.25 cottahs ≈ 151 sq.m.
Plot Size : 43´ X 45´ (A Quadrant Of 35´ Radius)
Plinth Area : 873 sft ≈ 81 sq.m.
Floor Area (In Each Floor) : 712 sft. ≈ 66 sq.m.

   The building plan is suitable for a family of middle income group. This is a two-storied building. The ground floor with one bed room, kitchen, dining hall, verandah and toilet may be rented with separate entrance as shown. The first verandah with drawing room at ground floor and the entire first floor can be used by the owner of the building. Offsets have been given following the land configuration, which help in giving access of air and light to each and every room with architectural effect in elevation. In ground floor one bed room with bath and toilet, kitchen and dining hall may be rented. The mezzanine room may be used as a guest room or recreation room.

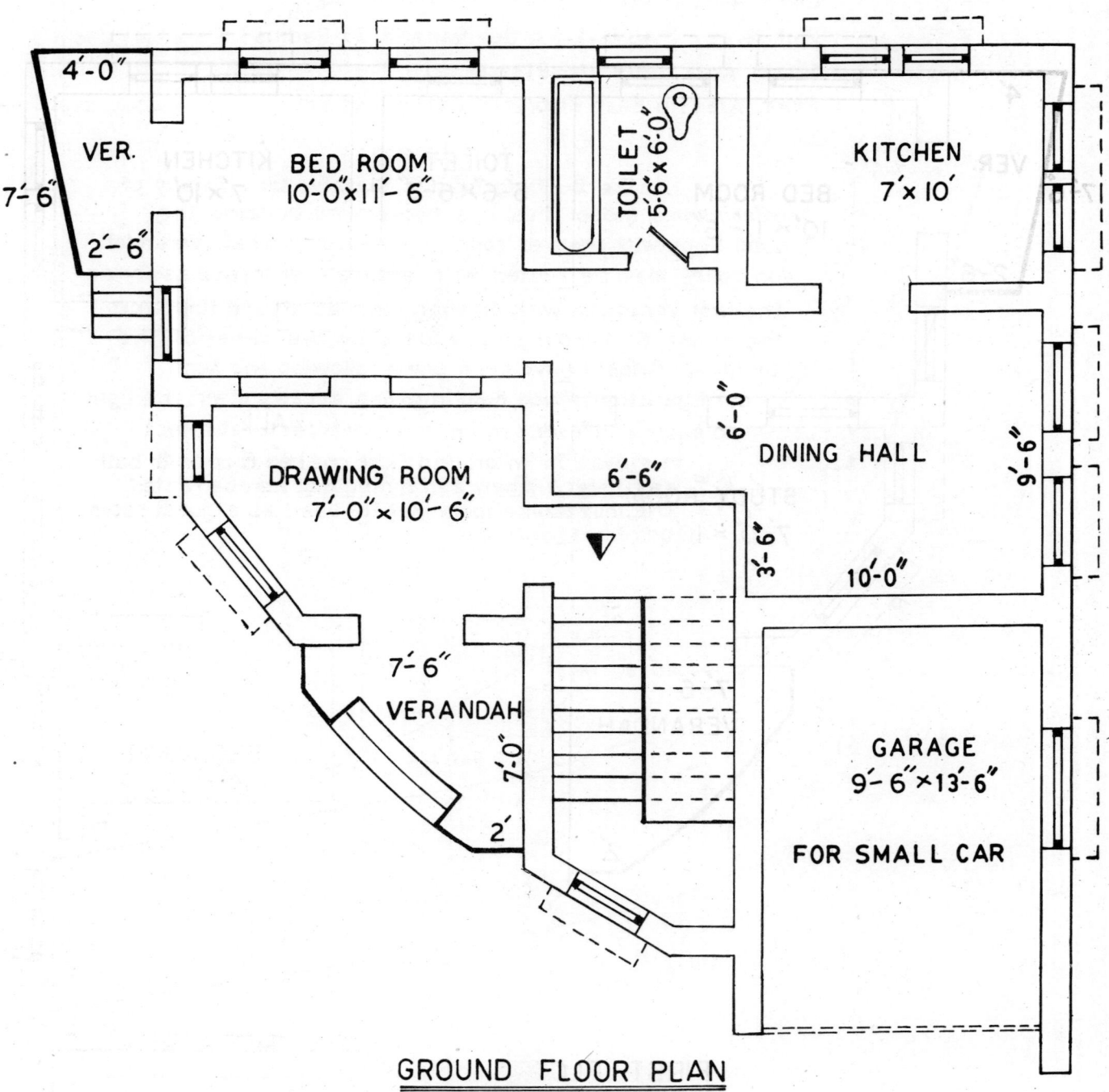

GROUND FLOOR PLAN

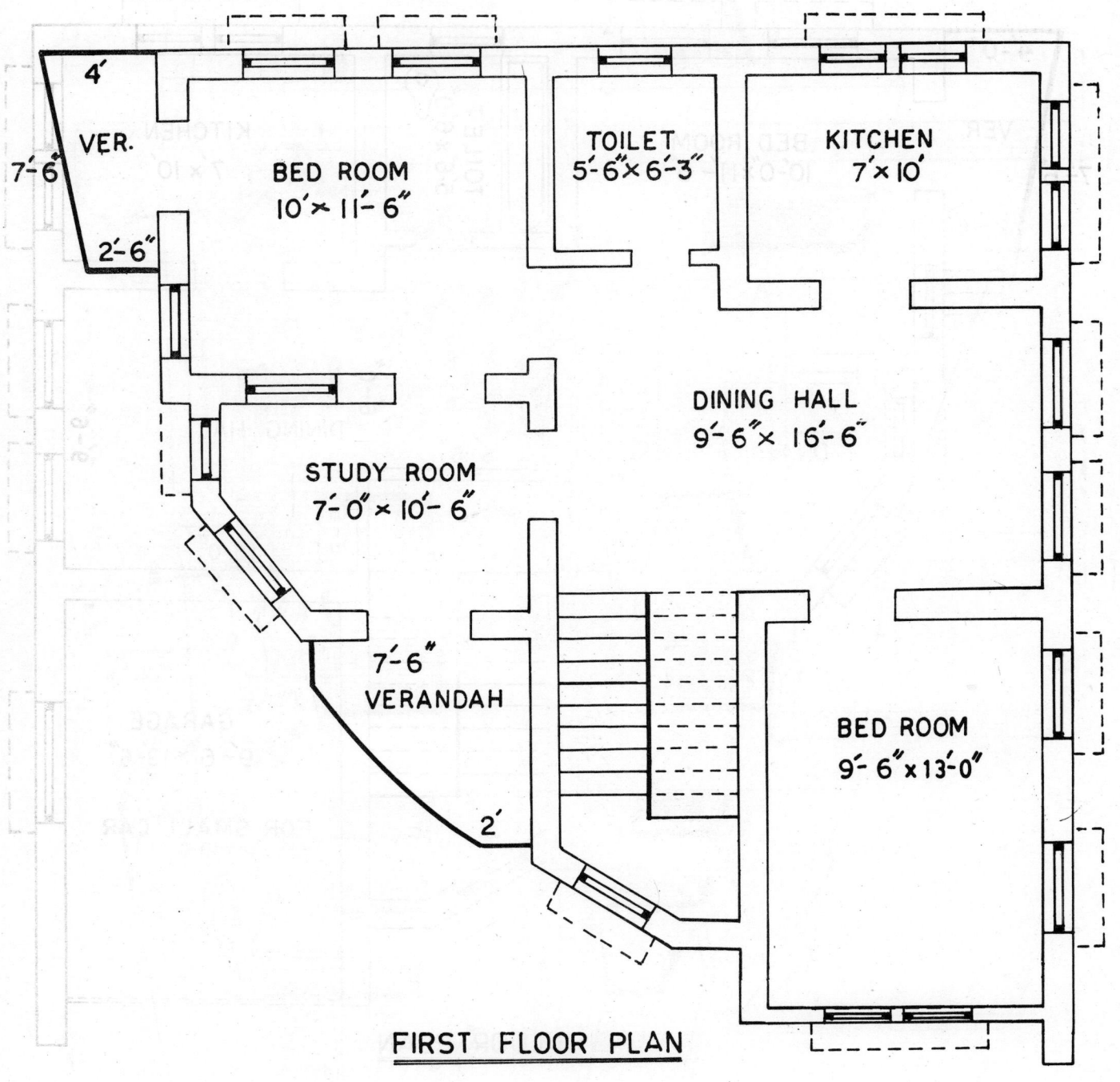

FIRST FLOOR PLAN

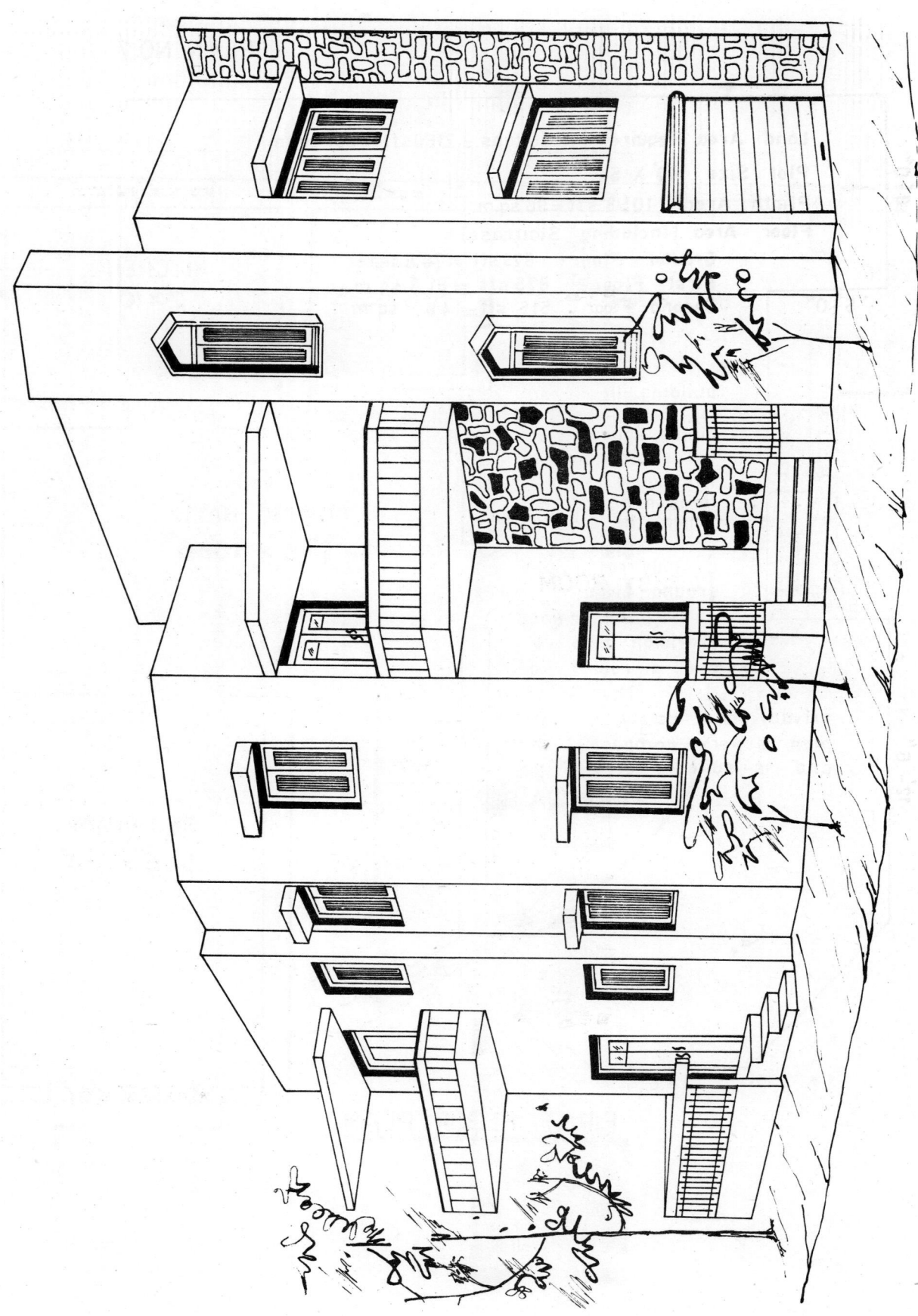

PLAN NO.7

32'- 0"

8'- 0"

5'- 0"

13'- 6"

5'- 0"

12'- 6"

27'- 0"

1'

3'- 6"

6'- 6"   1'

3'

2'-6"   3'

1'- 6"

6'

7'- 6"

Land Area Required : 3 cottahs = 2160 sft ≈ 201 sq.m.

Plot Size : 40' X 54'

Plinth Area : 1058 sft ≈ 98 sq.m.

Floor Area (Including Staircase) :
    Ground Floor : 823 sft = 76.5 sq.m.
    First Floor : 875 sft = 81.3 sq.m.
    Second Floor : 515 sft = 48 sq.m

       Planning has been done for a three-storied building. In ground floor, provisions have been made for two bed rooms with attached toilets, kitchen, dining space and a small drawing room with a verandah which can be used for rental purpose. The house owner will have a separate entrance through his drawing room in ground floor within which the staircase is housed.

       The first floor plan provides additionally a study room just above the drawing room in ground floor.

       The second floor plan facilitates provision of two rooms with attached toilets and a roof garden.

       The planning ensures comfort with privacy and safety and produces architectural effect. The plan is very compact with ventilation of air and light and is suitable for a joint family of 12 members.

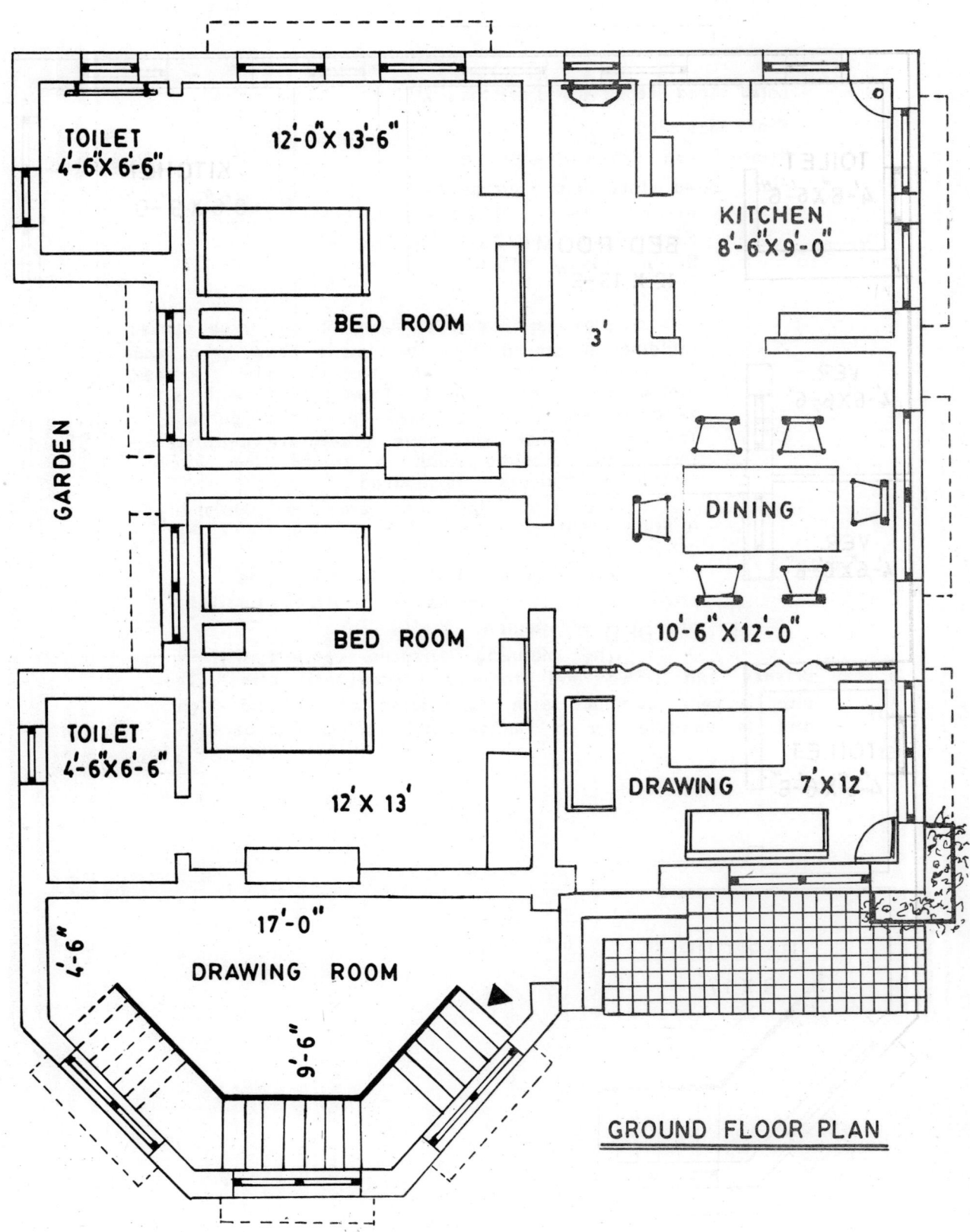

GROUND FLOOR PLAN

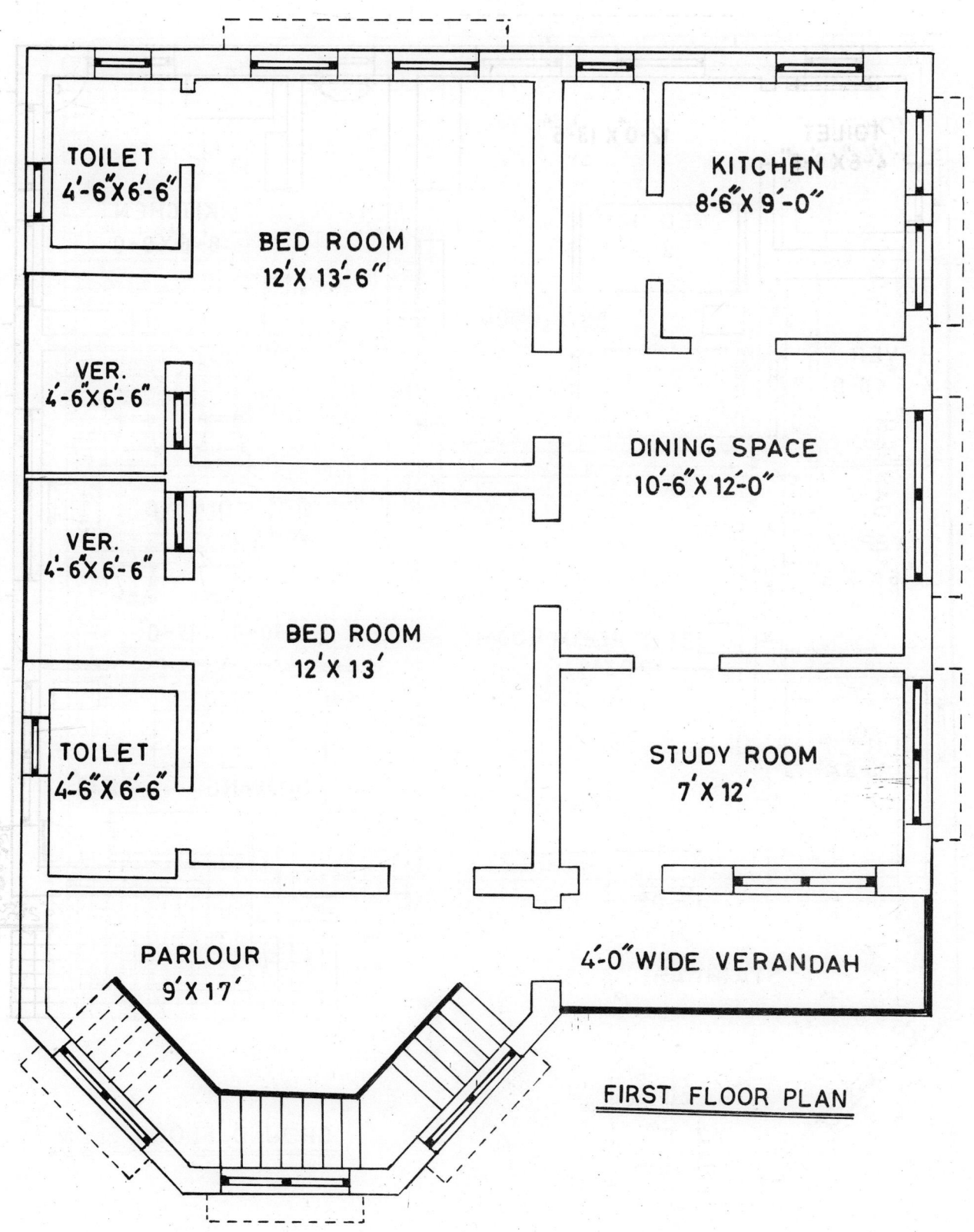

TOILET
4'-6"X6'-6"

BED ROOM
12'X 13'-6"

KITCHEN
8'-6"X9'-0"

VER.
4'-6"X6'-6"

DINING SPACE
10'-6"X12'-0"

VER.
4'-6"X6'-6"

BED ROOM
12'X 13'

TOILET
4'-6"X6'-6"

STUDY ROOM
7'X 12'

PARLOUR
9'X 17'

4'-0"WIDE VERANDAH

FIRST FLOOR PLAN

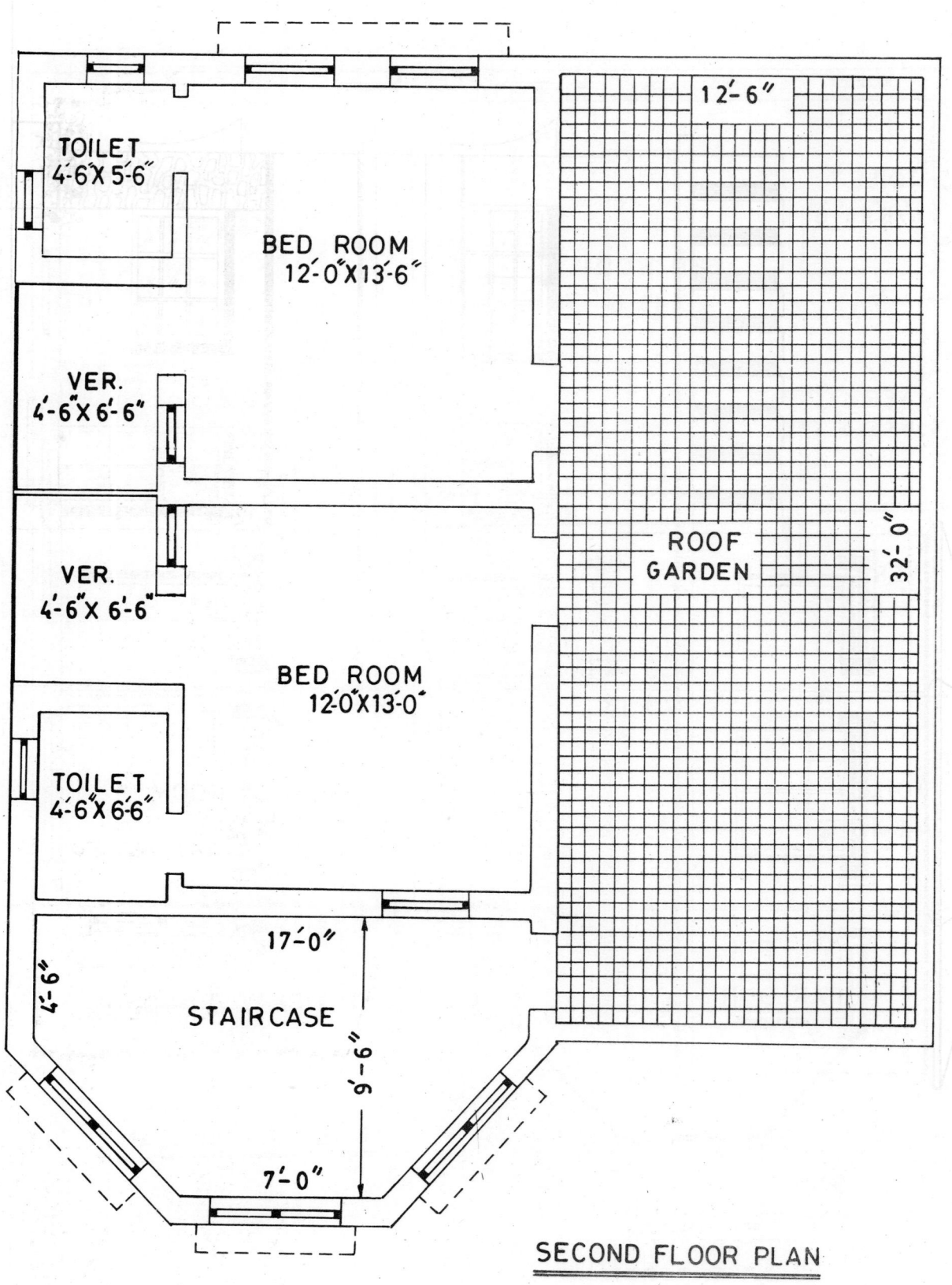

TOILET
4'-6"X 5'-6"

BED ROOM
12'-0"X13'-6"

12'-6"

VER.
4'-6"X 6'-6"

ROOF
GARDEN

32'-0"

VER.
4'-6"X 6'-6"

BED ROOM
12'-0"X13'-0"

TOILET
4'-6"X 6'-6"

17'-0"

4'-6"

STAIRCASE

9'-6"

7'-0"

SECOND FLOOR PLAN

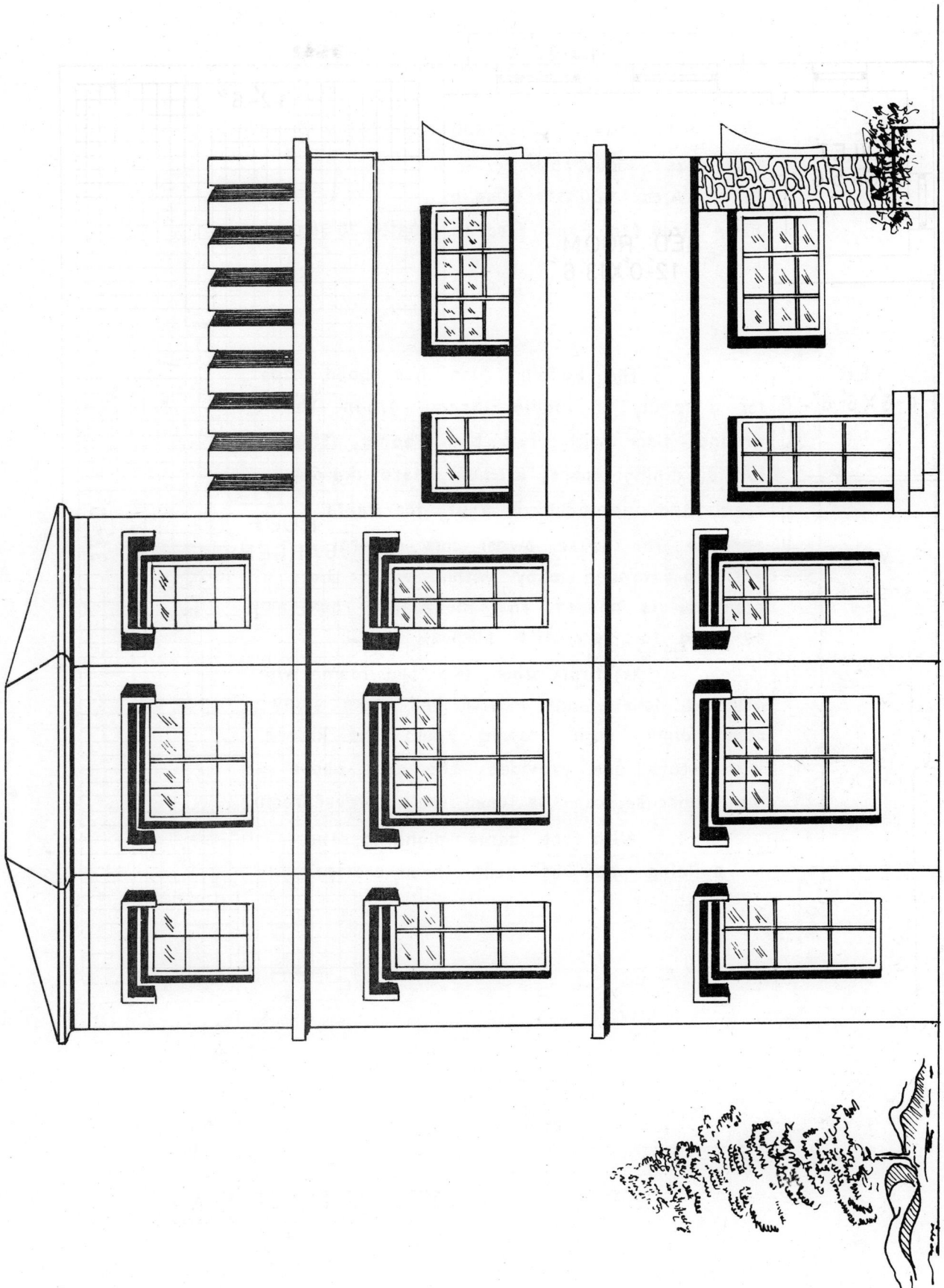

ELEVATION

PLAN NO. 8

26'-6"

11'-0"

Land Area Required: 2·75 cottahs = 1980 sft ≈ 184 sq.m.

Plot Size: 35'-0" X 56'-0"

Plinth Area: 1180 sft ≈ 110 sq.m.

Floor Area (In Each Floor): 810 sft ≈ 75·3 sq.m.

1'

34'-0"

23'-0"

The building plan has been made for a family of middle income group. The ground floor with two bed rooms, attached toilets, dining space, kitchen, store, drawing room and verandah is kept for rental purpose. The house owner has separate entrance through lobby within which the staircase is housed. The mezzanine room may be used for recreation purpose.

In first floor, two bed rooms with attached toilets, one master bed room, study room, dining cum drawing room and kitchen with store are provided. The attic above the staircase can be used as a prayer room.

With the same planning, the building may be built three-storied.

1'

3'-0"

1'

6'-0"

1'

12'-6"

2'

3'-6"

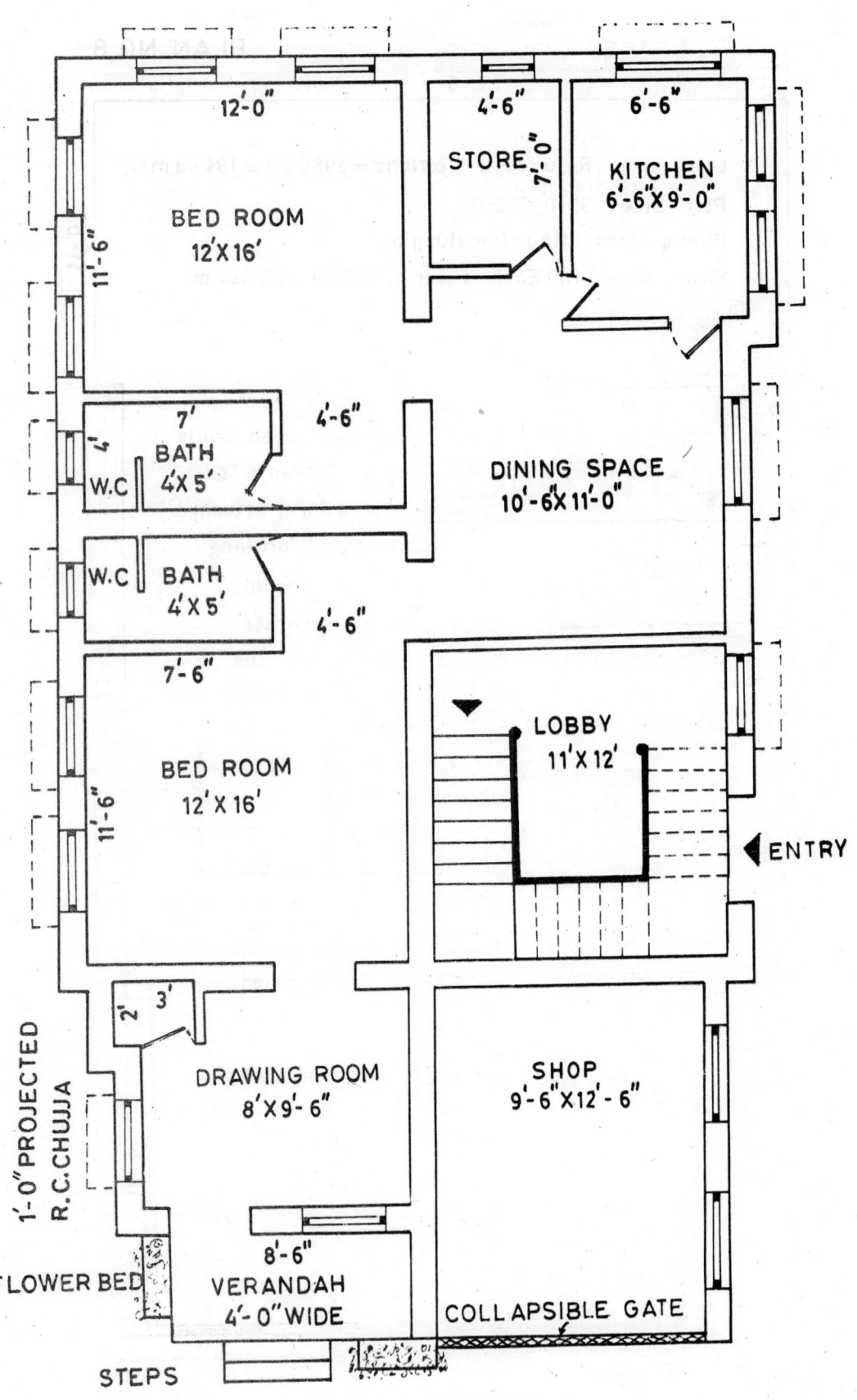

GROUND FLOOR PLAN

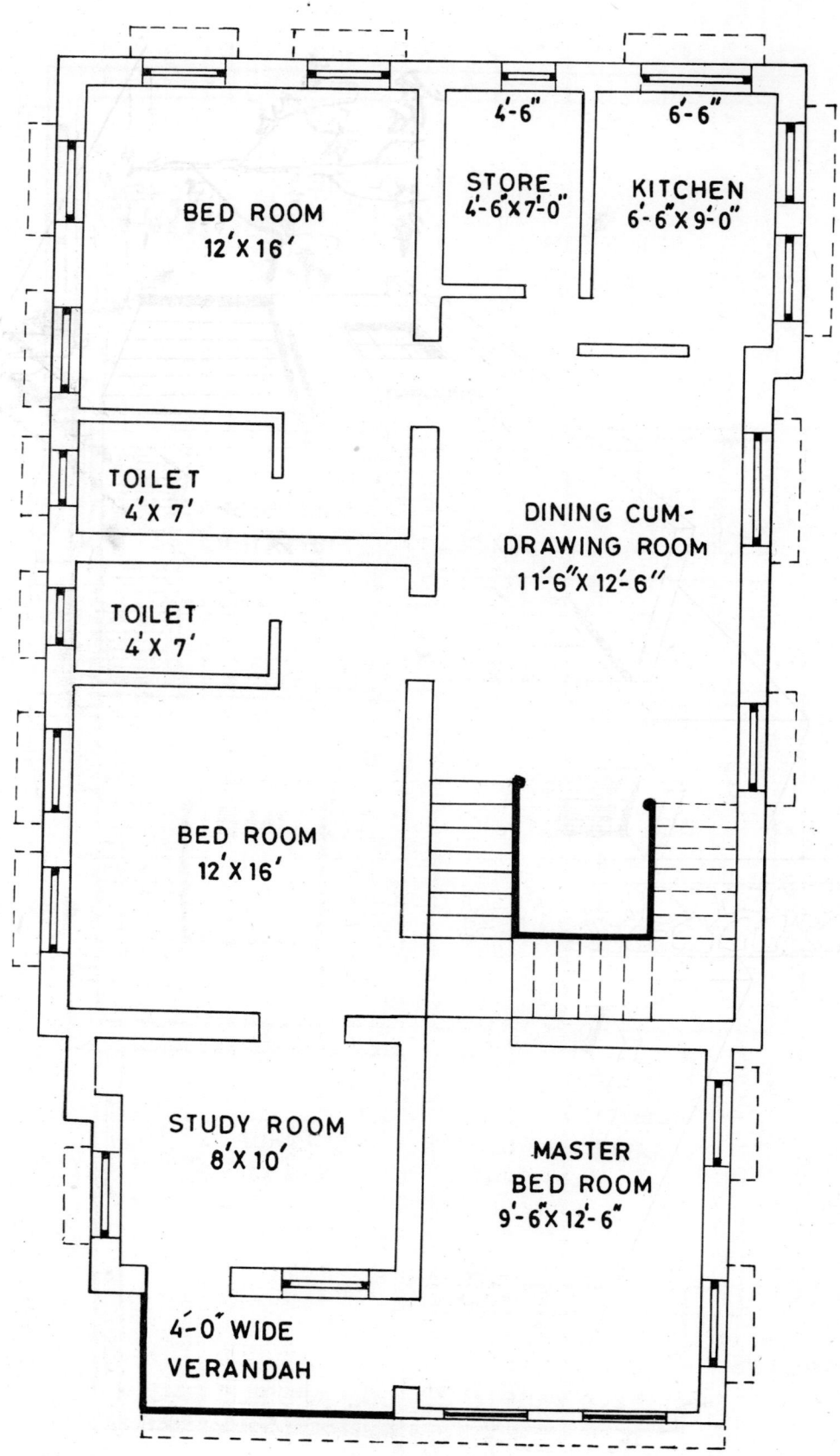

BED ROOM
12' X 16'

STORE
4'-6" X 7'-0"

4'-6"

KITCHEN
6'-6" X 9'-0"

6'-6"

TOILET
4' X 7'

DINING CUM-
DRAWING ROOM
11'-6" X 12'-6"

TOILET
4' X 7'

BED ROOM
12' X 16'

STUDY ROOM
8' X 10'

MASTER
BED ROOM
9'-6" X 12'-6"

4'-0" WIDE
VERANDAH

FIRST FLOOR PLAN

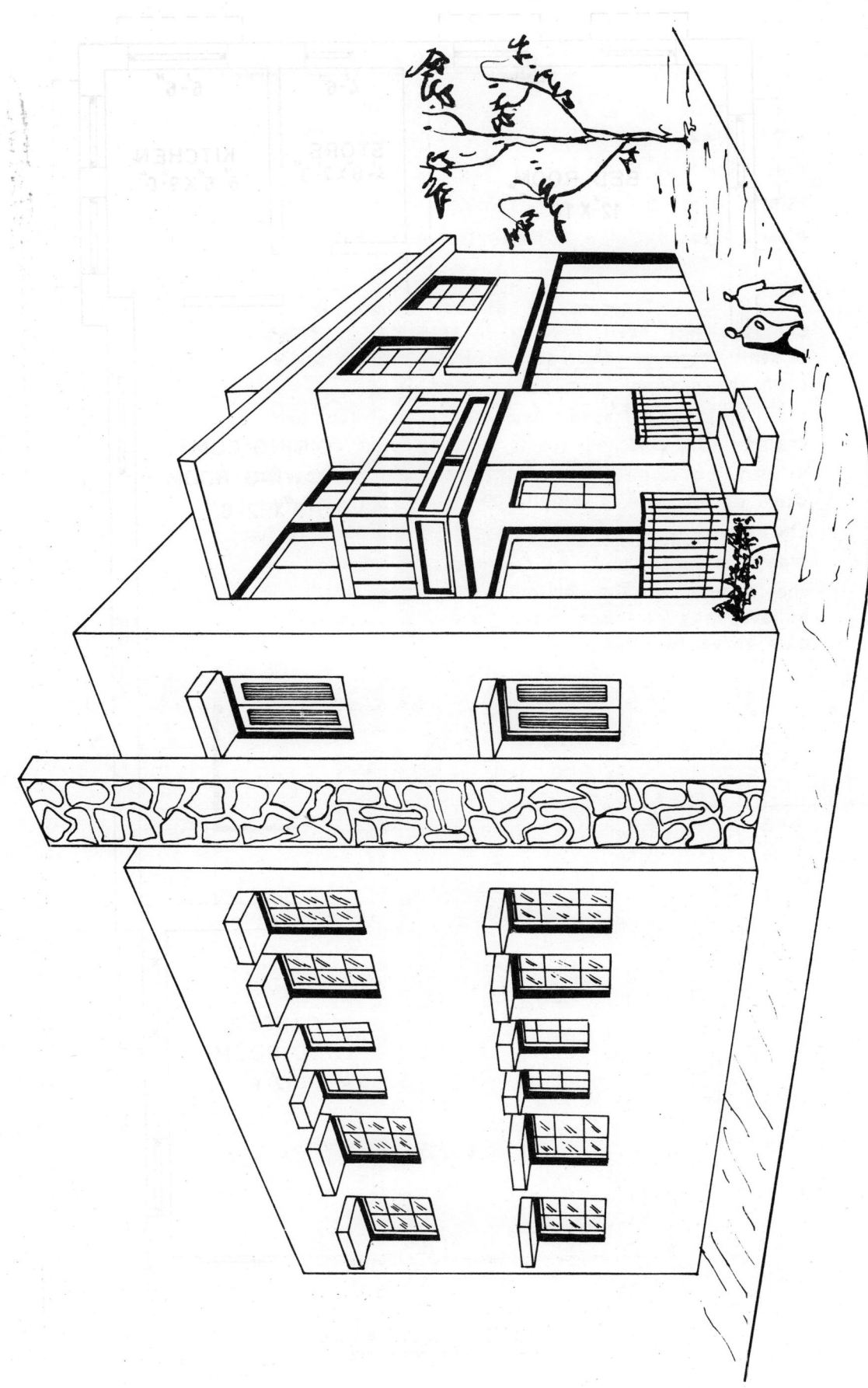

29'-6"

29'-6"

16'-6"

Land Area Required : 1872 sft = 2.6 cottahs ≈ 174 sq.m.

Plot Size : 41' X 46'

Plinth Area: 940 sft ≈ 87.4 sq.m.

Floor Area : 780 sft ≈ 72.5 sq.m.(in each floor)

        The plot is almost square in shape. Contrast has been brought in planning by providing a semi-circular staircase projected at an angle from the drawing cum dining room.

        In each floor, there are two bed rooms with attached toilets, drawing cum dining hall, kitchen, parlour and a front verandah. Each floor can accomodate a small family of 5 members. The staircase with window styles and wall texture, front verandah and R.C.C. sunshades bring the beauty of the building. If required, the building may be made three-storied to accomodate three families.

7'-6"

9'-6"

3'-6"

4'-0"

10'-0"

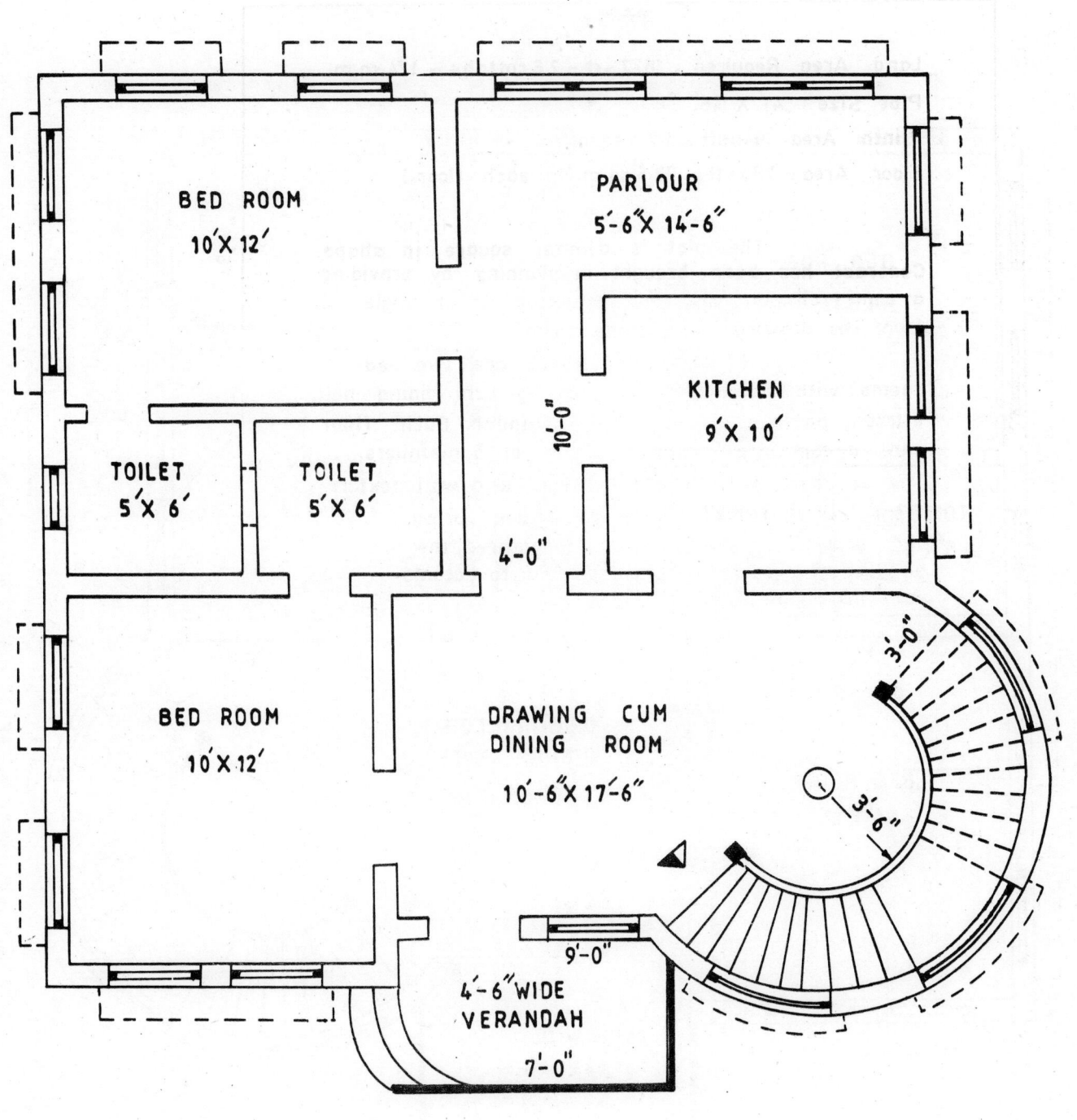

BED ROOM
10' X 12'

PARLOUR
5'-6" X 14'-6"

KITCHEN
9' X 10'

TOILET
5' X 6'

TOILET
5' X 6'

10'-0"

4'-0"

3'-0"

BED ROOM
10' X 12'

DRAWING CUM
DINING ROOM
10'-6" X 17'-6"

3'-6"

9'-0"

4'-6" WIDE
VERANDAH

7'-0"

GROUND FLOOR PLAN

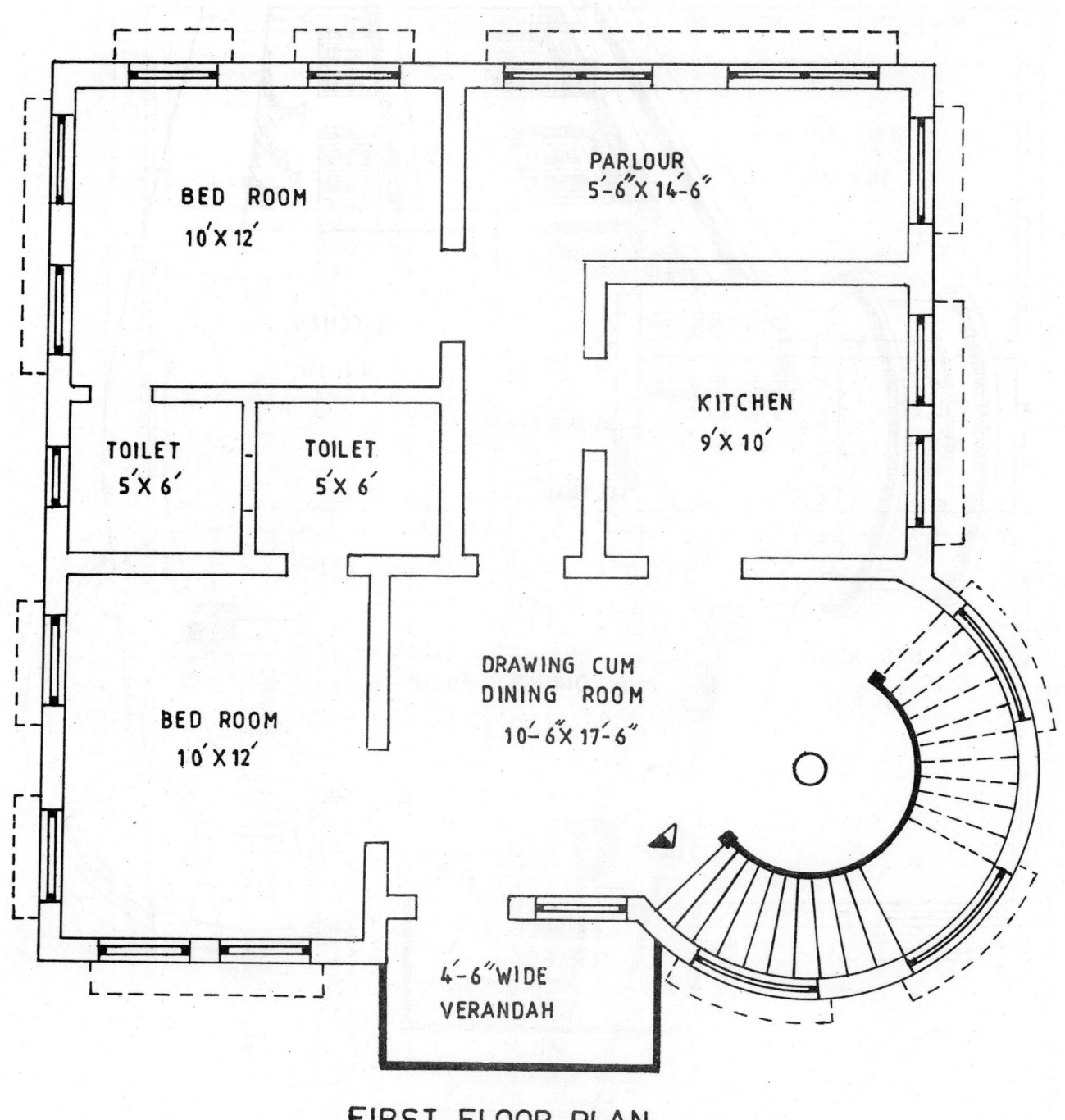

BED ROOM
10′ X 12′

PARLOUR
5′-6″ X 14′-6″

TOILET
5′ X 6′

TOILET
5′ X 6′

KITCHEN
9′ X 10′

BED ROOM
10′ X 12′

DRAWING CUM
DINING ROOM
10′-6″ X 17′-6″

4′-6″ WIDE
VERANDAH

FIRST FLOOR PLAN

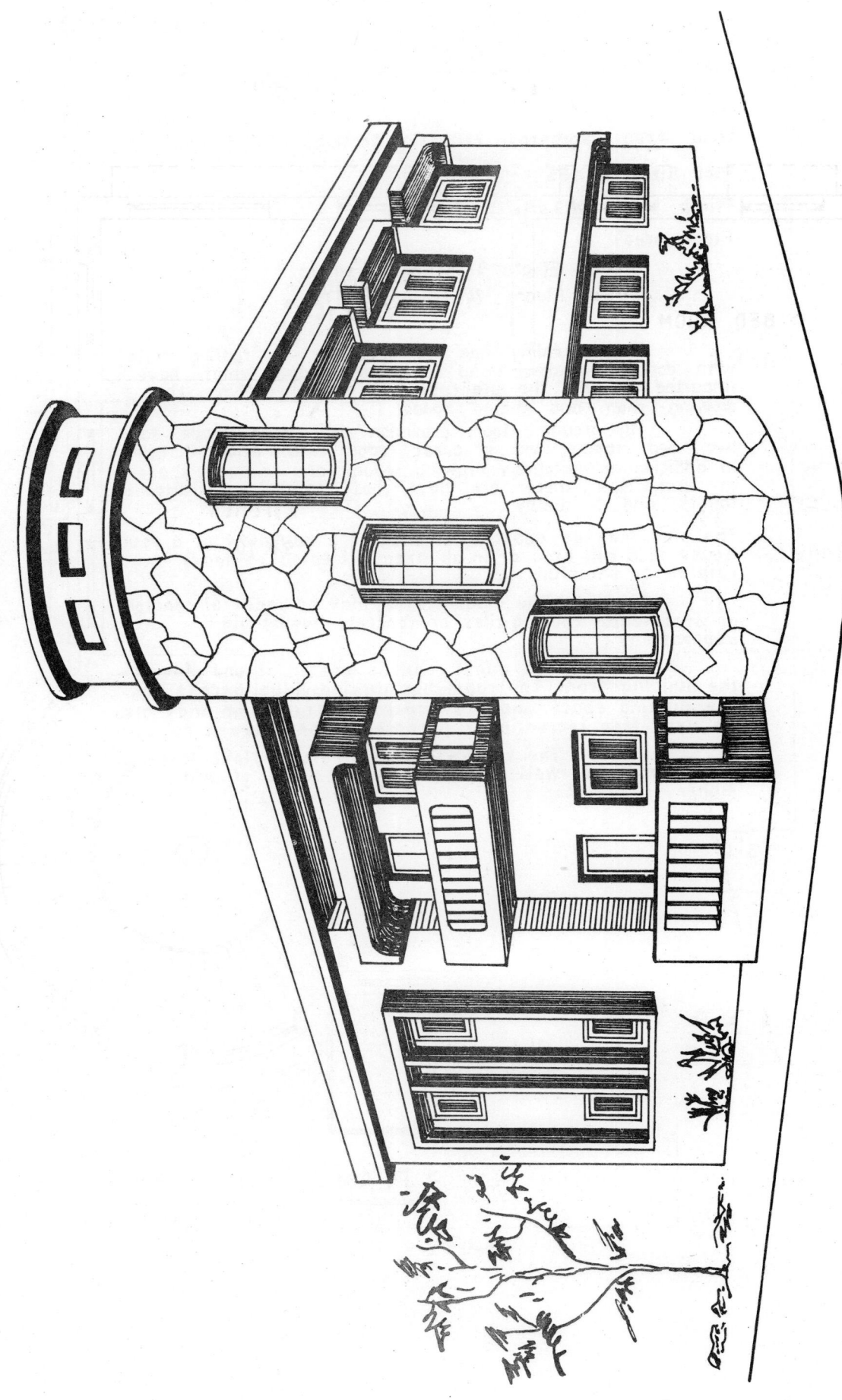

30'-0"

Land Area Required: 2200 sft ≈ 3 cottahs ≈ 204'5 sq.m.

Plot Size: 40'X55'

Plinth Area: 1239 sft ≈ 115 sq.m.

Floor Area:

     Ground Floor: 973 sft ≈ 90'4 sq.m.
     Firest Floor: 749 sft ≈ 69'6 sq.m.

     This building has its speciality in roofing style with dormer windows and a gable window, which have imparted beauty. The staircase is housed within the drawing room and dining space.
     In ground floor, provision has been made for two bed rooms and a guest room with attached toilets in addition to kitchen, drawing room and dining space. In first floor, there are three bed rooms with attached toilets and a lobby

The plan may be adopted for making a bungalow or a rest house in a hill area or in an area subjected to heavy rainfall or snowfall.

     The sloping roof may be made of concrete or ornamental roofing tiles or concrete having tile impressions.

     The guest room is located at one side of the building front. The room has an individual access to the dinning space and drawing room. This room may also be used for recreational purposes when there is no guest.

     The staircase ends at first floor level. The rooms are well-ventilated with access of natural light.

31'-0"

34'-0"

5'-0"

7'-0"

16'-0"

13'-0"

18'-0"

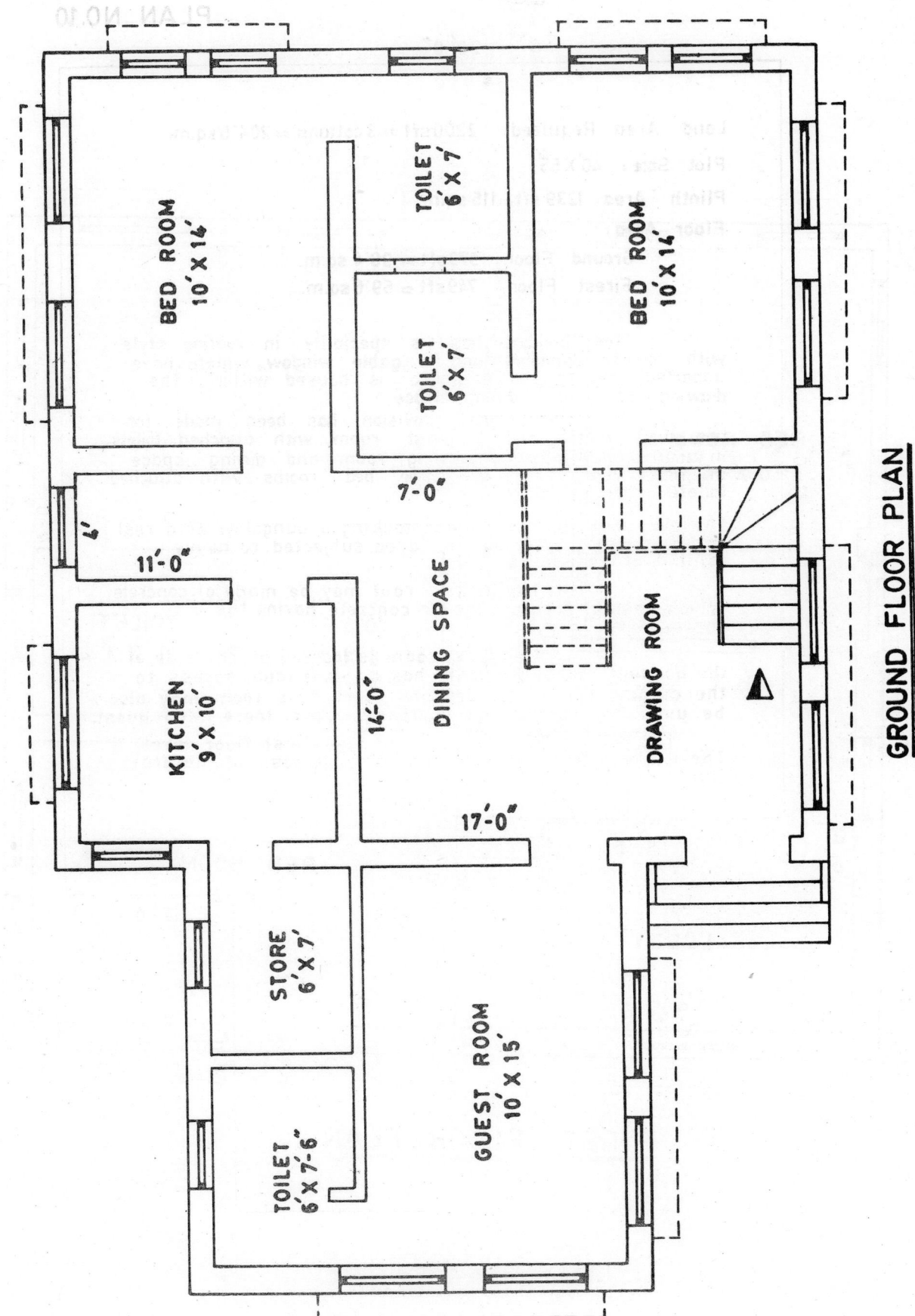

GROUND FLOOR PLAN

BED ROOM
10' X 14'

TOILET
6' X 7'

BED ROOM
10' X 14'

TOILET
6' X 7'

KITCHEN
9' X 10'

11'-0"

7'-0"

DINING SPACE

14'-0"

DRAWING ROOM

17'-0"

STORE
6' X 7'

TOILET
6' X 7'-6"

GUEST ROOM
10' X 15'

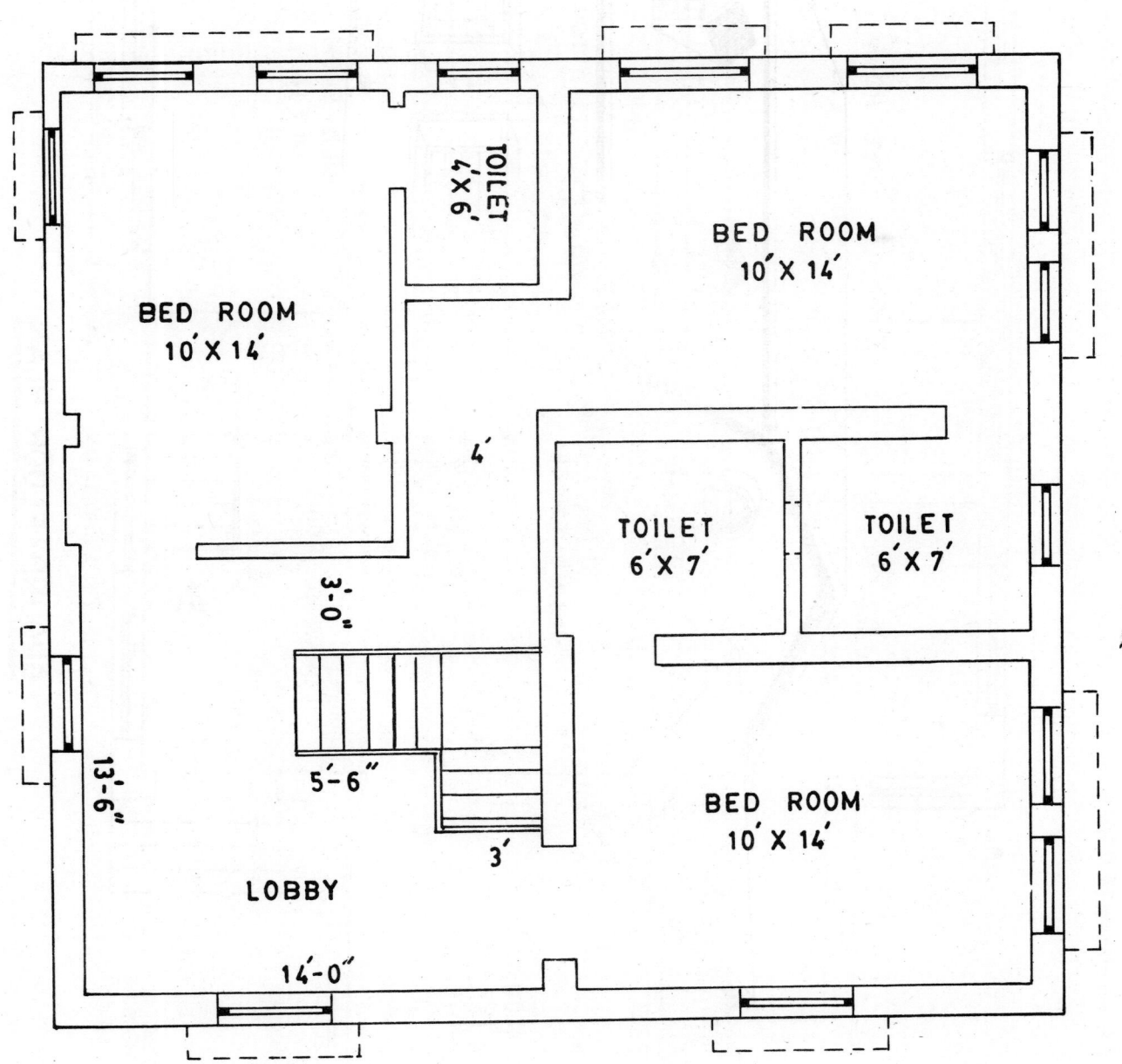

FIRST FLOOR PLAN

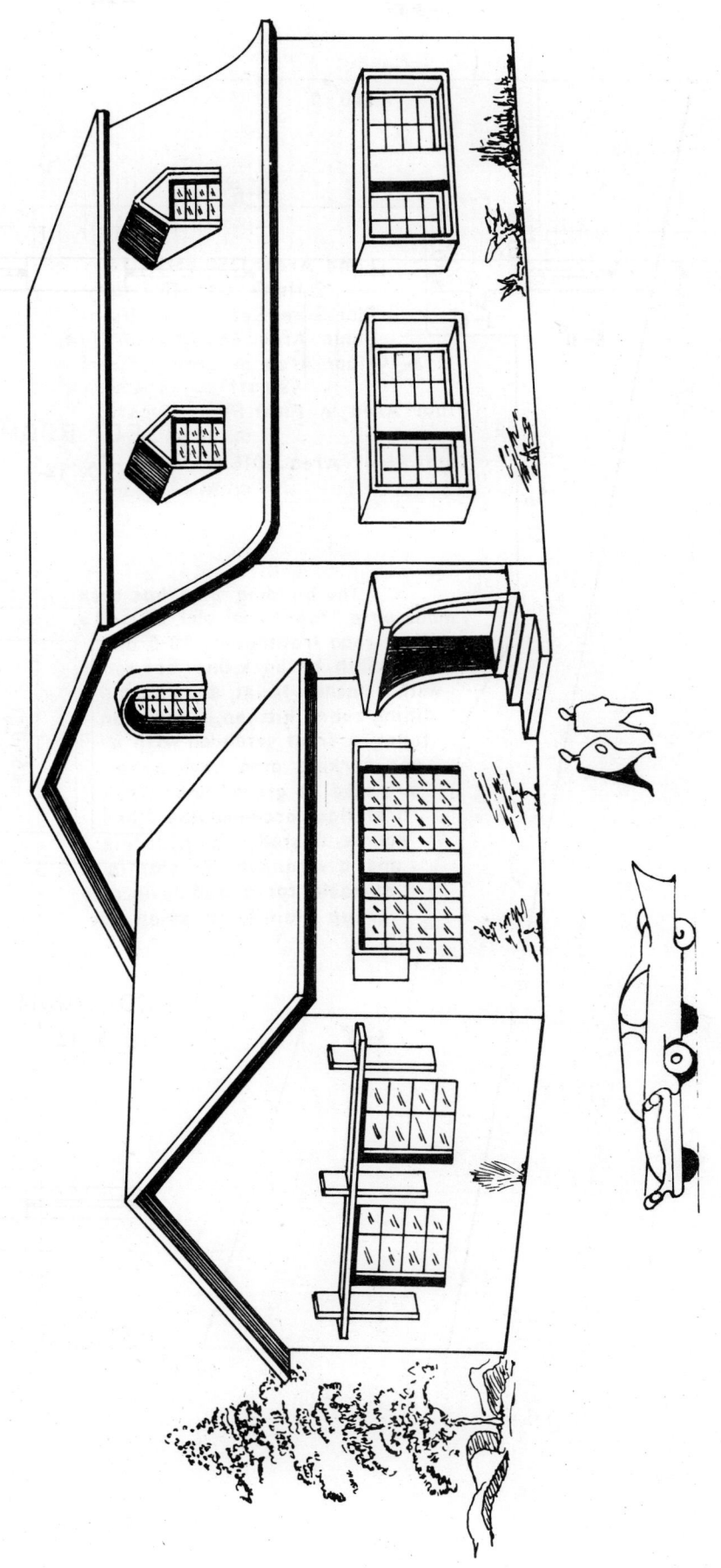

PLAN NO. 11

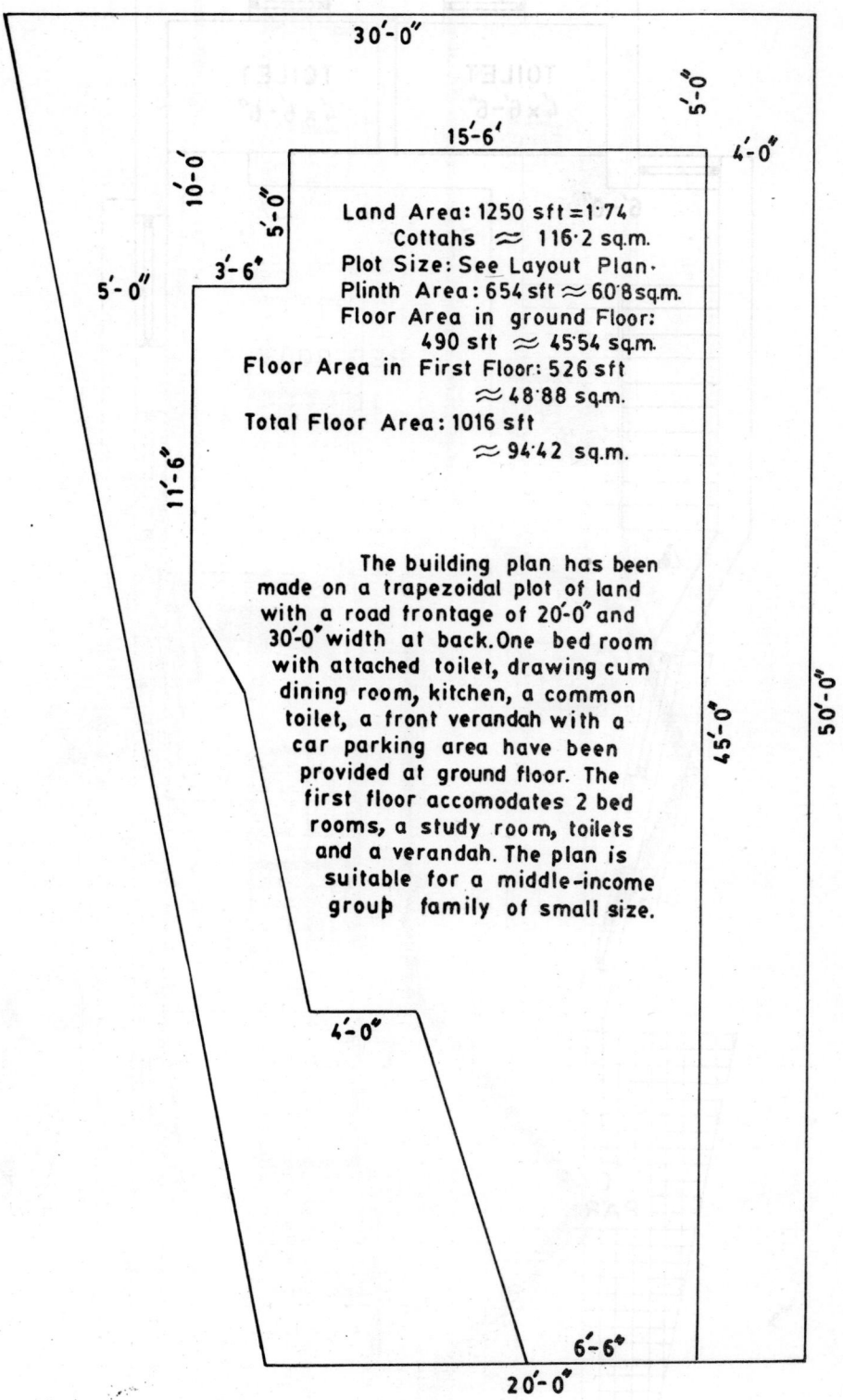

30'-0"

15'-6'

5'-0"

4'-0"

10'-0'

5'-0"

3'-6"

5'-0"

Land Area: 1250 sft = 1.74
Cottahs ≈ 116.2 sq.m.
Plot Size: See Layout Plan.
Plinth Area: 654 sft ≈ 60.8 sq.m.
Floor Area in ground Floor:
490 sft ≈ 45.54 sq.m.
Floor Area in First Floor: 526 sft
≈ 48.88 sq.m.
Total Floor Area: 1016 sft
≈ 94.42 sq.m.

11'-6"

The building plan has been
made on a trapezoidal plot of land
with a road frontage of 20'-0" and
30'-0" width at back. One bed room
with attached toilet, drawing cum
dining room, kitchen, a common
toilet, a front verandah with a
car parking area have been
provided at ground floor. The
first floor accomodates 2 bed
rooms, a study room, toilets
and a verandah. The plan is
suitable for a middle-income
group family of small size.

45'-0"

50'-0"

4'-0"

6'-6"

20'-0"

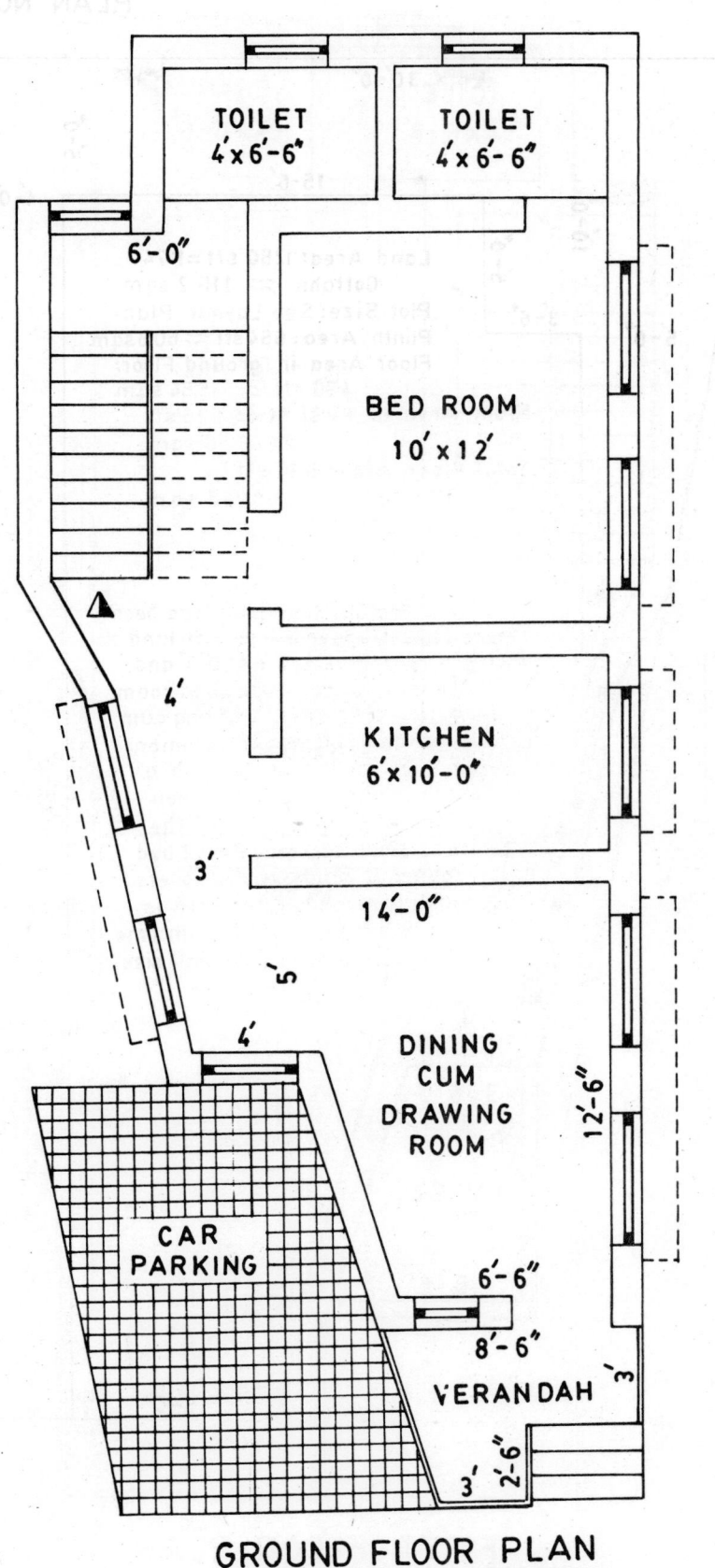

GROUND FLOOR PLAN

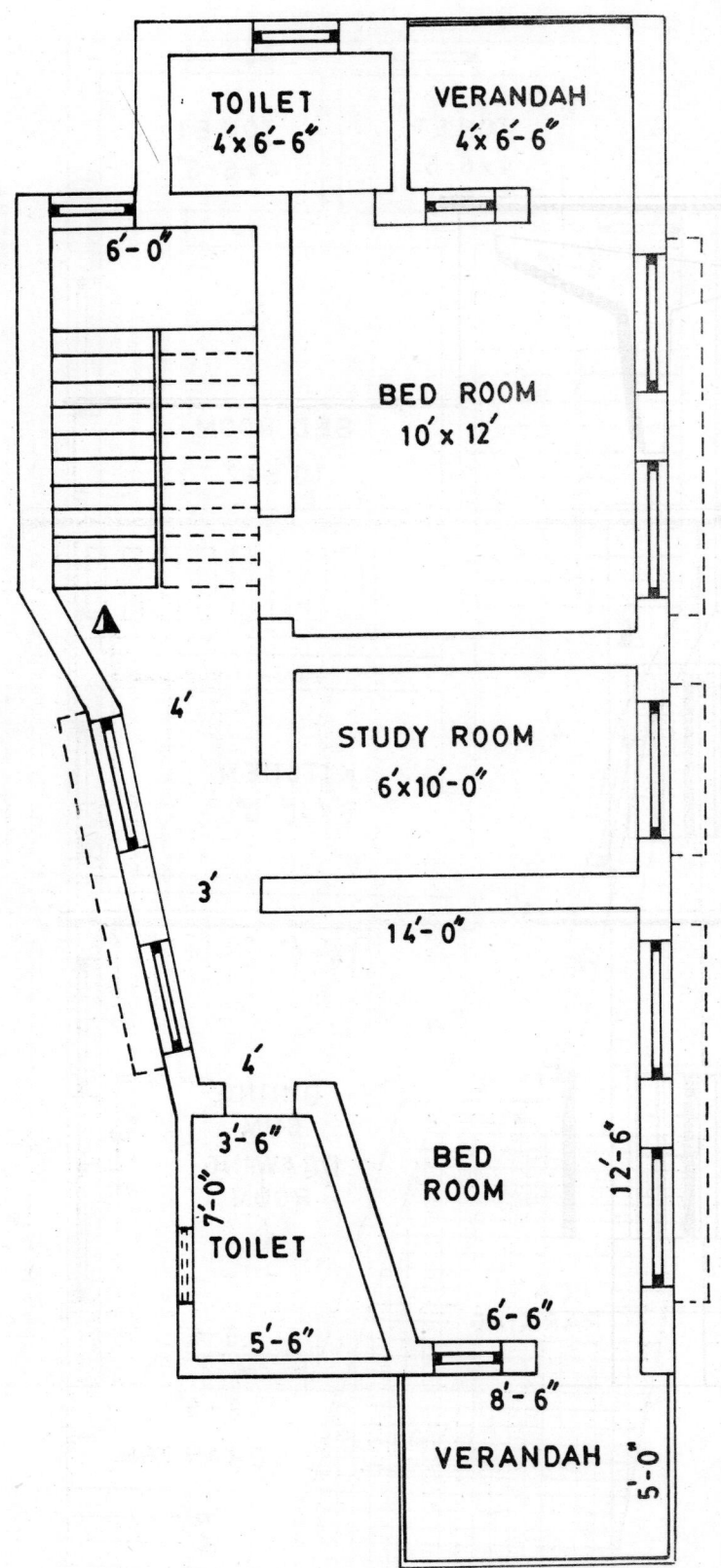

TOILET
4' x 6'-6"

VERANDAH
4' x 6'-6"

6'-0"

BED ROOM
10' x 12'

STUDY ROOM
6' x 10'-0"

4'

3'

14'-0"

4'

3'-6"

7'-0"

TOILET

BED ROOM

12'-6"

5'-6"

6'-6"

8'-6"

VERANDAH

5'-0"

FIRST FLOOR PLAN

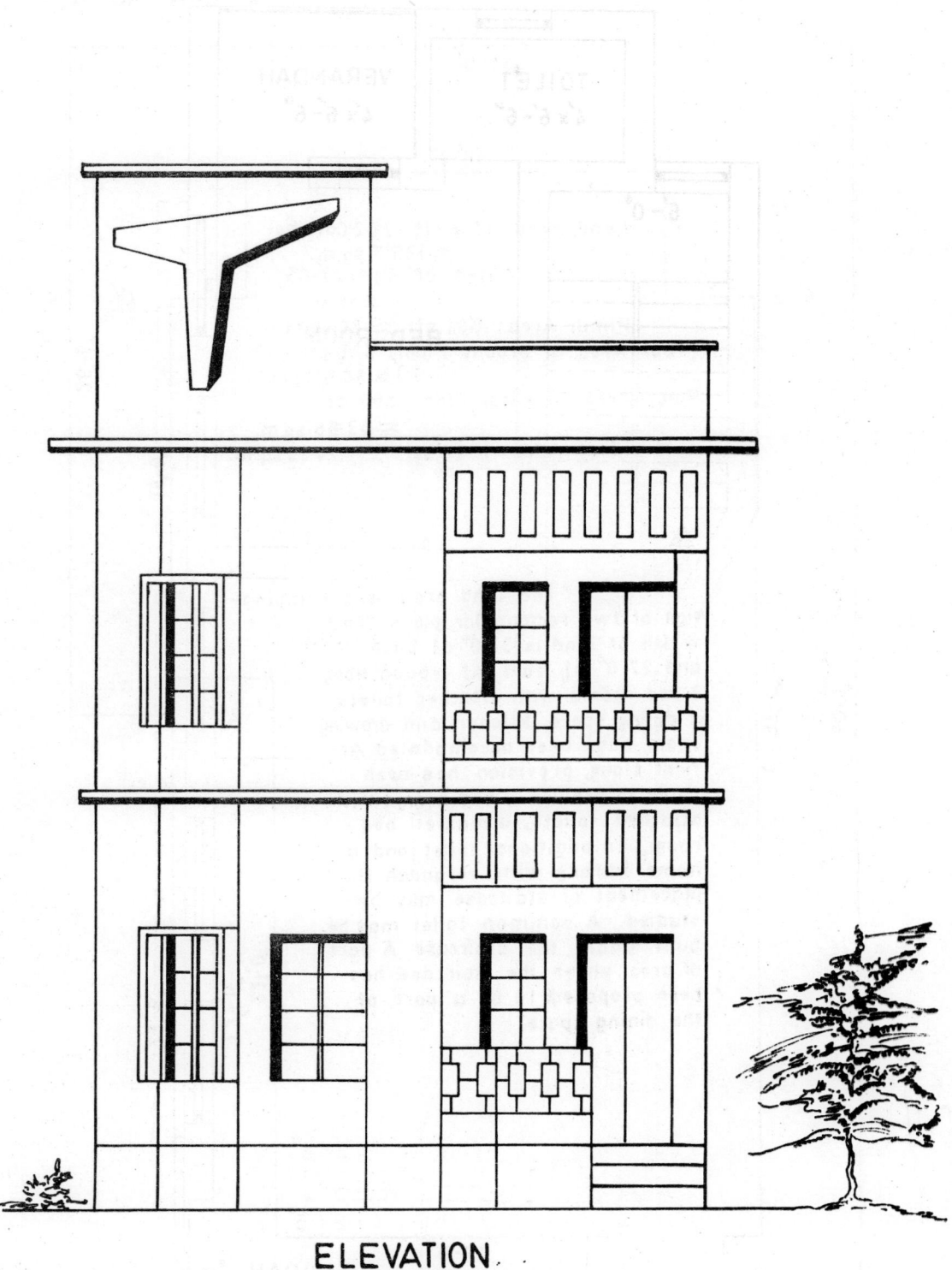

ELEVATION

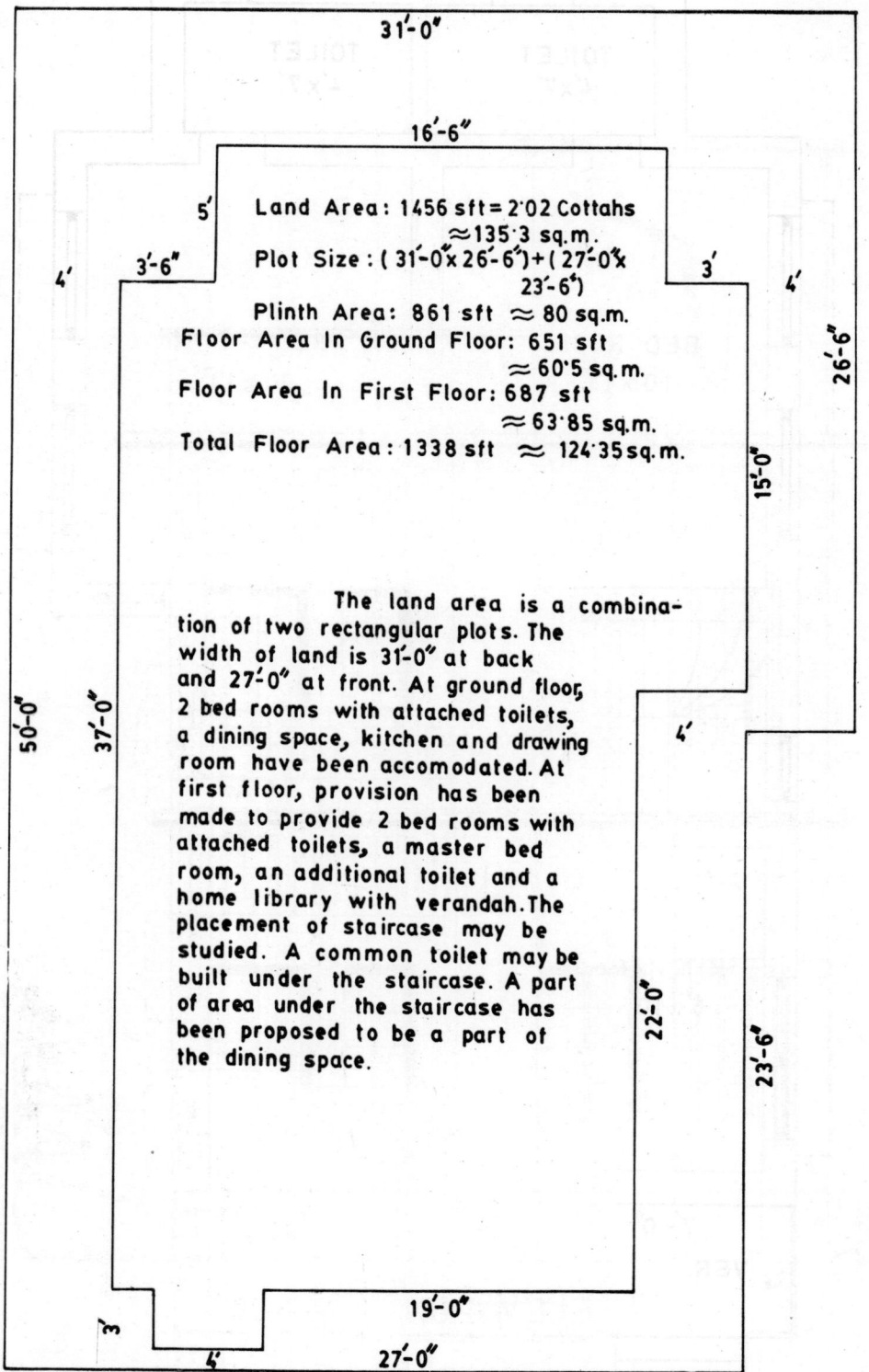

31'-0"

16'-6"

5'

Land Area : 1456 sft = 2'02 Cottahs
≈ 135'3 sq.m.
Plot Size : (31'-0"x 26'-6") + (27'-0"x
23'-6")
Plinth Area : 861 sft ≈ 80 sq.m.
Floor Area In Ground Floor : 651 sft
≈ 60'5 sq.m.
Floor Area In First Floor : 687 sft
≈ 63'85 sq.m.
Total Floor Area : 1338 sft ≈ 124'35 sq.m.

4'

3'-6"

3'

4'

26'-6"

15'-0"

     The land area is a combina-
tion of two rectangular plots. The
width of land is 31'-0" at back
and 27'-0" at front. At ground floor,
2 bed rooms with attached toilets,
a dining space, kitchen and drawing
room have been accomodated. At
first floor, provision has been
made to provide 2 bed rooms with
attached toilets, a master bed
room, an additional toilet and a
home library with verandah. The
placement of staircase may be
studied. A common toilet may be
built under the staircase. A part
of area under the staircase has
been proposed to be a part of
the dining space.

4'

50'-0"

37'-0"

22'-0"

23'-6"

3'

19'-0"

4'

27'-0"

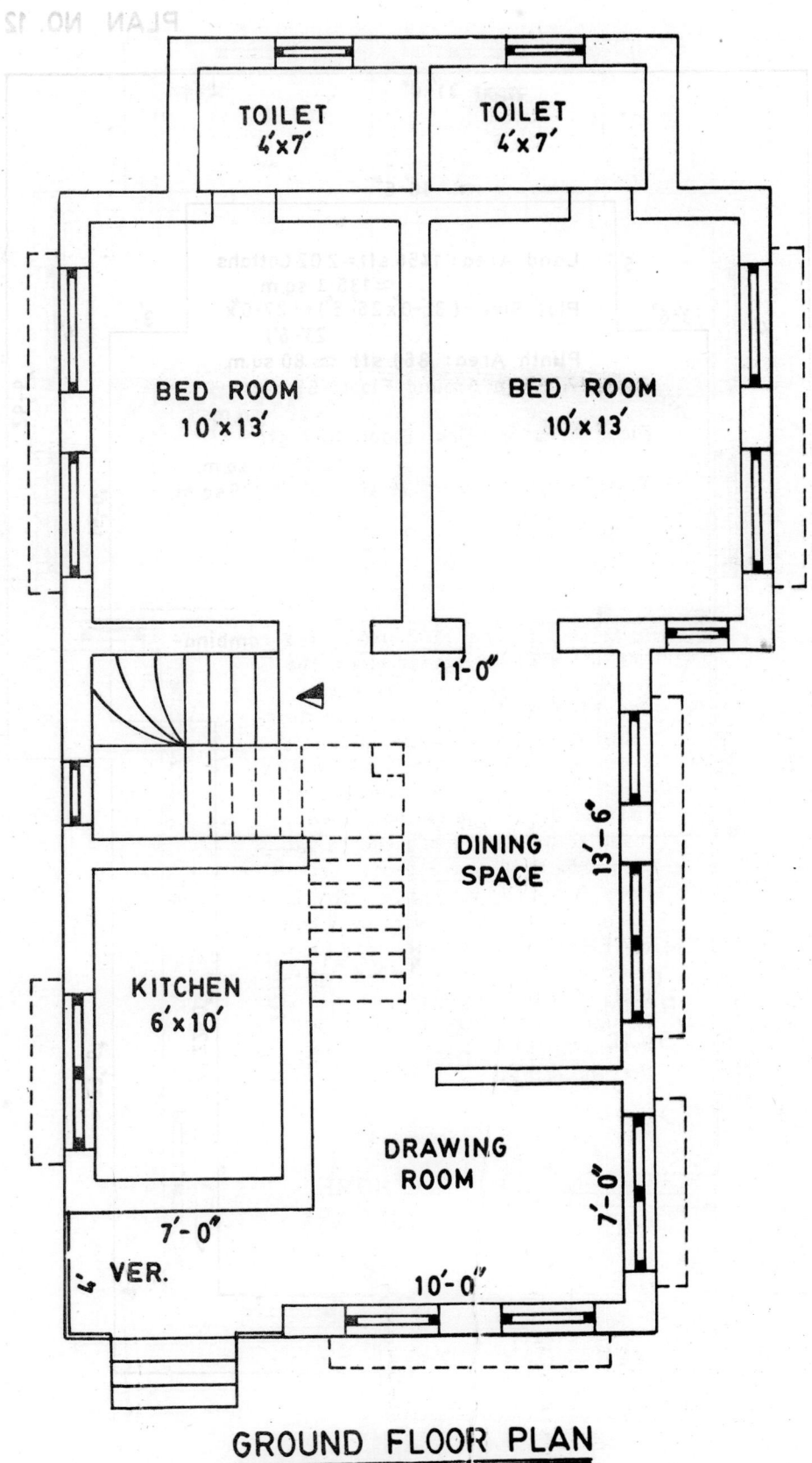

GROUND FLOOR PLAN

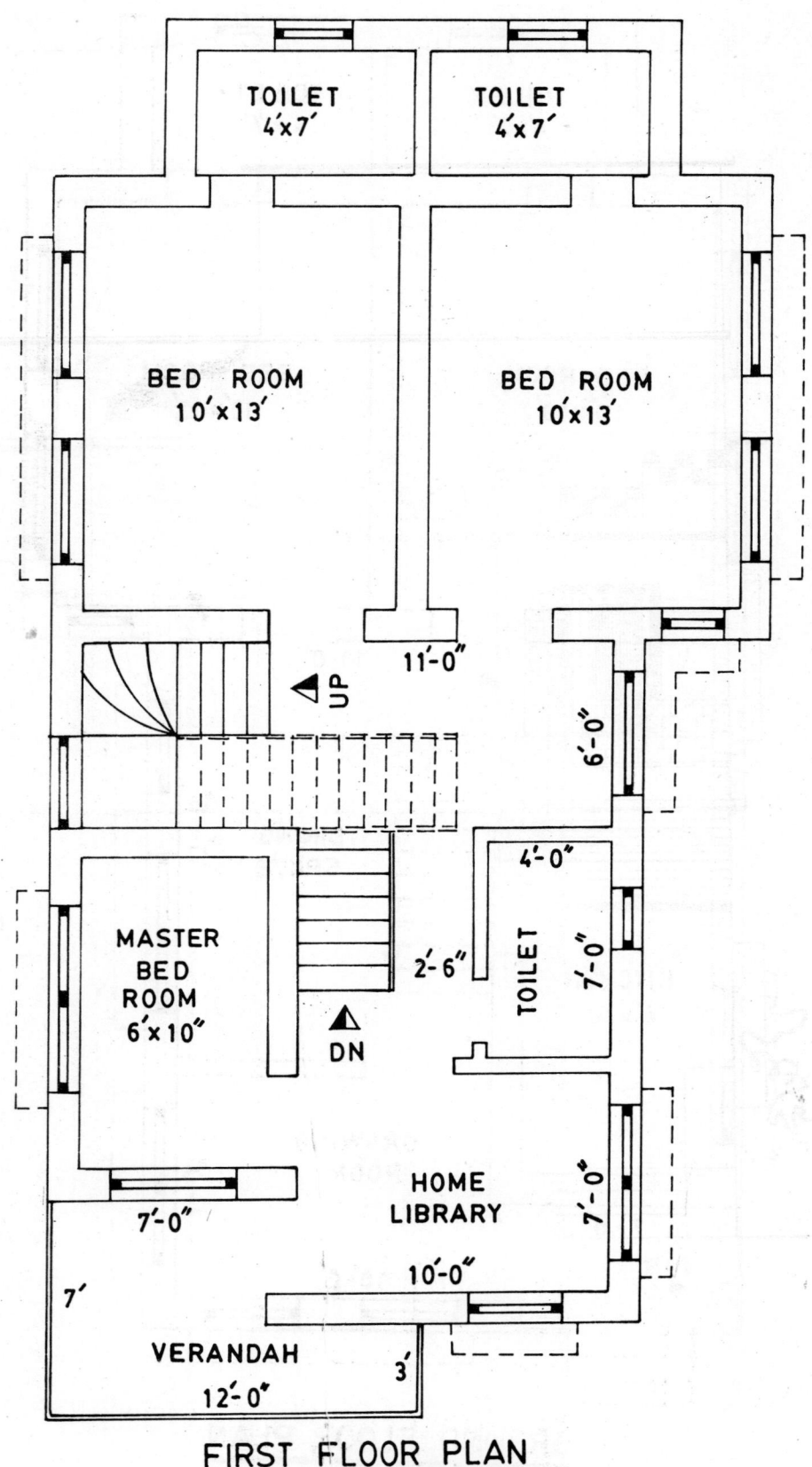

TOILET
4'x7'

TOILET
4'x7'

BED ROOM
10'x13'

BED ROOM
10'x13'

11'-0"

UP

6'-0"

MASTER
BED
ROOM
6'x10'

2'-6"

DN

TOILET

4'-0"

7'-0"

7'-0"

HOME
LIBRARY

7'-0"

10'-0"

7'-0"

7'

VERANDAH

3'

12'-0"

FIRST FLOOR PLAN

ELEVATION

PLAN NO. 13

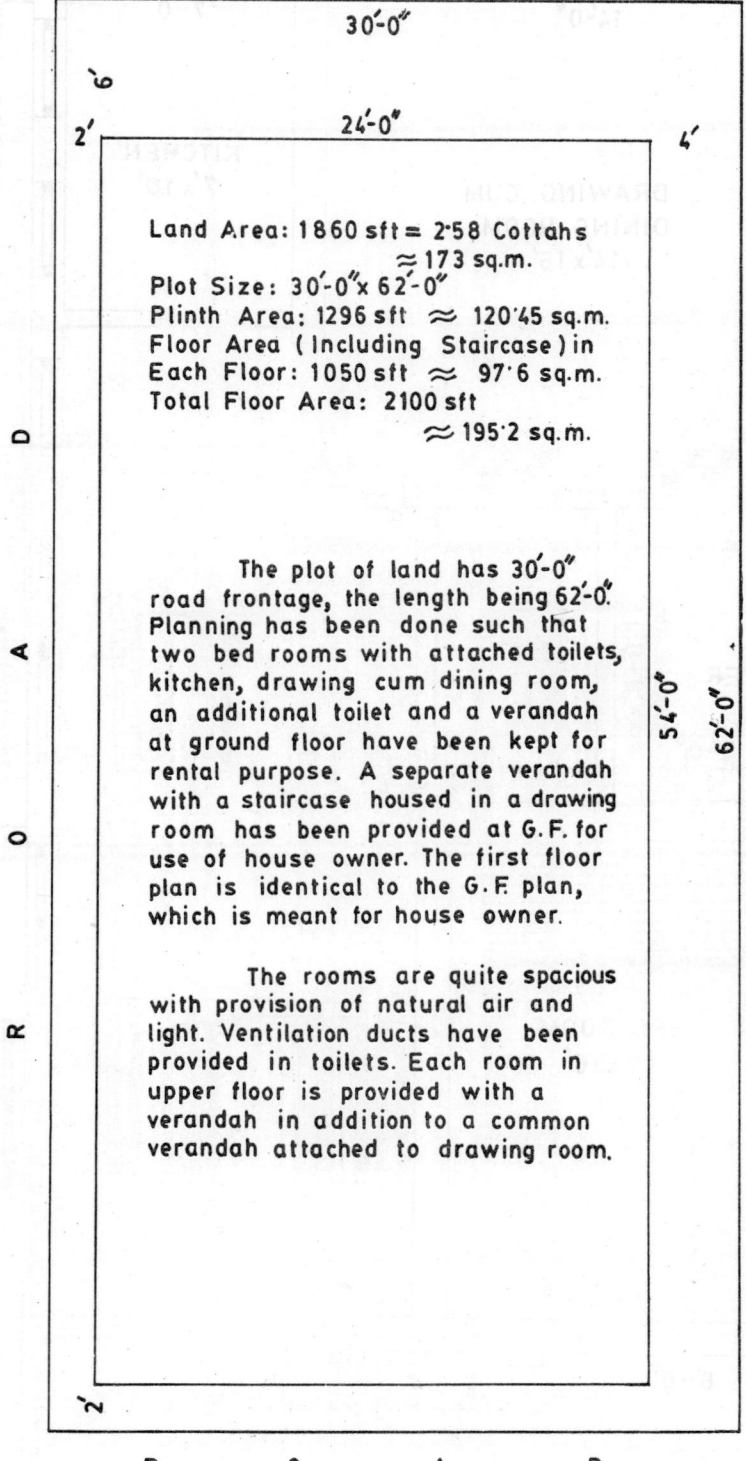

Land Area: 1860 sft = 2·58 Cottahs
                    ≈ 173 sq.m.
Plot Size: 30'-0"x 62'-0"
Plinth Area: 1296 sft ≈ 120·45 sq.m.
Floor Area (Including Staircase) in
Each Floor: 1050 sft ≈ 97·6 sq.m.
Total Floor Area: 2100 sft
                        ≈ 195·2 sq.m.

The plot of land has 30'-0" road frontage, the length being 62'-0". Planning has been done such that two bed rooms with attached toilets, kitchen, drawing cum dining room, an additional toilet and a verandah at ground floor have been kept for rental purpose. A separate verandah with a staircase housed in a drawing room has been provided at G.F. for use of house owner. The first floor plan is identical to the G.F. plan, which is meant for house owner.

The rooms are quite spacious with provision of natural air and light. Ventilation ducts have been provided in toilets. Each room in upper floor is provided with a verandah in addition to a common verandah attached to drawing room.

R   O   A   D

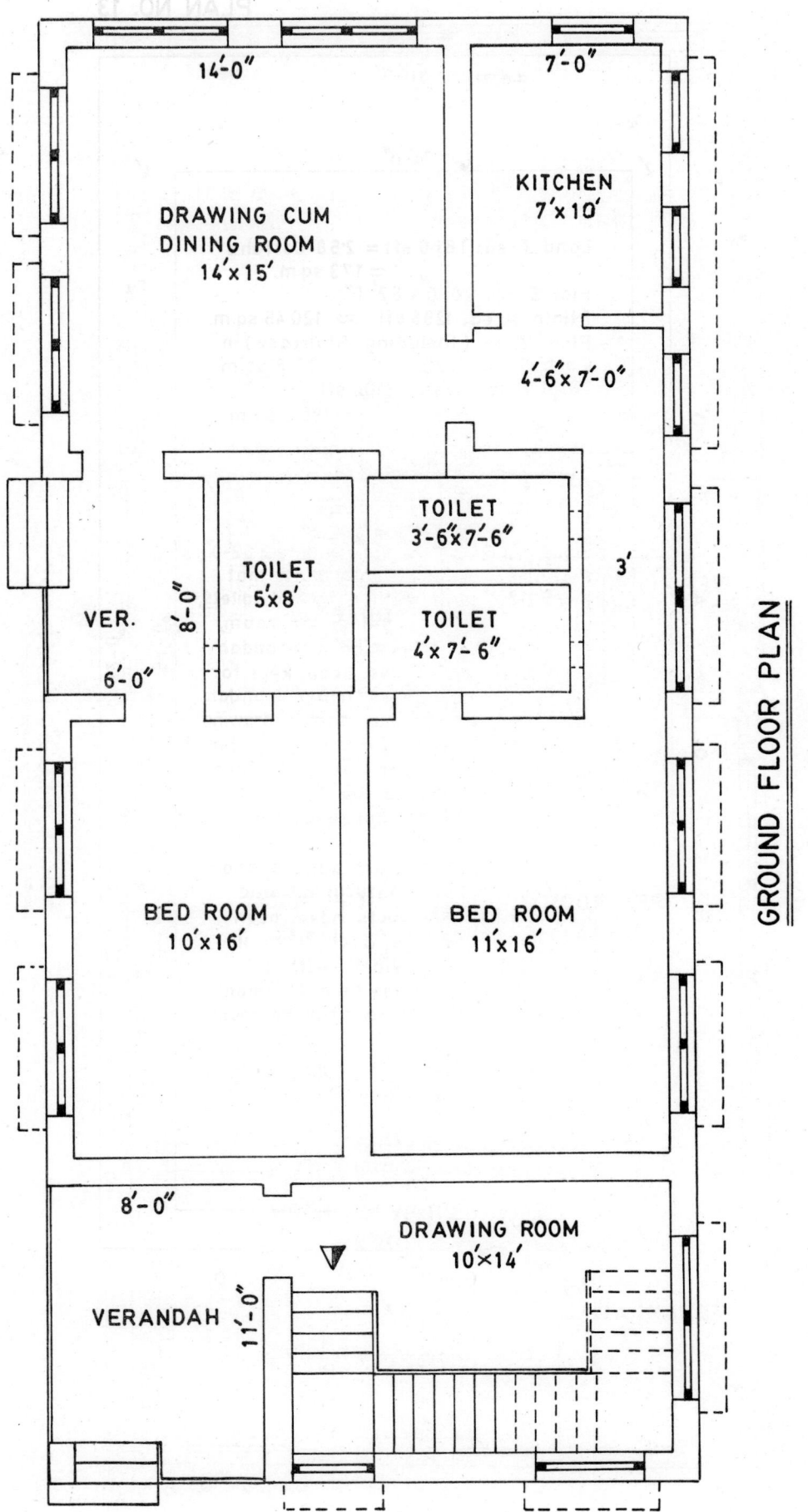

DRAWING CUM
DINING ROOM
14'x15'

14'-0"

KITCHEN
7'x10'

7'-0"

4'-6"x7'-0"

TOILET
3'-6"x7'-6"

TOILET
5'x8'

TOILET
4'x7'-6"

3'

8'-0"

VER.

6'-0"

BED ROOM
10'x16'

BED ROOM
11'x16'

GROUND FLOOR PLAN

8'-0"

DRAWING ROOM
10'x14'

11'-0"

VERANDAH

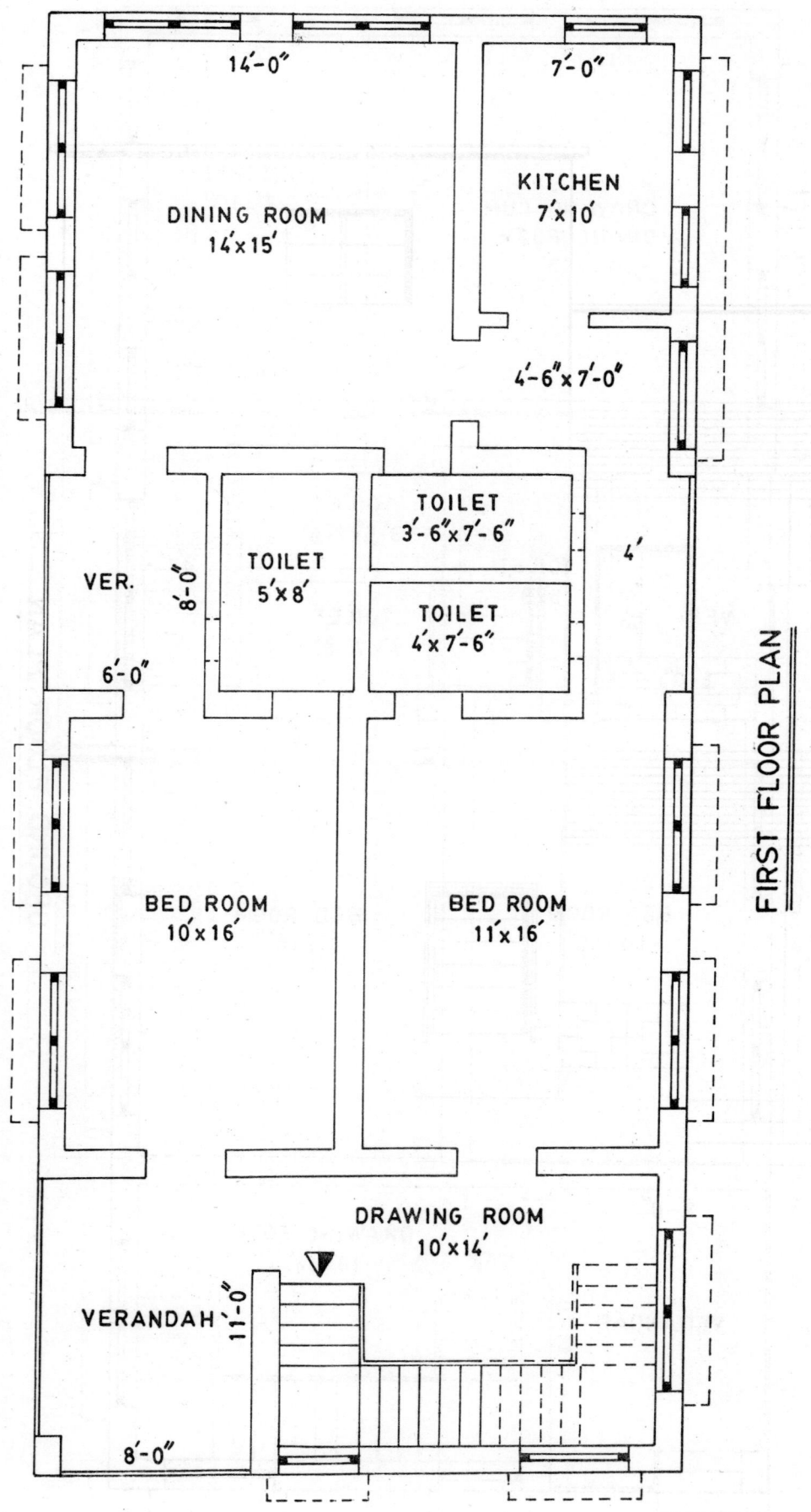

14'-0"

7'-0"

KITCHEN
7'x10'

DINING ROOM
14'x15'

4'-6"x7'-0"

TOILET
3'-6"x7'-6"

8'-0"

VER.

TOILET
5'x8'

TOILET
4'x7'-6"

4'

6'-0"

BED ROOM
10'x16'

BED ROOM
11'x16'

FIRST FLOOR PLAN

DRAWING ROOM
10'x14'

VERANDAH

11'-0"

8'-0"

ELEVATION

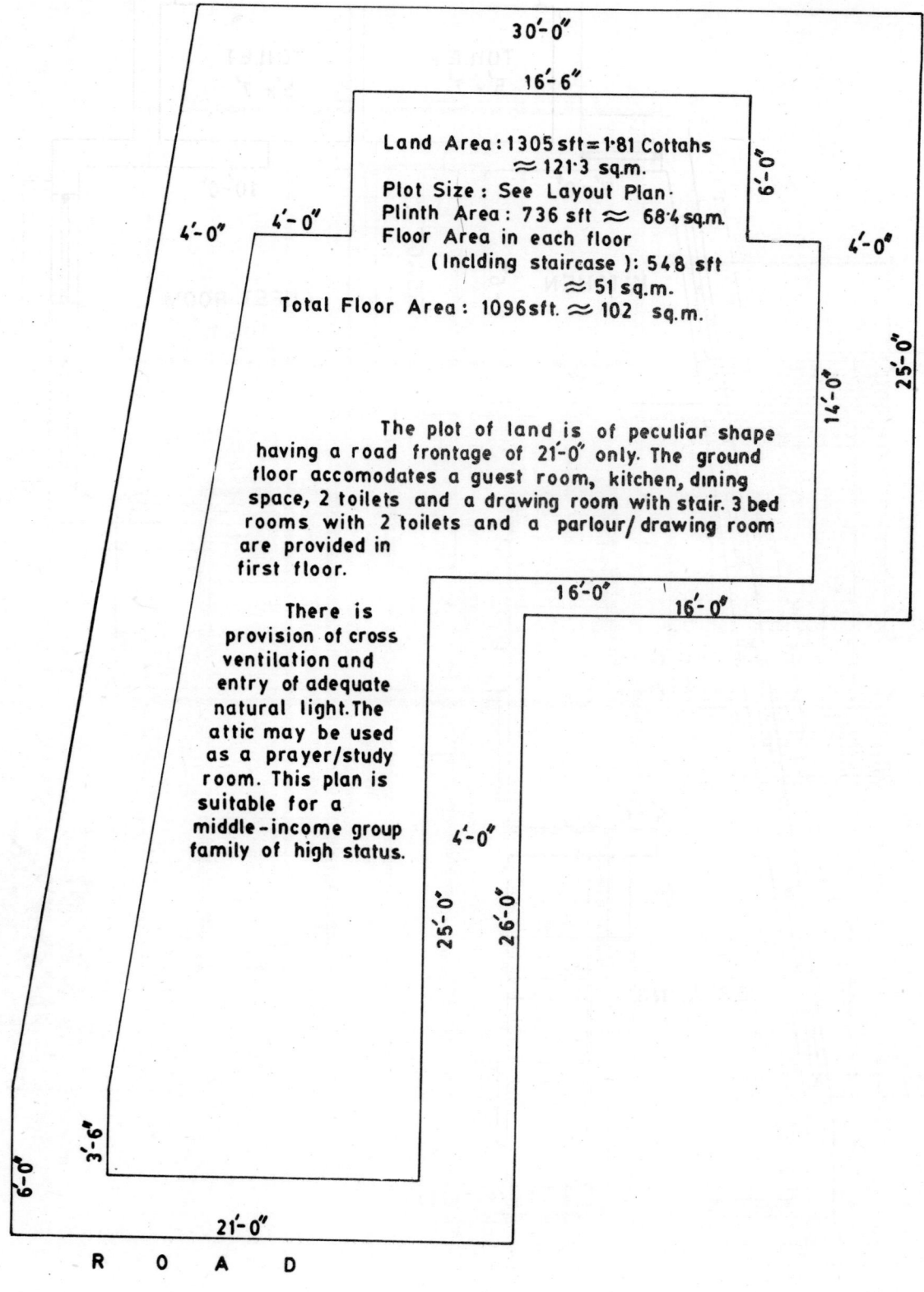

30'-0"

16'-6"

6'-0"

4'-0"   4'-0"

4'-0"

25'-0"

14'-0"

Land Area:1305 sft=1·81 Cottahs
≈ 121·3 sq.m.
Plot Size: See Layout Plan·
Plinth Area: 736 sft ≈ 68·4 sq.m.
Floor Area in each floor
(Inclding staircase): 548 sft
≈ 51 sq.m.
Total Floor Area: 1096 sft. ≈ 102 sq.m.

The plot of land is of peculiar shape
having a road frontage of 21'-0" only. The ground
floor accomodates a guest room, kitchen, dining
space, 2 toilets and a drawing room with stair. 3 bed
rooms with 2 toilets and a parlour/drawing room
are provided in
first floor.

There is
provision of cross
ventilation and
entry of adequate
natural light. The
attic may be used
as a prayer/study
room. This plan is
suitable for a
middle-income group
family of high status.

16'-0"   16'-0"

4'-0"

25'-0"   26'-0"

6'-0"

3'-6"

21'-0"

R   O   A   D

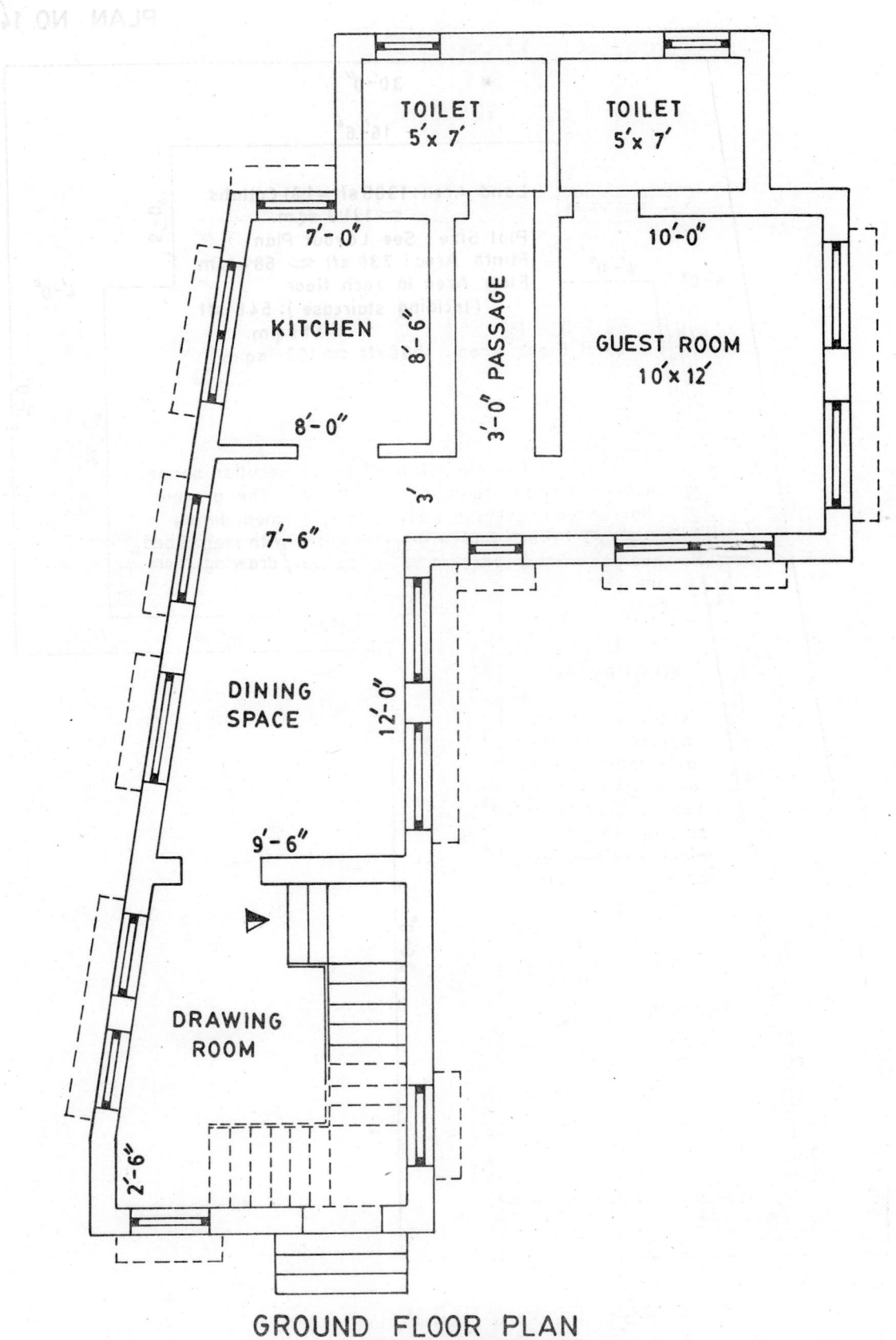

GROUND FLOOR PLAN

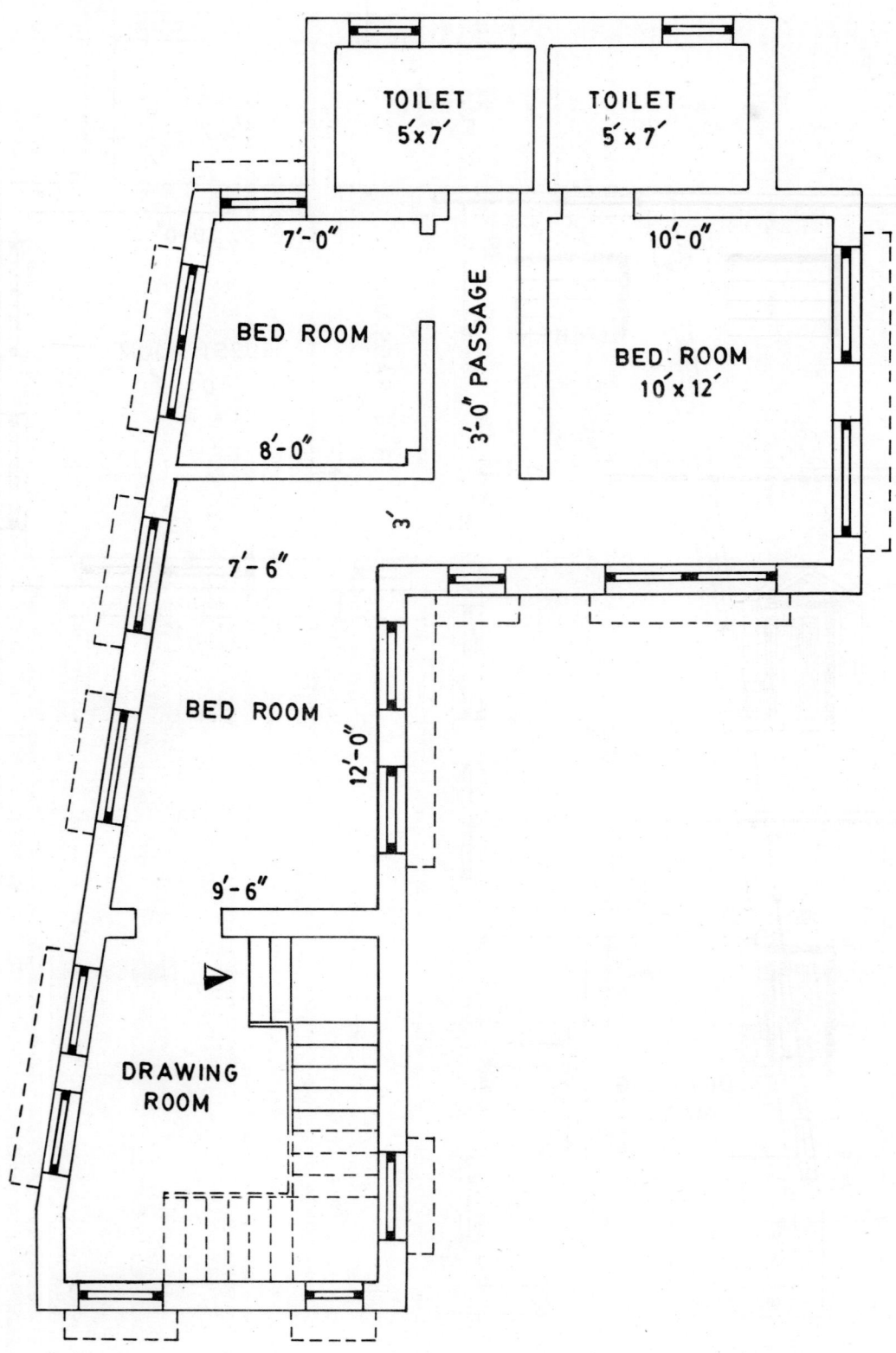

TOILET
5′ x 7′

TOILET
5′ x 7′

7′-0″

10′-0″

BED ROOM

3′-0″ PASSAGE

BED ROOM
10′ x 12′

8′-0″

3′

7′-6″

BED ROOM

12′-0″

9′-6″

DRAWING
ROOM

FIRST FLOOR PLAN

ELEVATION

PLAN NO. 15

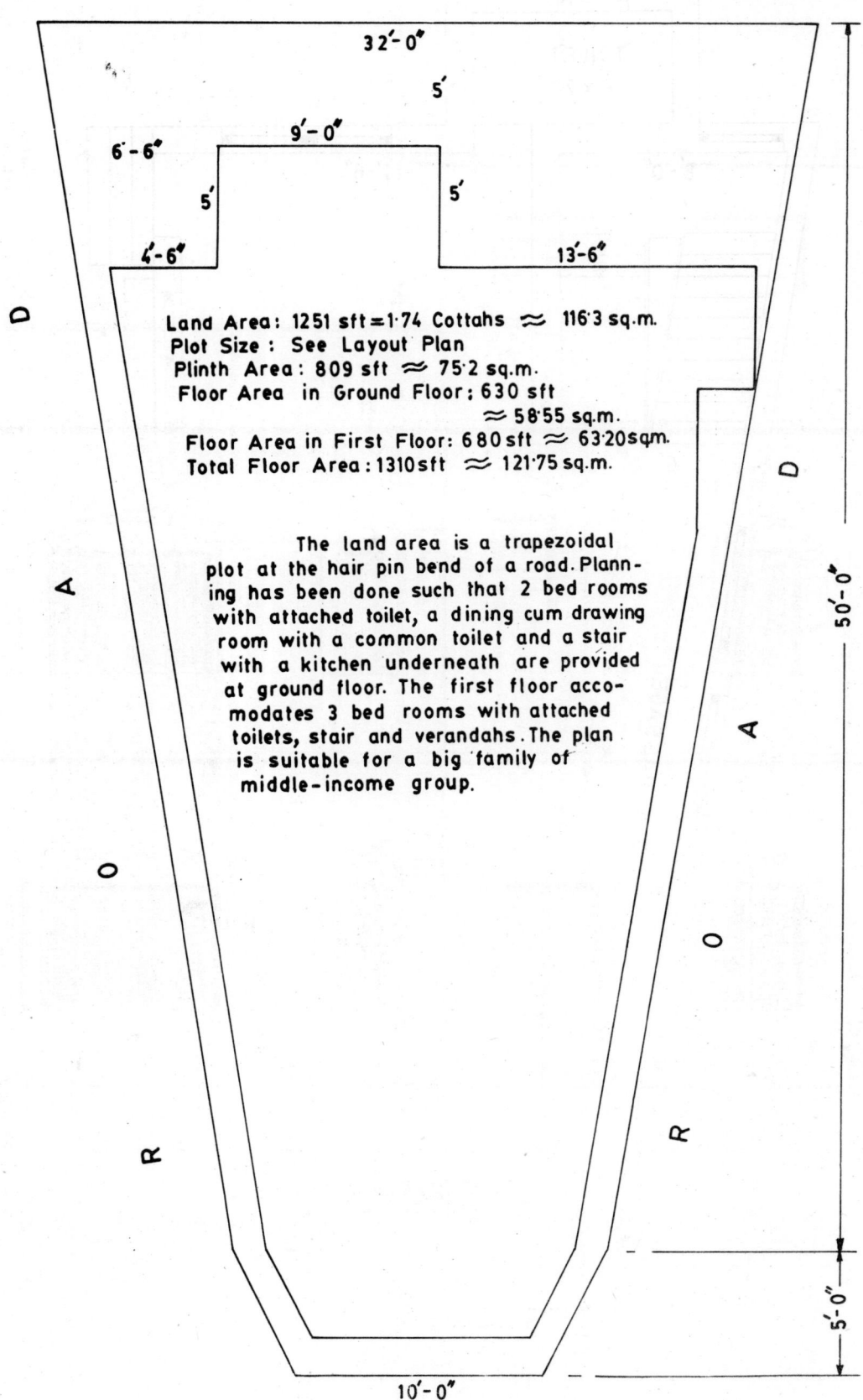

32'-0"

5'

9'-0"

6'-6"

5'   5'

4'-6"   13'-6"

Land Area: 1251 sft = 1·74 Cottahs ≈ 116·3 sq.m.
Plot Size : See Layout Plan
Plinth Area: 809 sft ≈ 75·2 sq.m.
Floor Area in Ground Floor: 630 sft
                              ≈ 58·55 sq.m.
Floor Area in First Floor: 680 sft ≈ 63·20 sq.m.
Total Floor Area : 1310 sft ≈ 121·75 sq.m.

The land area is a trapezoidal
plot at the hair pin bend of a road. Planning has been done such that 2 bed rooms
with attached toilet, a dining cum drawing
room with a common toilet and a stair
with a kitchen underneath are provided
at ground floor. The first floor accomodates 3 bed rooms with attached
toilets, stair and verandahs. The plan
is suitable for a big family of
middle-income group.

D A O R

D A O R

50'-0"

5'-0"

10'-0"

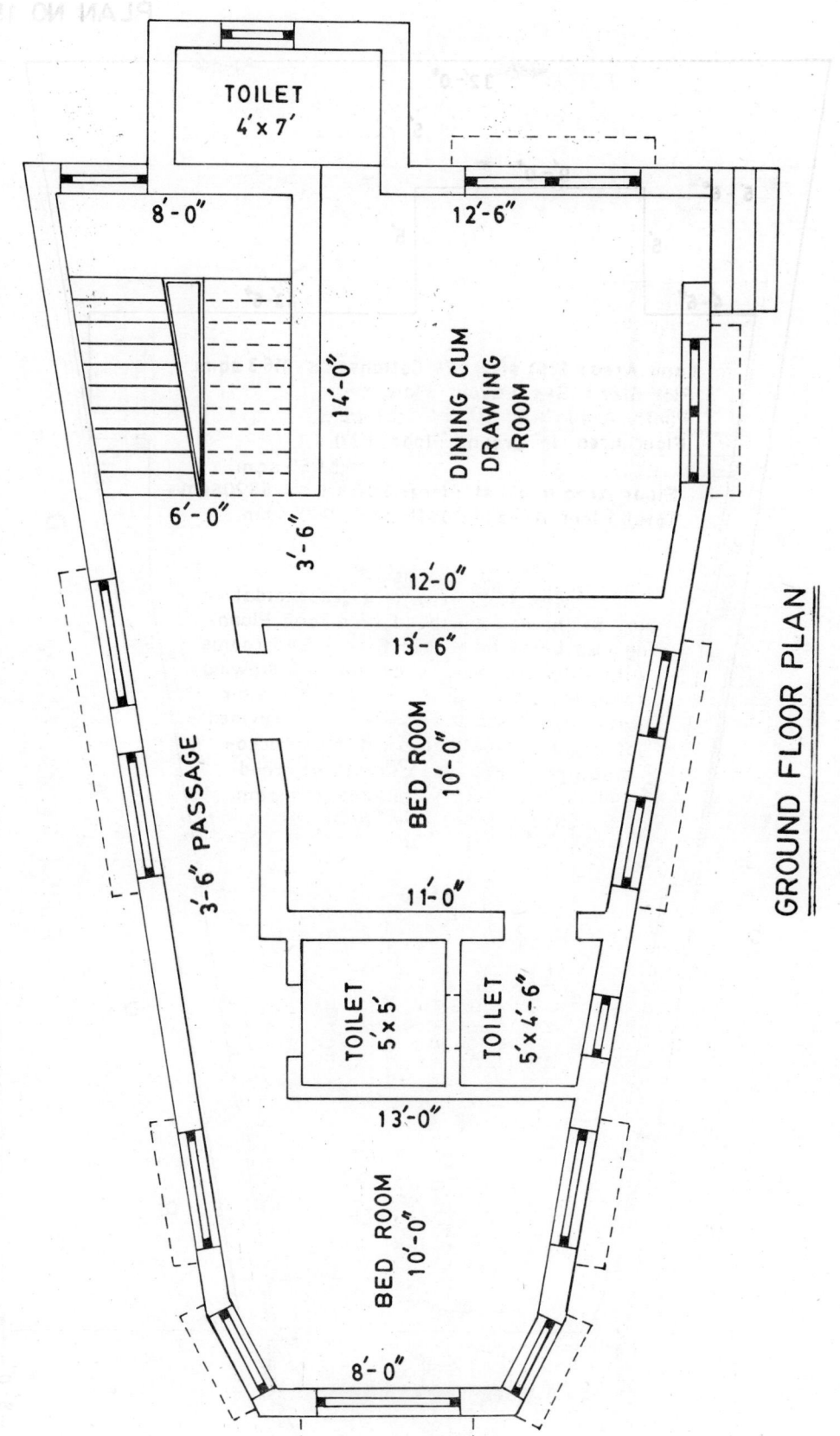

TOILET
4' x 7'

8'-0"

12'-6"

14'-0"

DINING CUM
DRAWING
ROOM

6'-0"

3'-6"

12'-0"

13'-6"

3'-6" PASSAGE

BED ROOM
10'-0"

11'-0"

TOILET
5' x 5'

TOILET
5' x 4'-6"

13'-0"

BED ROOM
10'-0"

8'-0"

GROUND FLOOR PLAN

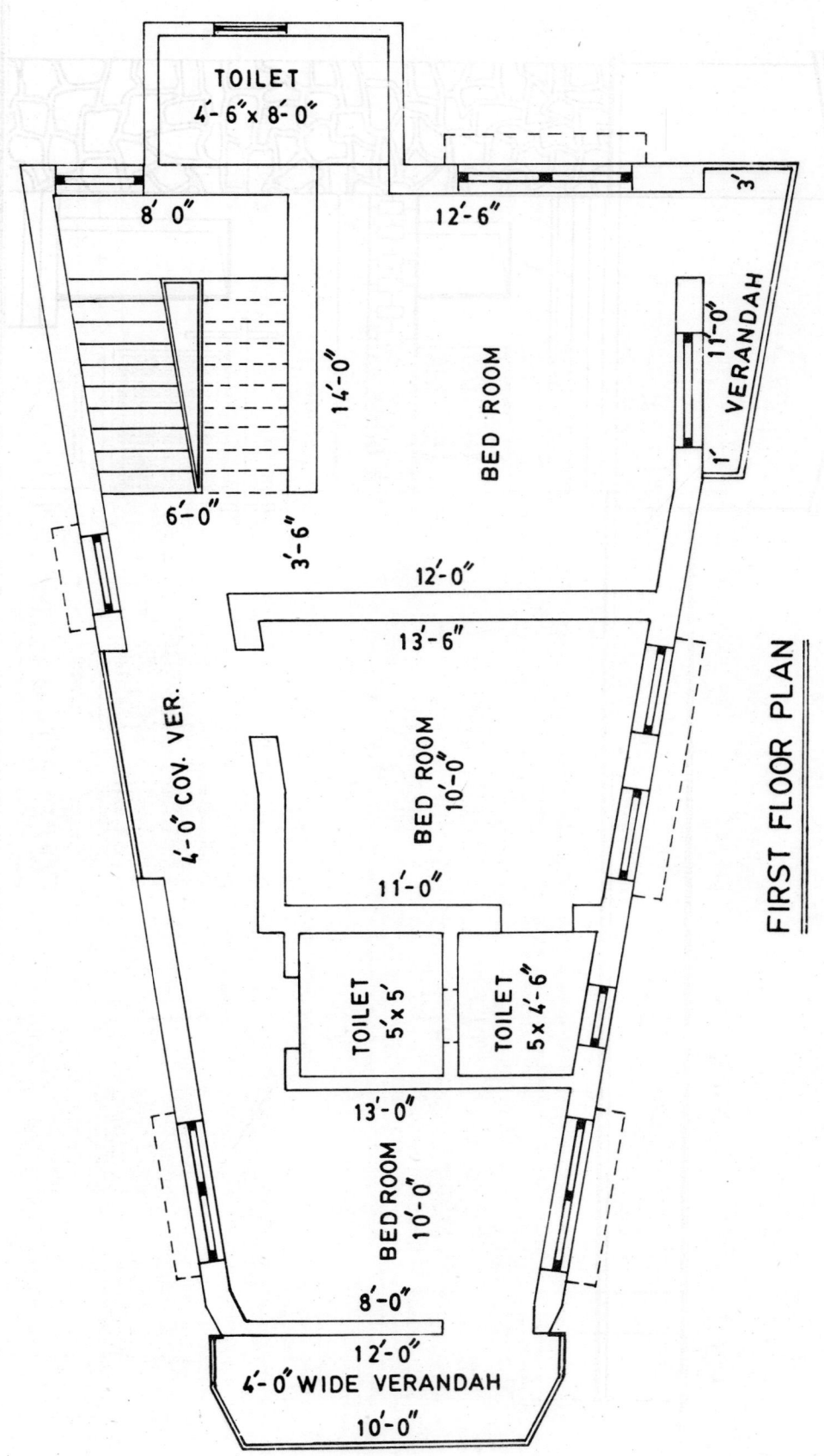

FIRST FLOOR PLAN

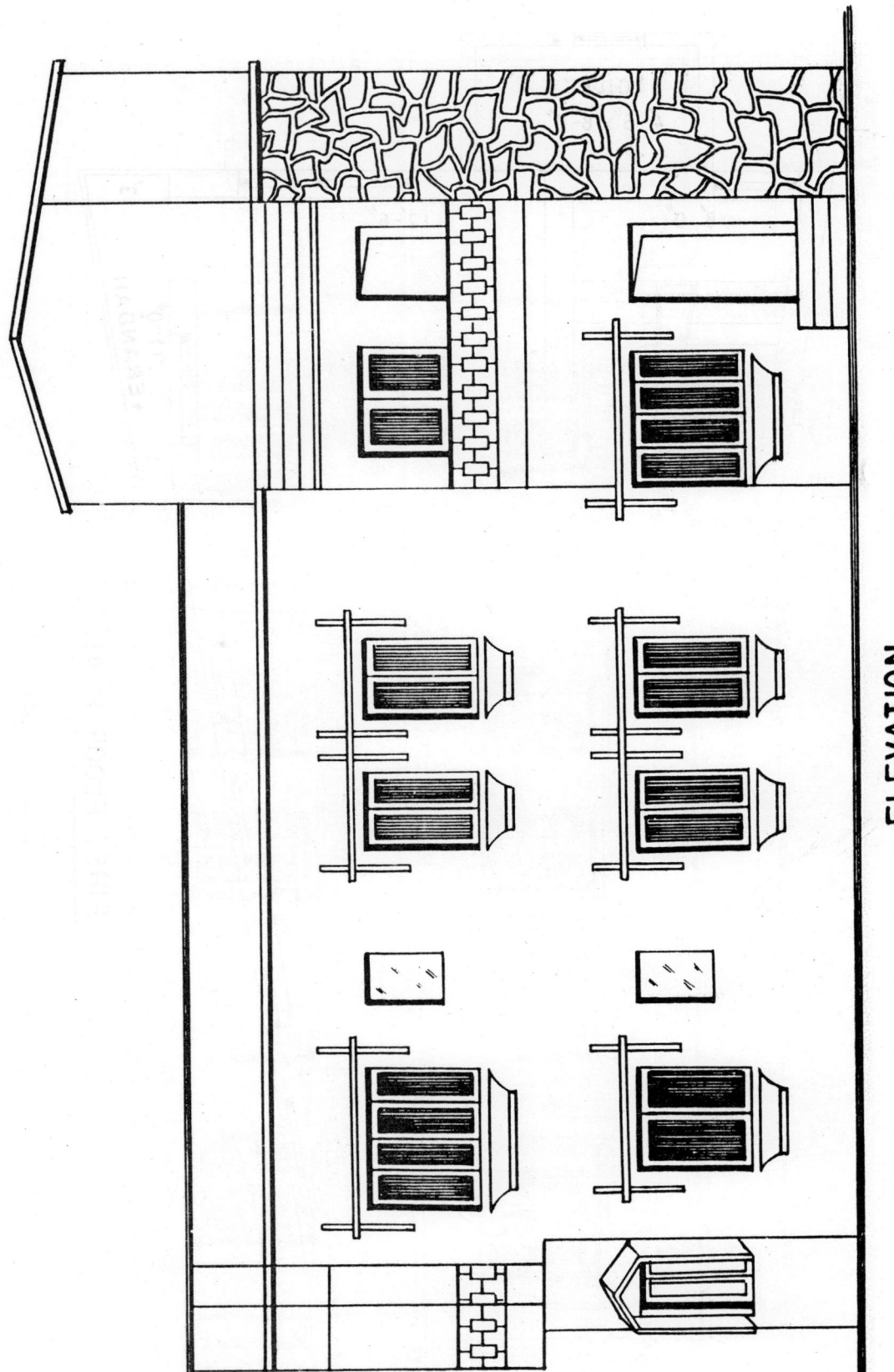

ELEVATION

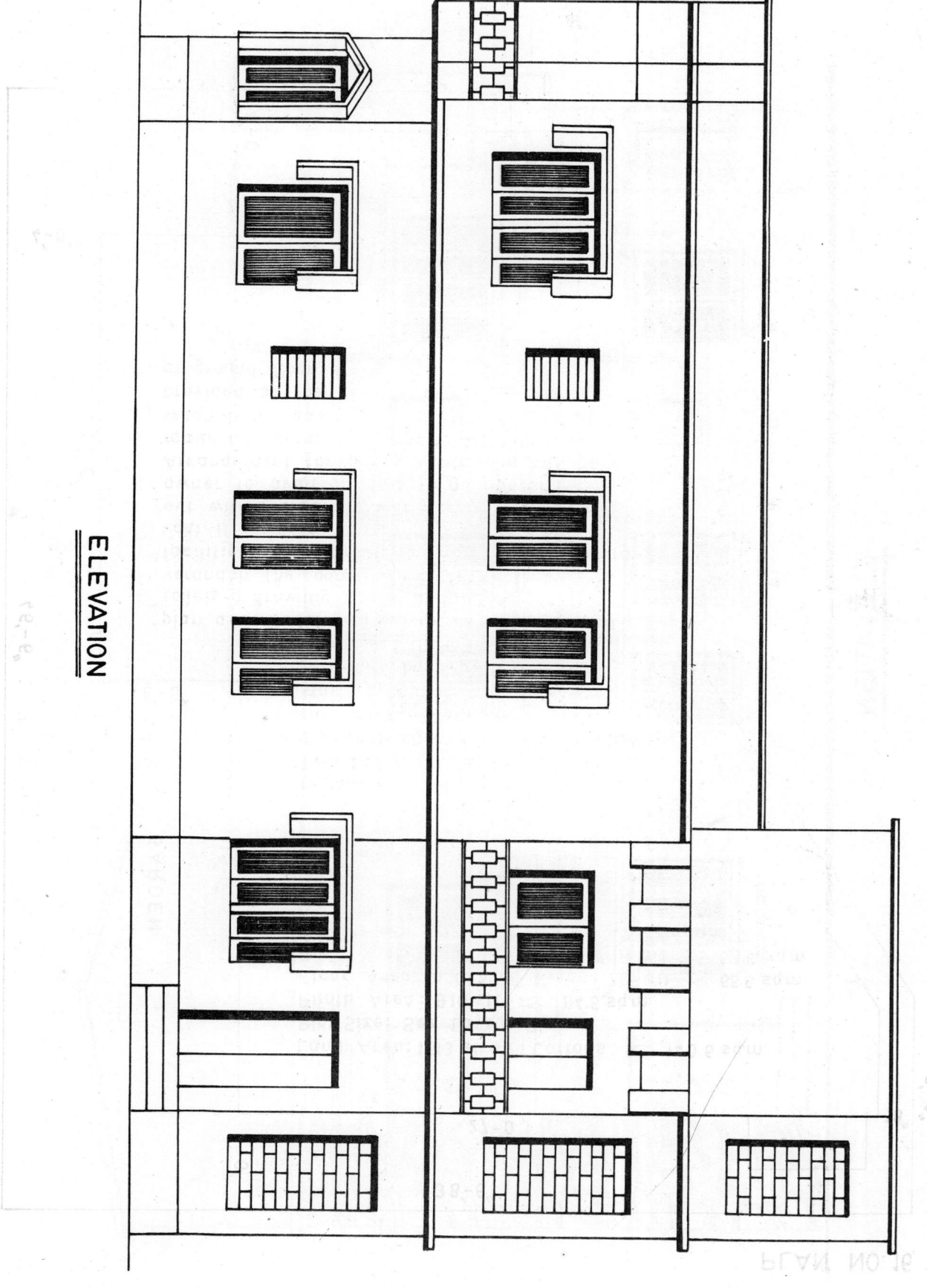

ELEVATION

PLAN NO. 16

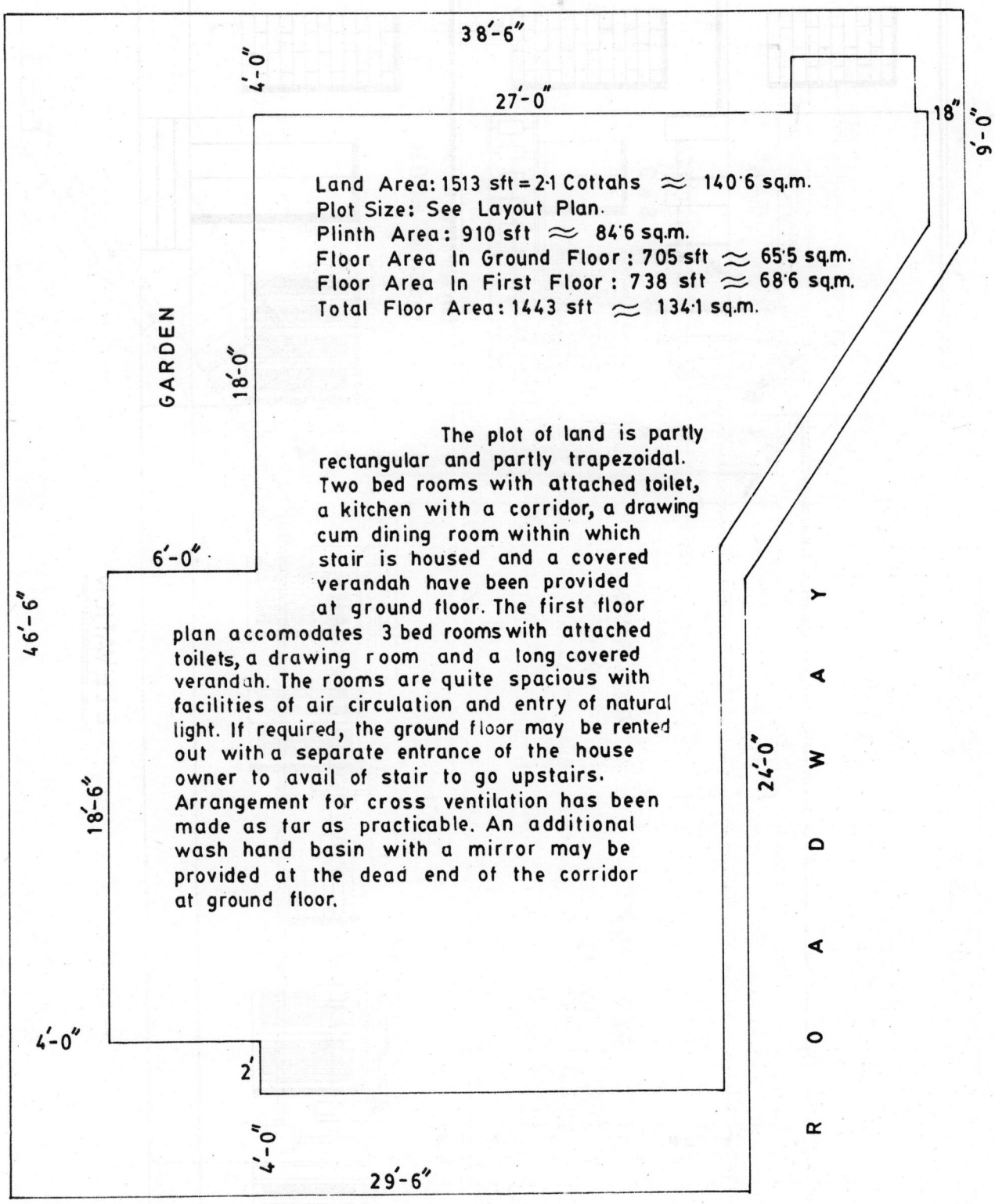

38'-6"

4'-0"

27'-0"

18"

6'-0"

Land Area: 1513 sft = 2·1 Cottahs ≈ 140·6 sq.m.
Plot Size: See Layout Plan.
Plinth Area: 910 sft ≈ 84·6 sq.m.
Floor Area In Ground Floor : 705 sft ≈ 65·5 sq.m.
Floor Area In First Floor : 738 sft ≈ 68·6 sq.m.
Total Floor Area: 1443 sft ≈ 134·1 sq.m.

GARDEN

18'-0"

6'-0"

46'-6"

18'-6"

         The plot of land is partly
rectangular and partly trapezoidal.
Two bed rooms with attached toilet,
a kitchen with a corridor, a drawing
cum dining room within which
stair is housed and a covered
verandah have been provided
at ground floor. The first floor
plan accomodates 3 bed rooms with attached
toilets, a drawing room and a long covered
verandah. The rooms are quite spacious with
facilities of air circulation and entry of natural
light. If required, the ground floor may be rented
out with a separate entrance of the house
owner to avail of stair to go upstairs.
Arrangement for cross ventilation has been
made as far as practicable. An additional
wash hand basin with a mirror may be
provided at the dead end of the corridor
at ground floor.

R O A D W A Y

24'-0"

4'-0"

2'

4'-0"

29'-6"

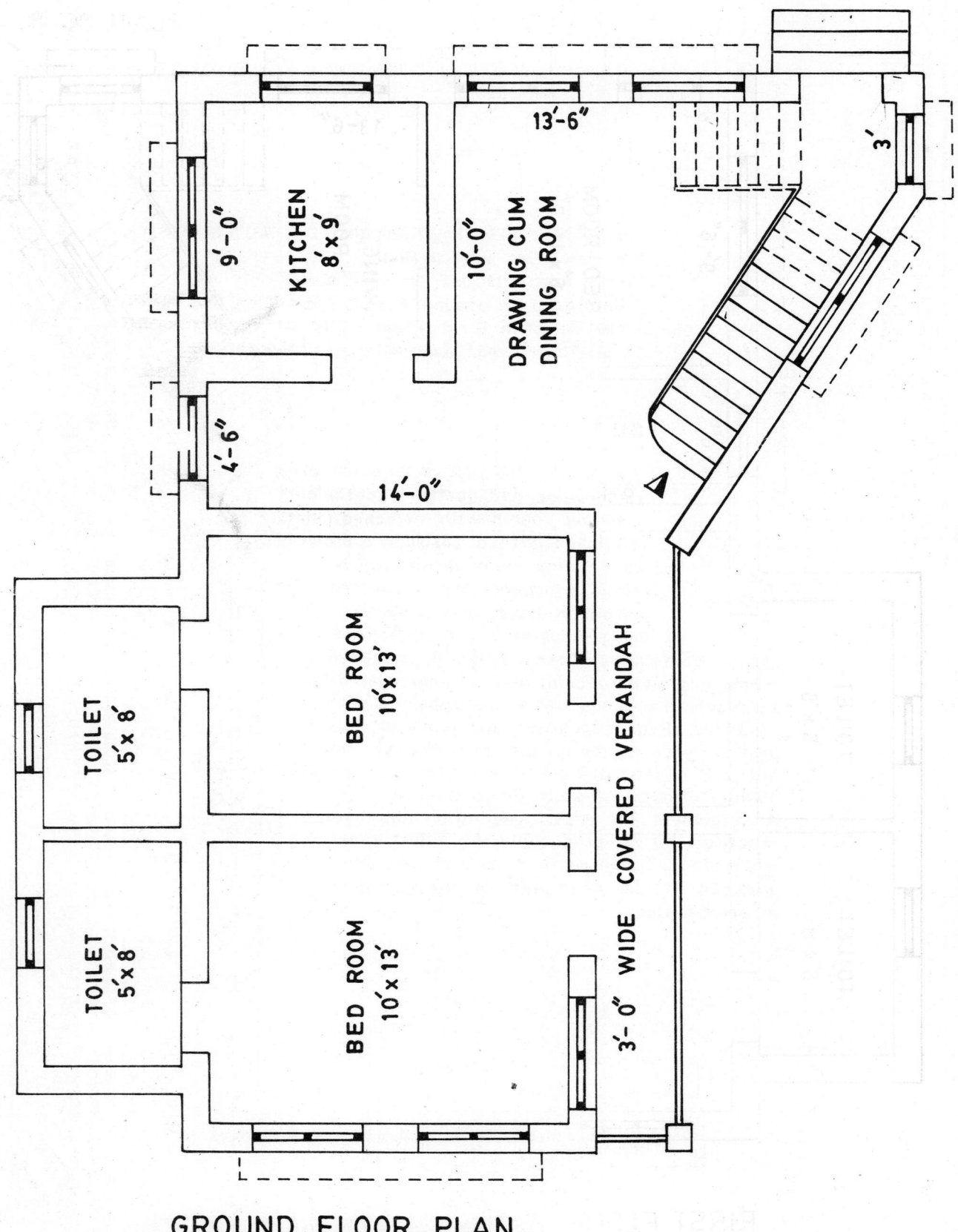

9'-0"

KITCHEN
8' x 9'

13'-6"

10'-0"

DRAWING CUM
DINING ROOM

3'

4'-6"

14'-0"

TOILET
5' x 8'

BED ROOM
10' x 13'

3'-0" WIDE COVERED VERANDAH

TOILET
5' x 8'

BED ROOM
10' x 13'

GROUND FLOOR PLAN

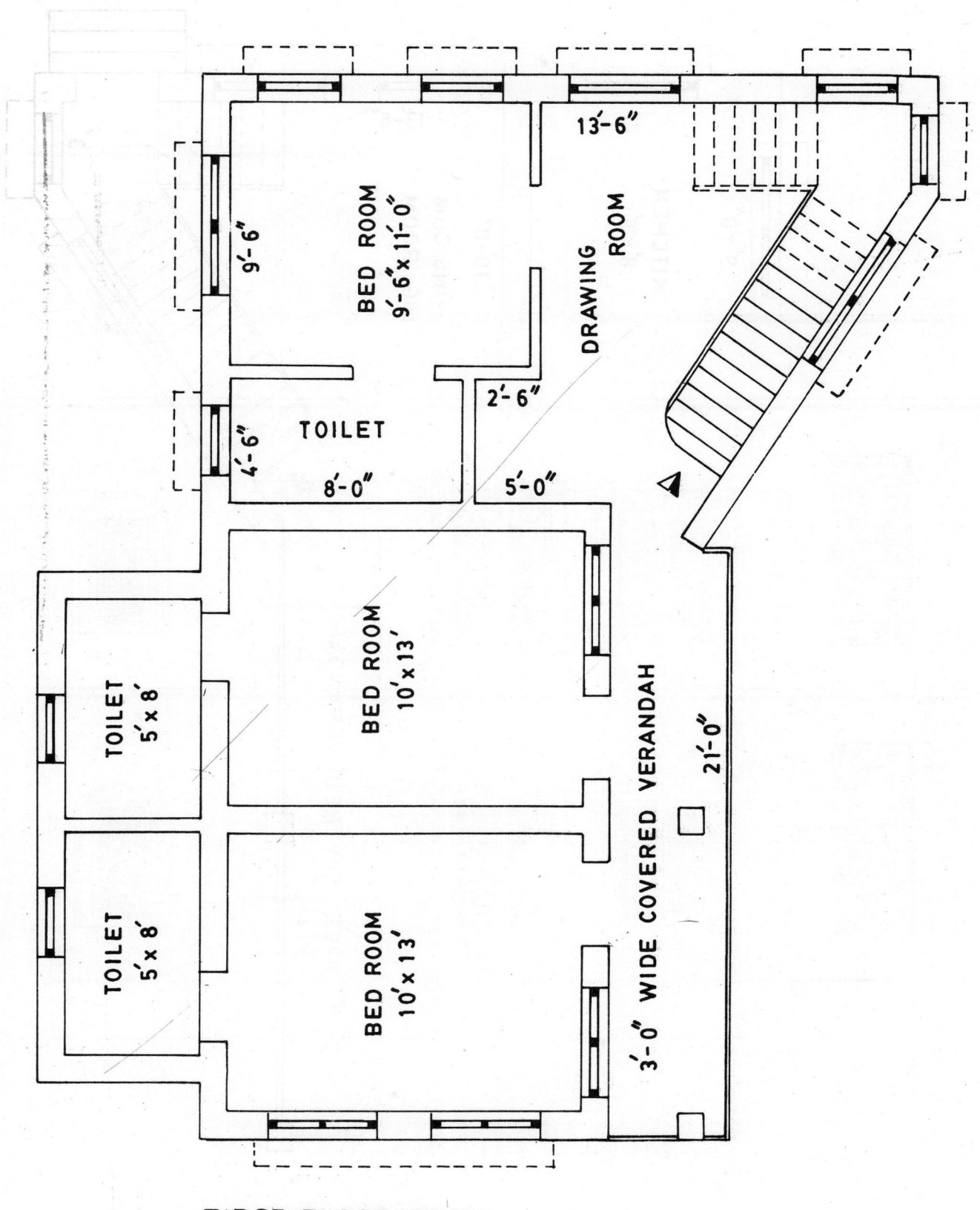

FIRST FLOOR PLAN

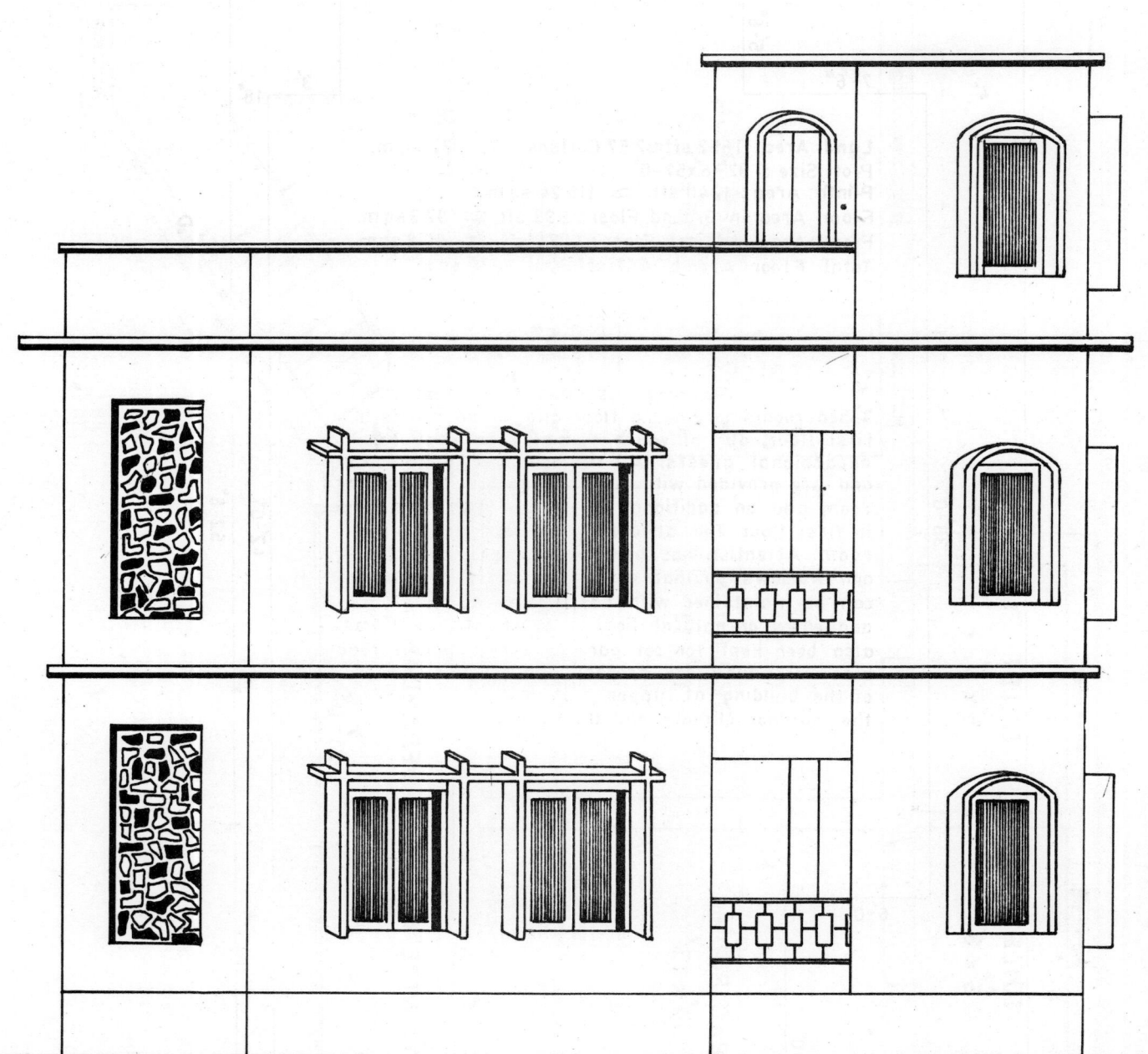

ELEVATION

PLAN NO. 17

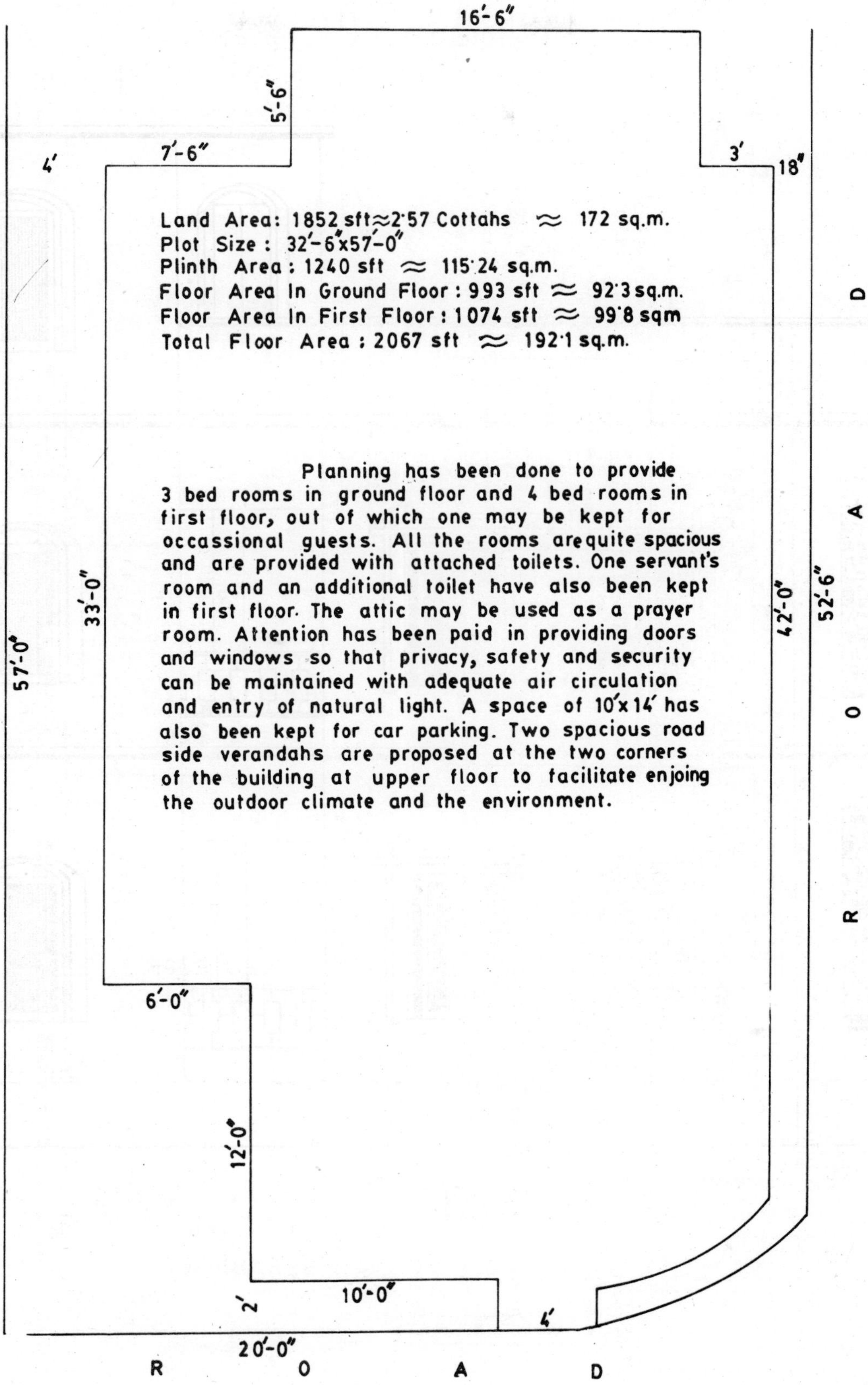

16'-6"

5'-6"

7'-6"   3'   18"

4'

Land Area: 1852 sft ≈ 2·57 Cottahs ≈ 172 sq.m.
Plot Size : 32'-6"x57'-0"
Plinth Area : 1240 sft ≈ 115·24 sq.m.
Floor Area In Ground Floor : 993 sft ≈ 92·3 sq.m.
Floor Area In First Floor : 1074 sft ≈ 99·8 sqm
Total Floor Area : 2067 sft ≈ 192·1 sq.m.

D

A

Planning has been done to provide
3 bed rooms in ground floor and 4 bed rooms in
first floor, out of which one may be kept for
occassional guests. All the rooms are quite spacious
and are provided with attached toilets. One servant's
room and an additional toilet have also been kept
in first floor. The attic may be used as a prayer
room. Attention has been paid in providing doors
and windows so that privacy, safety and security
can be maintained with adequate air circulation
and entry of natural light. A space of 10'x 14' has
also been kept for car parking. Two spacious road
side verandahs are proposed at the two corners
of the building at upper floor to facilitate enjoing
the outdoor climate and the environment.

O

42'-0"   52'-6"

33'-0"

57'-0"

R

6'-0"

12'-0"

2'   10'-0"   4'

20'-0"

R   O   A   D

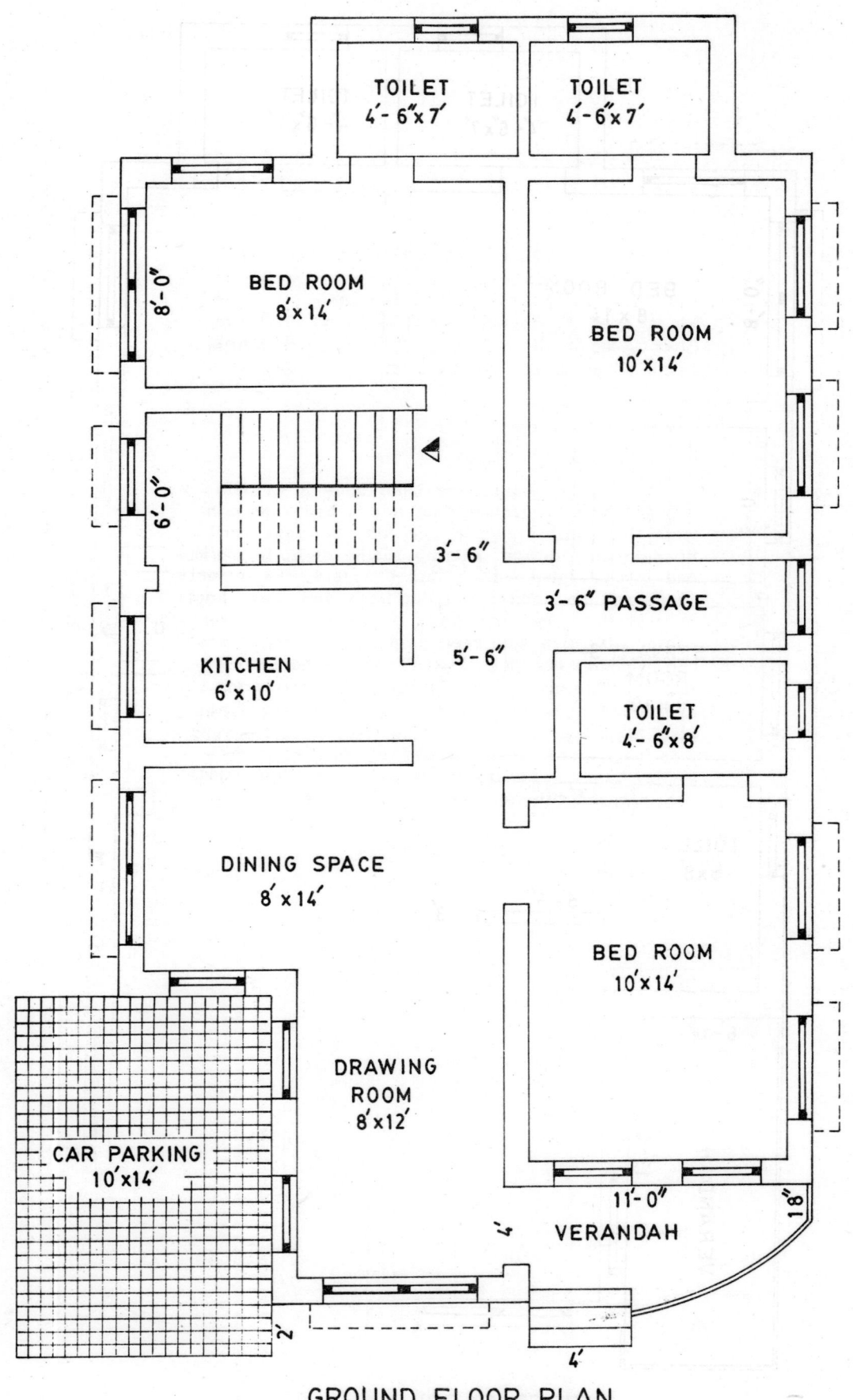

GROUND FLOOR PLAN

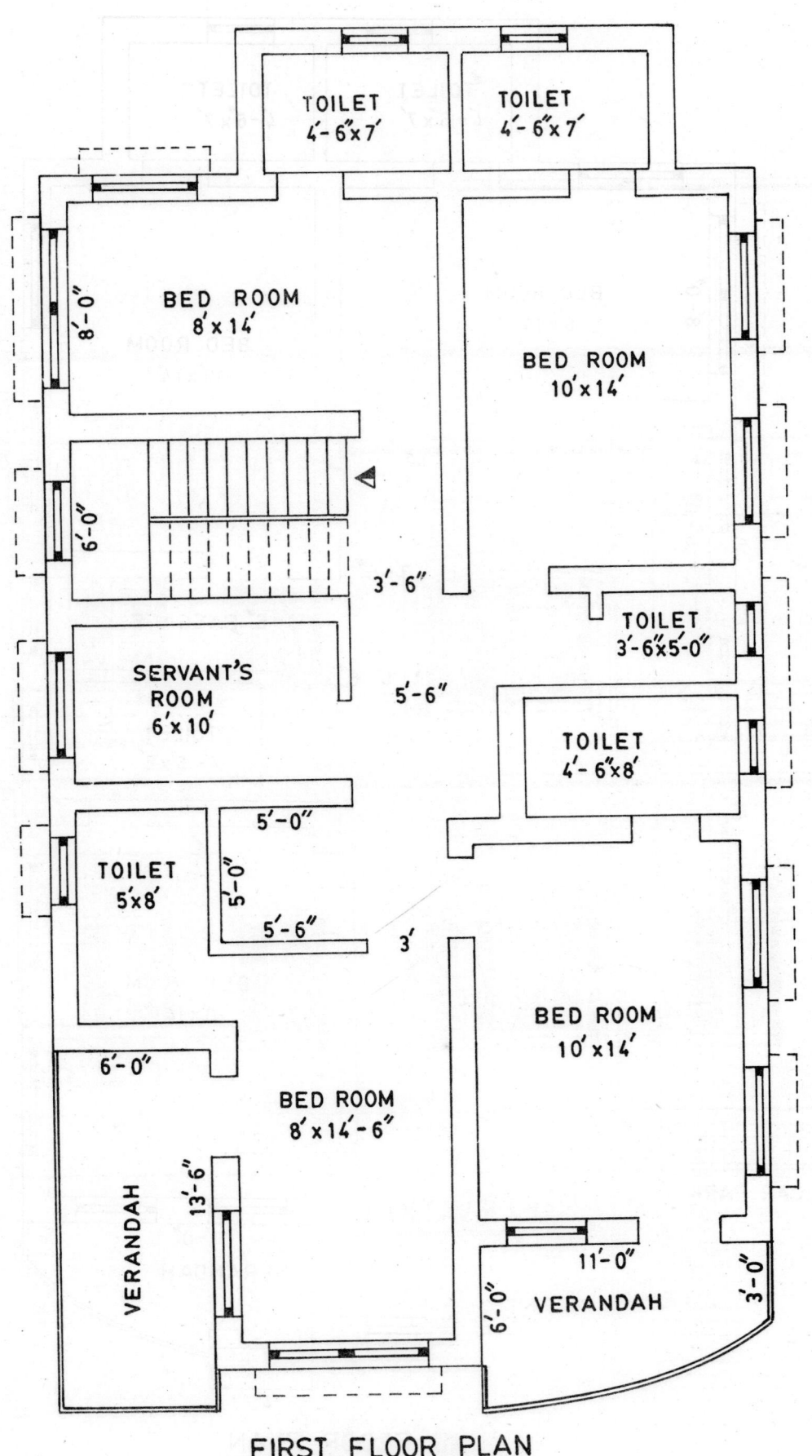

FIRST FLOOR PLAN

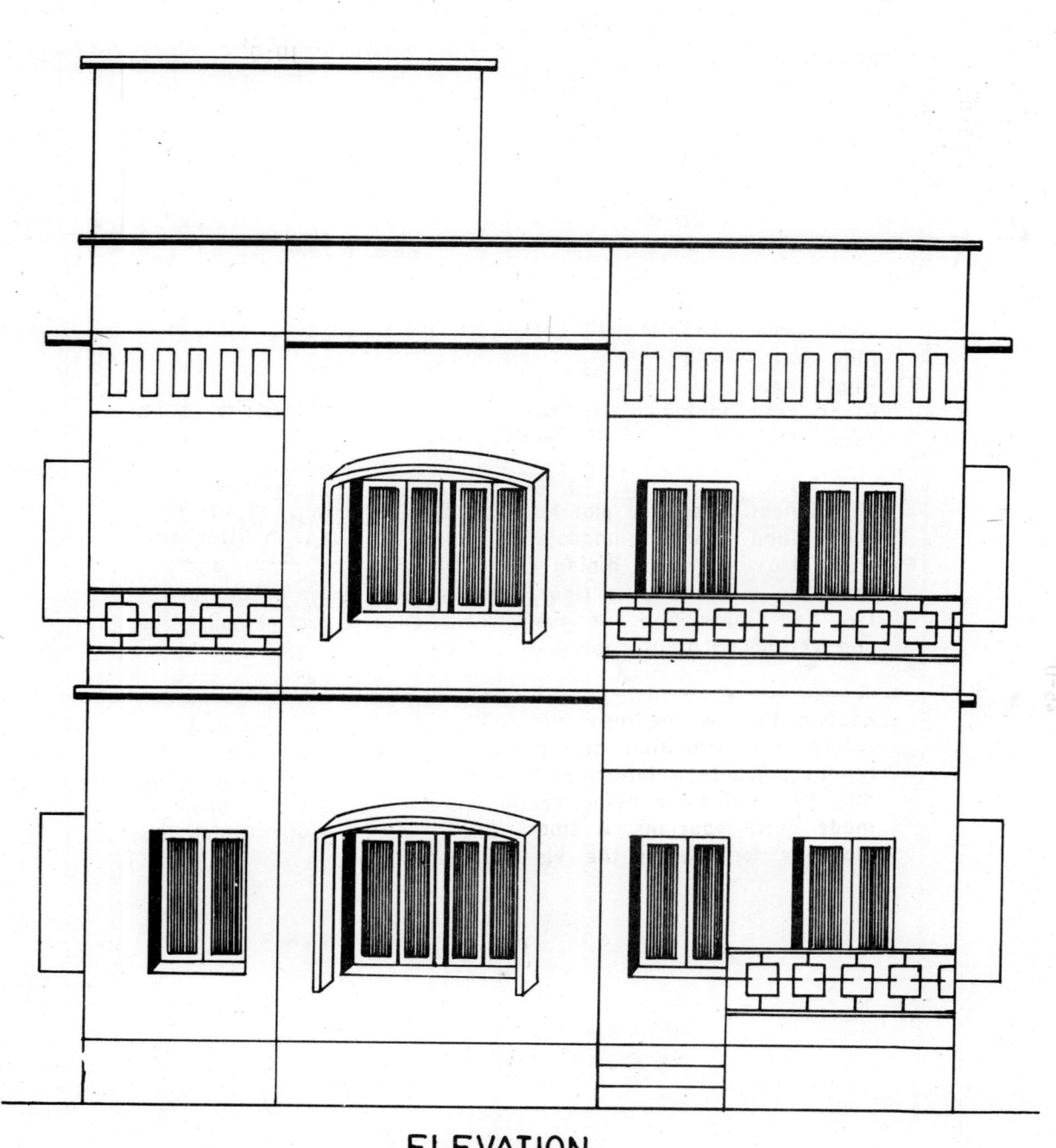

ELEVATION

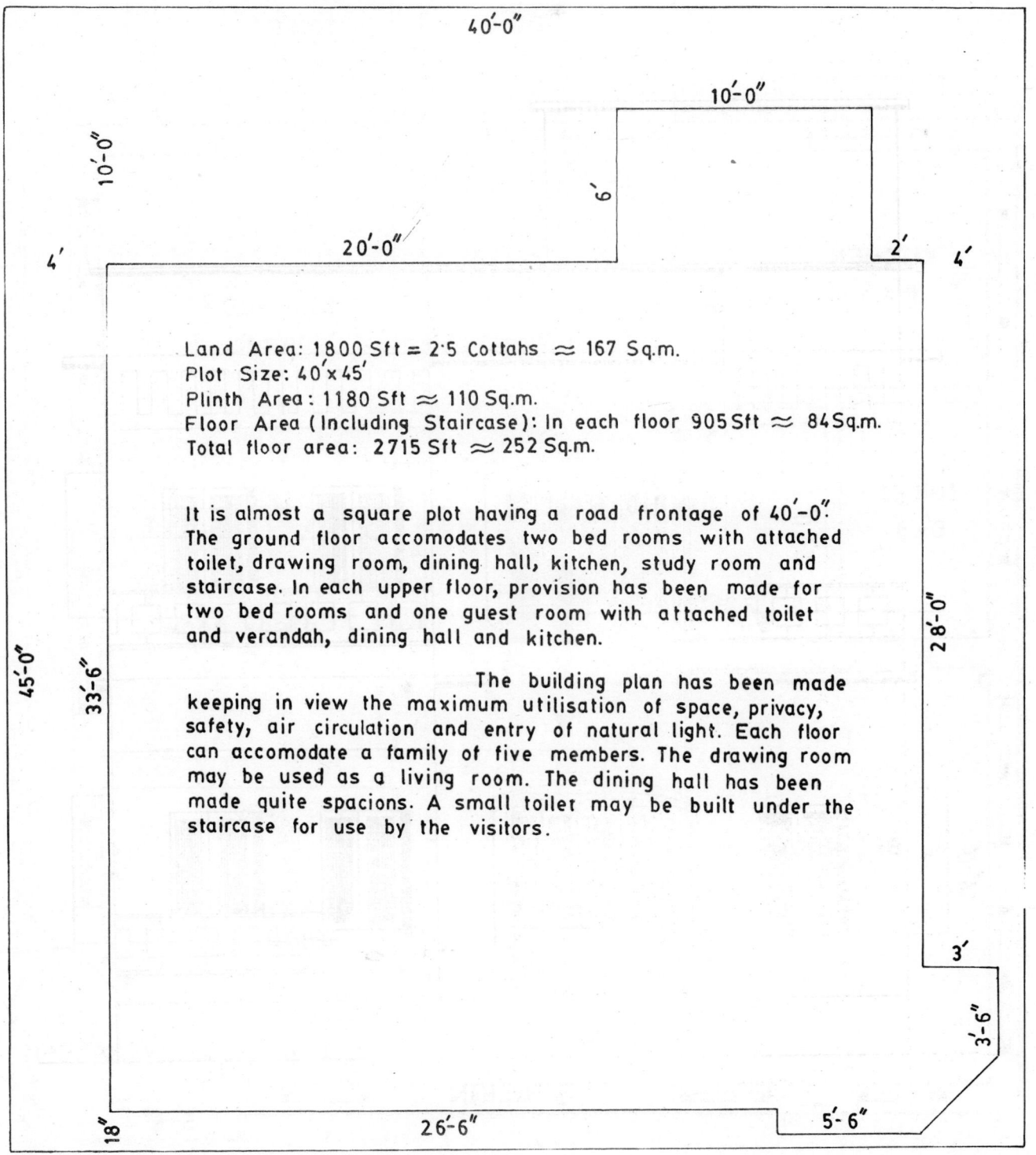

Land Area: 1800 Sft = 2·5 Cottahs ≈ 167 Sq.m.
Plot Size: 40'x45'
Plinth Area: 1180 Sft ≈ 110 Sq.m.
Floor Area (Including Staircase): In each floor 905 Sft ≈ 84 Sq.m.
Total floor area: 2715 Sft ≈ 252 Sq.m.

It is almost a square plot having a road frontage of 40'-0". The ground floor accomodates two bed rooms with attached toilet, drawing room, dining hall, kitchen, study room and staircase. In each upper floor, provision has been made for two bed rooms and one guest room with attached toilet and verandah, dining hall and kitchen.

The building plan has been made keeping in view the maximum utilisation of space, privacy, safety, air circulation and entry of natural light. Each floor can accomodate a family of five members. The drawing room may be used as a living room. The dining hall has been made quite spacions. A small toilet may be built under the staircase for use by the visitors.

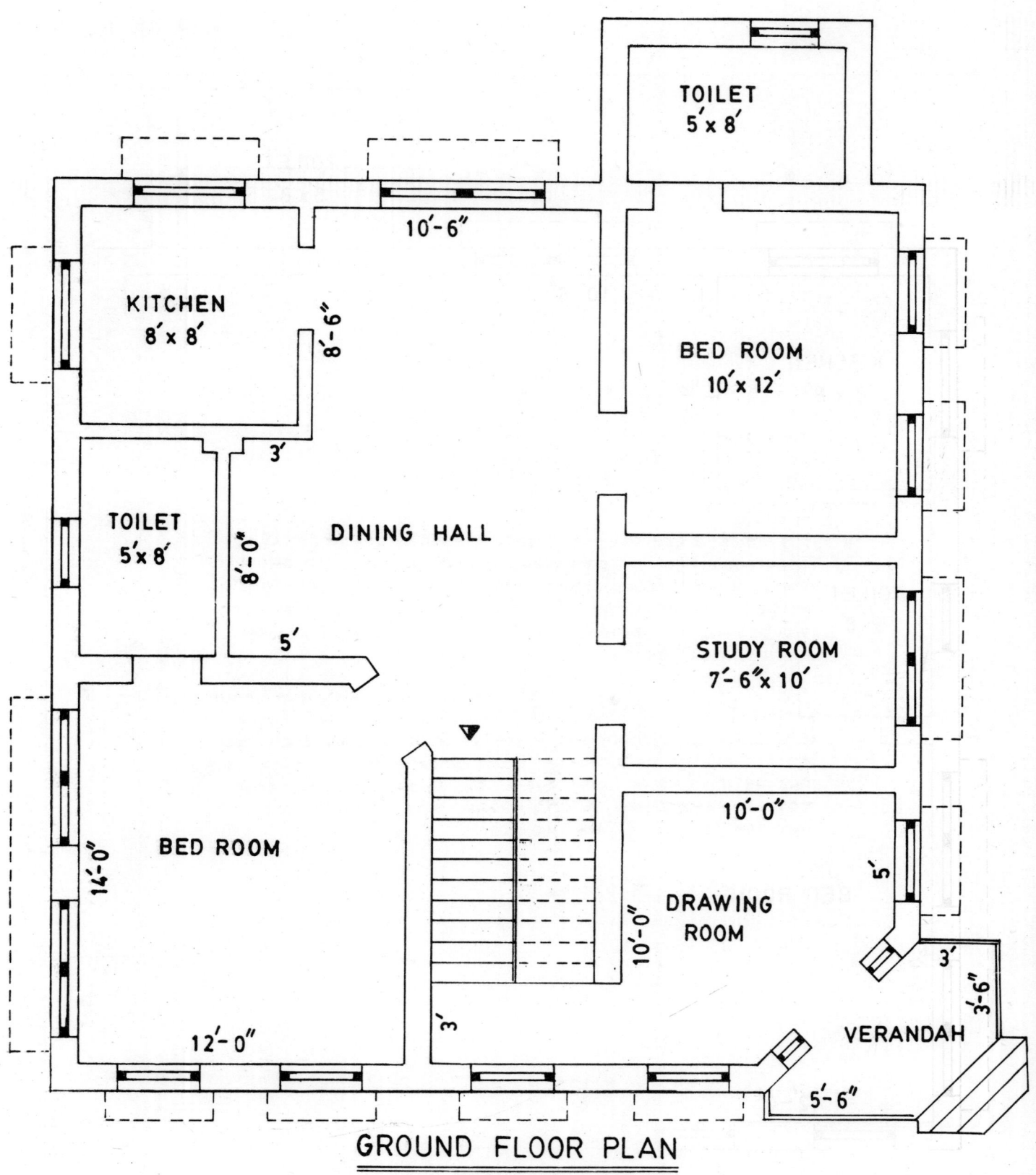

TOILET
5' x 8'

KITCHEN
8' x 8'

10'-6"

8'-6"

BED ROOM
10' x 12'

3'

TOILET
5' x 8'

8'-0"

DINING HALL

STUDY ROOM
7'-6" x 10'

5'

10'-0"

14'-0"

BED ROOM

DRAWING
ROOM

10'-0"

5'

3'

VERANDAH

3'

3'-6"

12'-0"

3'

5'-6"

GROUND FLOOR PLAN

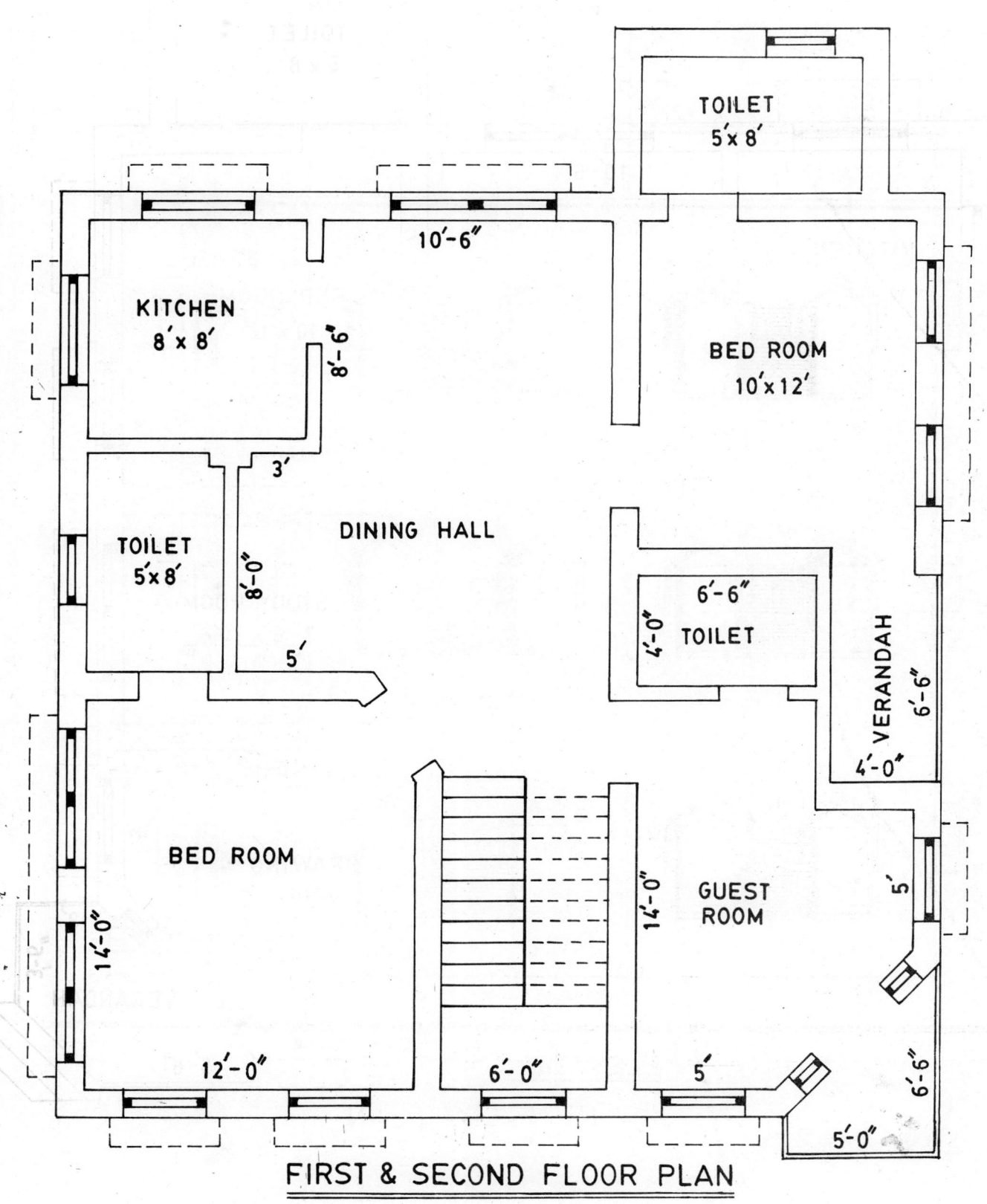

TOILET
5'x 8'

KITCHEN
8' x 8'

8'-6"

3'

DINING HALL

10'-6"

BED ROOM
10'x12'

TOILET
5'x 8'

8'-0"

5'

6'-6"
4'-0"
TOILET

VERANDAH
6'-6"
4'-0"

BED ROOM

14'-0"

GUEST
ROOM

14'-0"

5'

5'

12'-0"

6'-0"

5'

6'-6"

5'-0"

FIRST & SECOND FLOOR PLAN

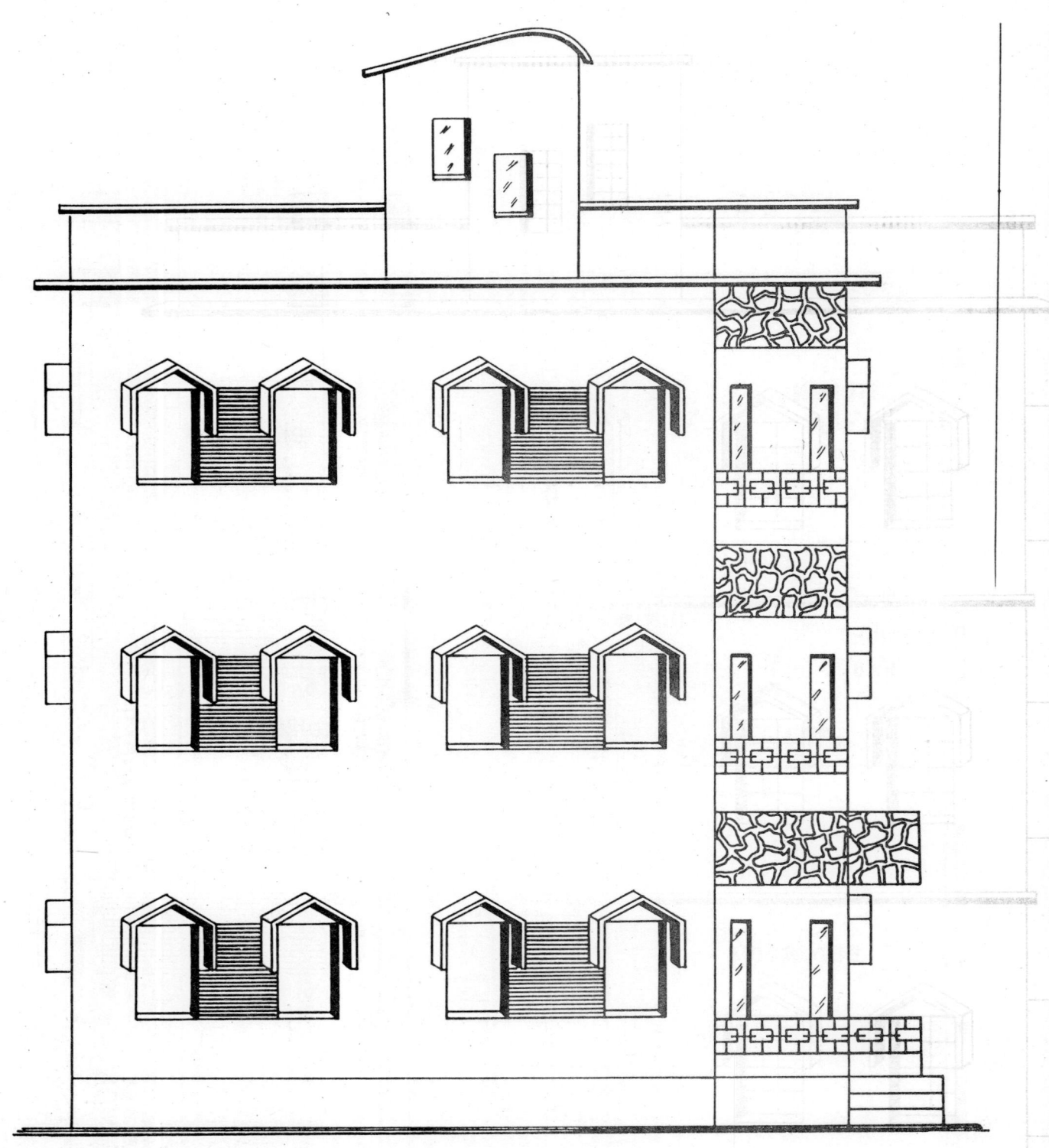

ELEVATION

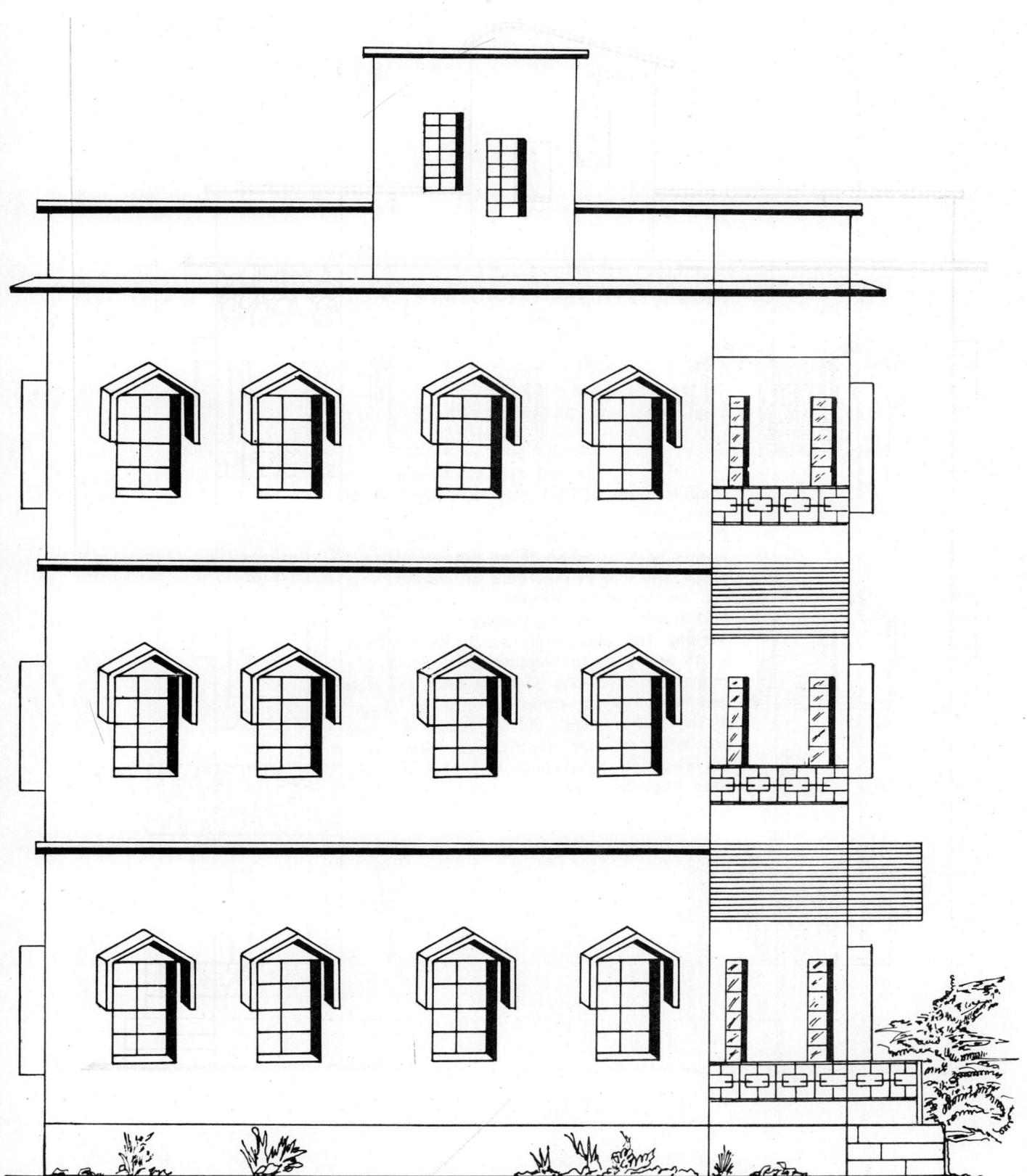

ELEVATION

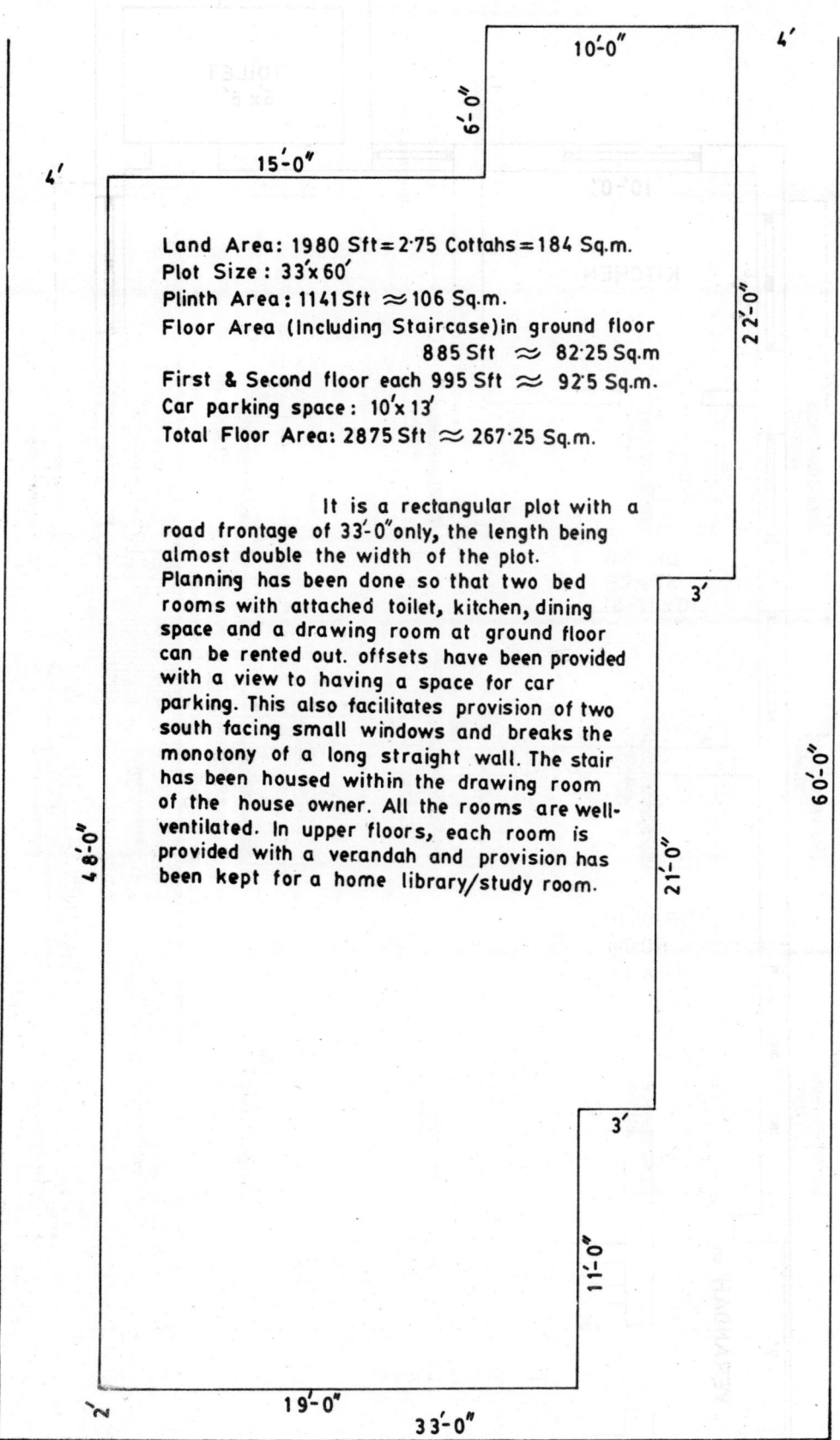

Land Area: 1980 Sft = 2·75 Cottahs = 184 Sq.m.
Plot Size: 33′ x 60′
Plinth Area: 1141 Sft ≈ 106 Sq.m.
Floor Area (Including Staircase) in ground floor
                                885 Sft ≈ 82·25 Sq.m
First & Second floor each 995 Sft ≈ 92·5 Sq.m.
Car parking space: 10′ x 13′
Total Floor Area: 2875 Sft ≈ 267·25 Sq.m.

It is a rectangular plot with a road frontage of 33′-0″ only, the length being almost double the width of the plot. Planning has been done so that two bed rooms with attached toilet, kitchen, dining space and a drawing room at ground floor can be rented out. offsets have been provided with a view to having a space for car parking. This also facilitates provision of two south facing small windows and breaks the monotony of a long straight wall. The stair has been housed within the drawing room of the house owner. All the rooms are well-ventilated. In upper floors, each room is provided with a verandah and provision has been kept for a home library/study room.

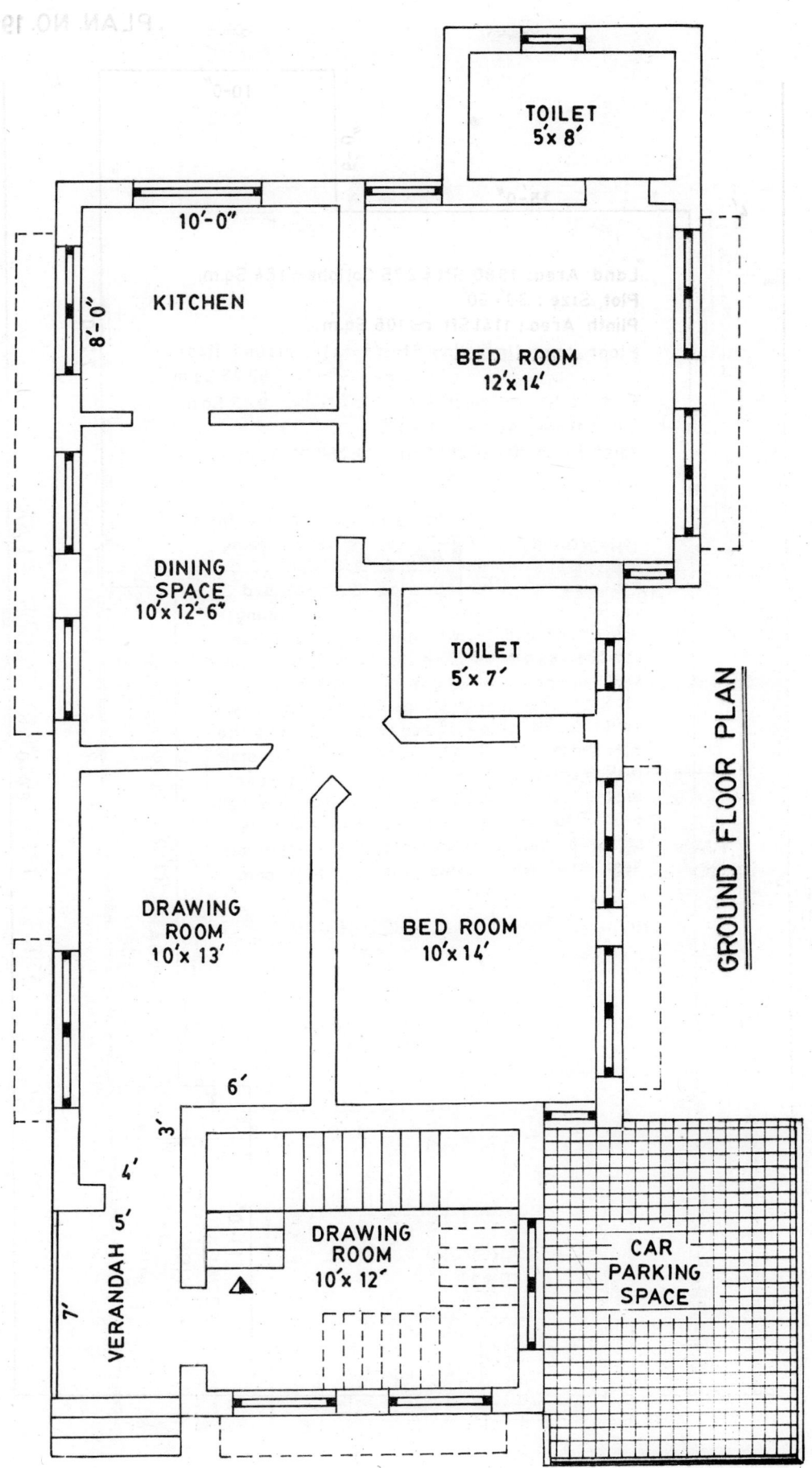

GROUND FLOOR PLAN

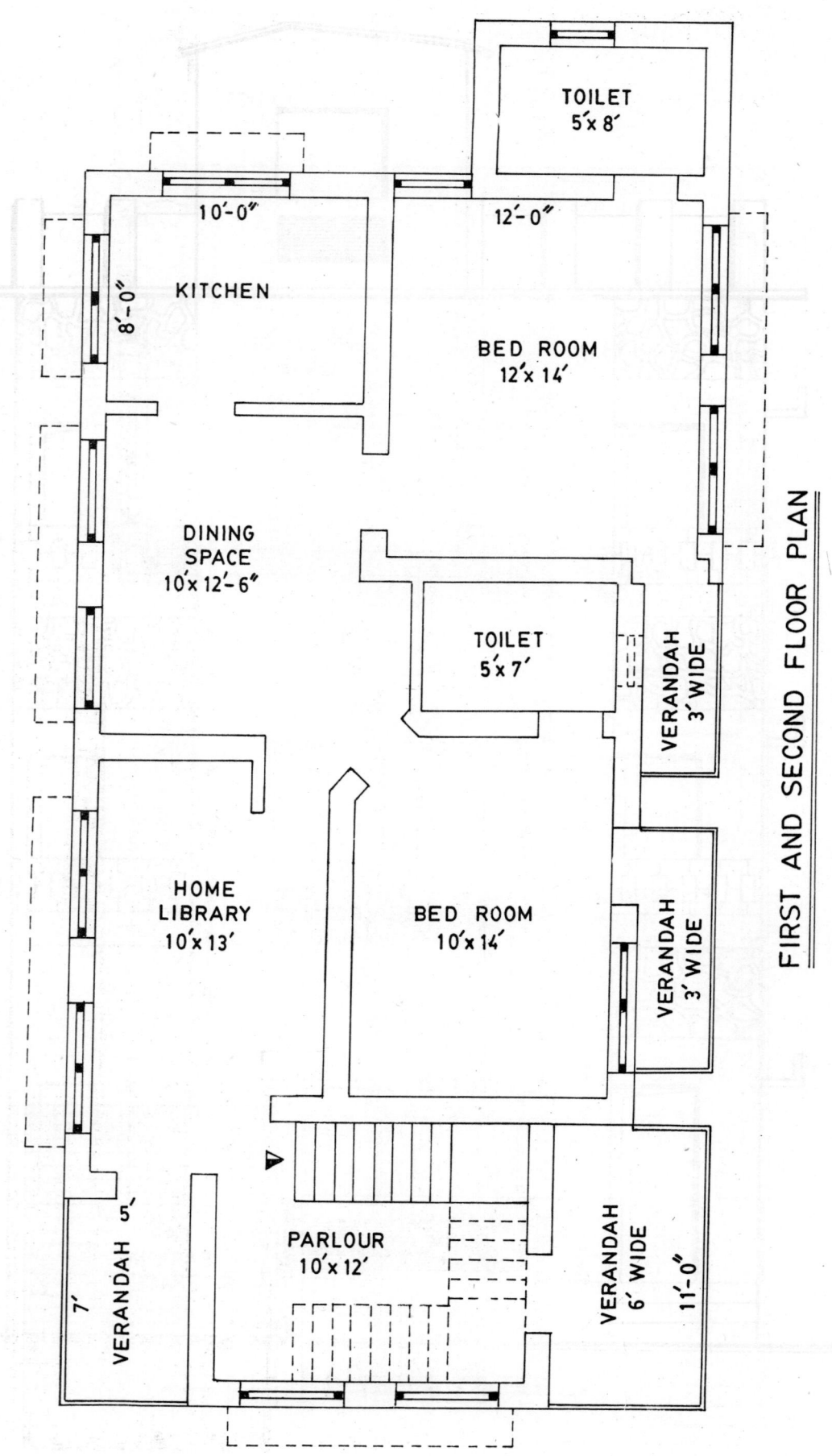

TOILET
5' x 8'

12'-0"

10'-0"

8'-0"

KITCHEN

BED ROOM
12' x 14'

DINING
SPACE
10' x 12'-6"

TOILET
5' x 7'

VERANDAH
3' WIDE

HOME
LIBRARY
10' x 13'

BED ROOM
10' x 14'

VERANDAH
3' WIDE

VERANDAH 5'

7'

PARLOUR
10' x 12'

VERANDAH
6' WIDE

11'-0"

FIRST AND SECOND FLOOR PLAN

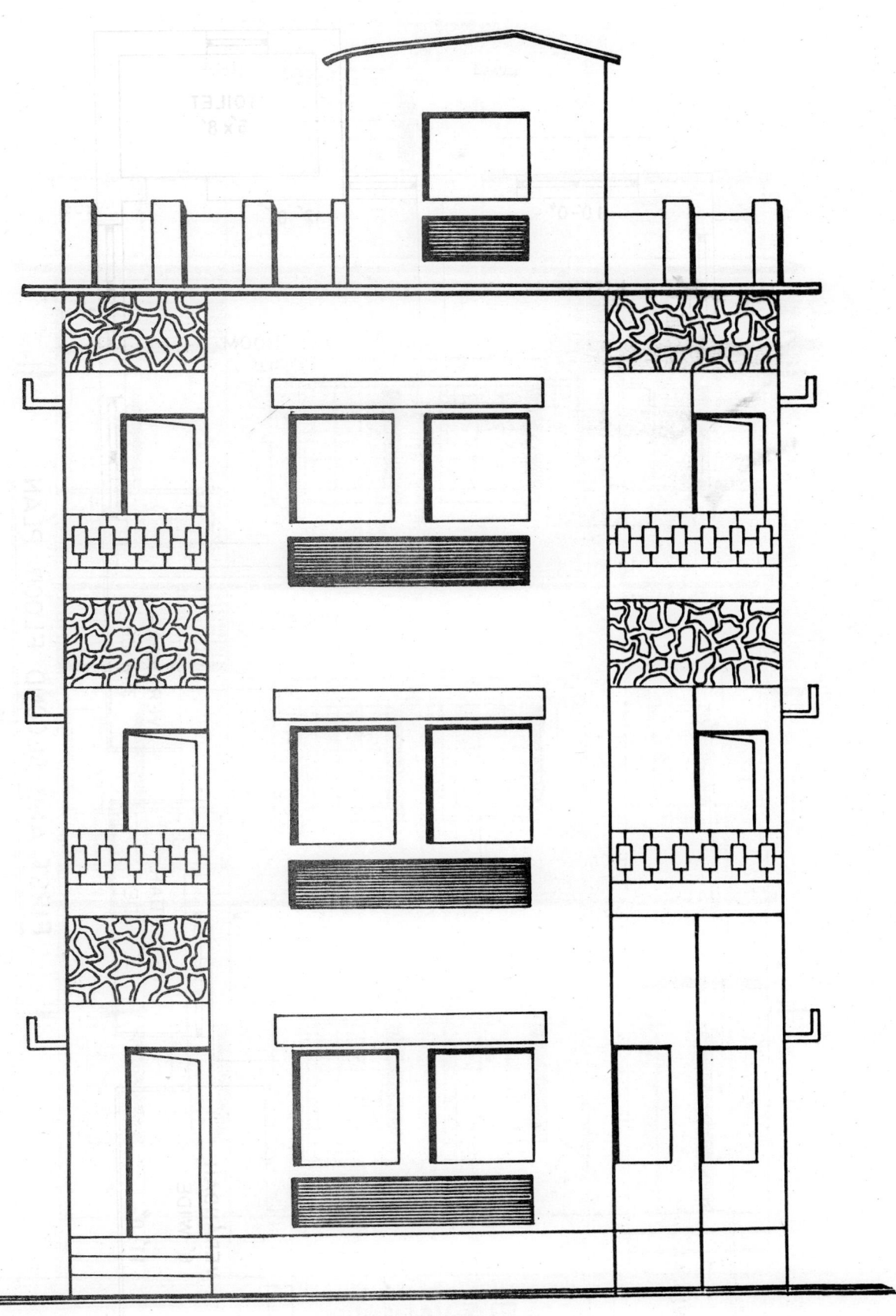

ELEVATION

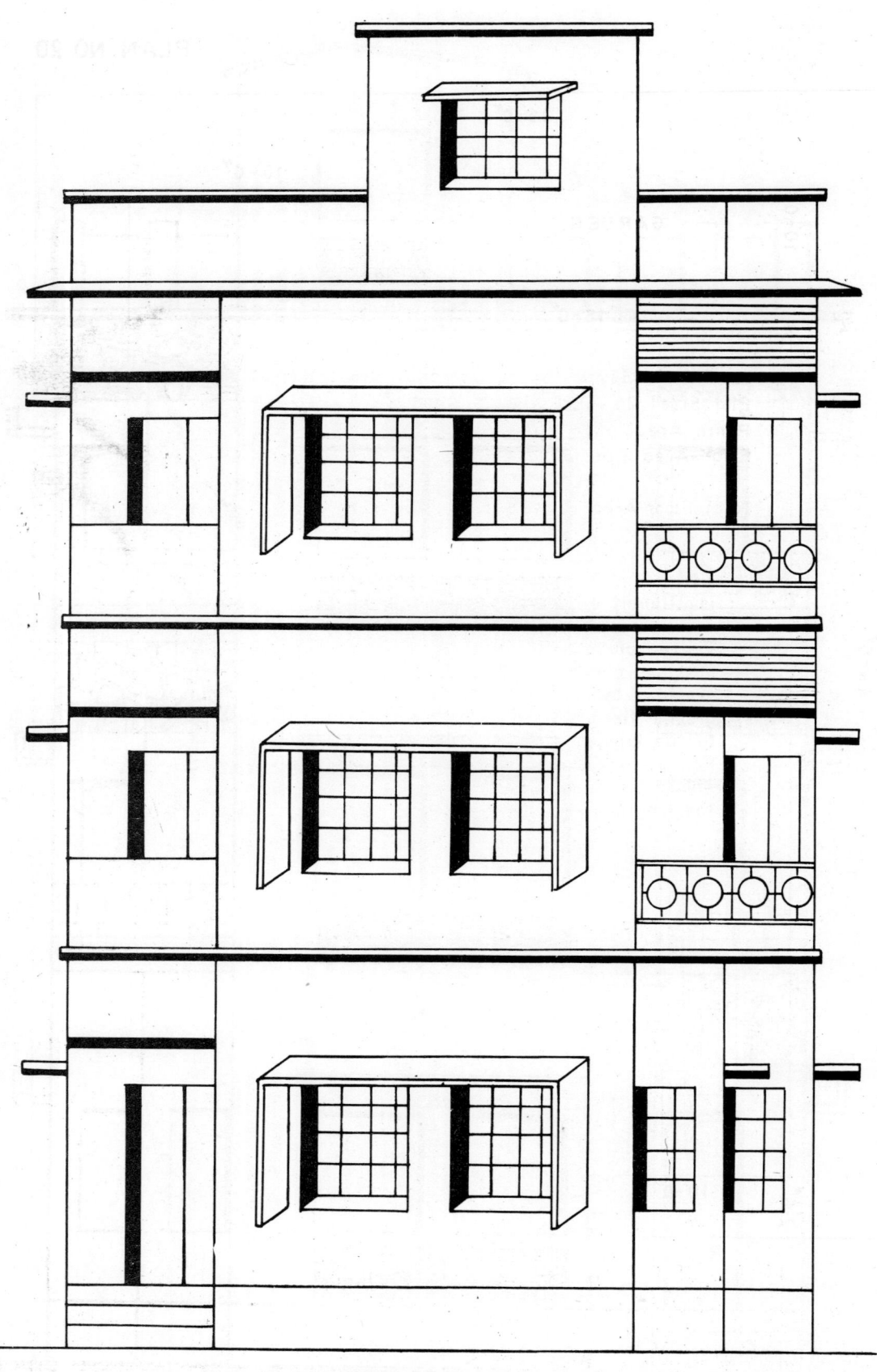

ELEVATION

PLAN NO. 20

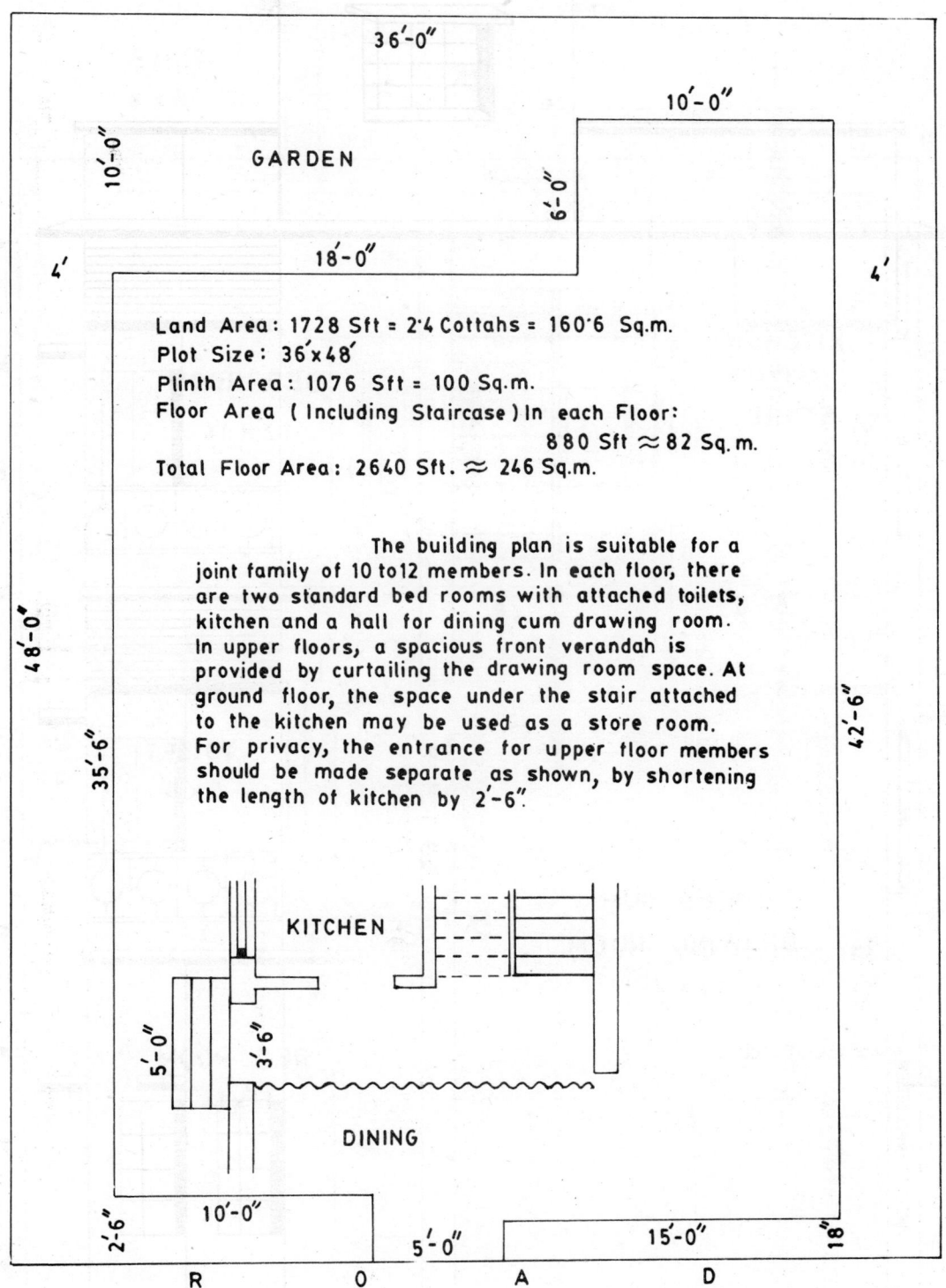

GARDEN

36'-0"

10'-0"

10'-0"

6'-0"

18'-0"

4'

4'

48'-0"

35'-6"

42'-6"

Land Area : 1728 Sft = 2·4 Cottahs = 160·6 Sq.m.
Plot Size : 36'x48'
Plinth Area : 1076 Sft = 100 Sq.m.
Floor Area (Including Staircase) In each Floor :
880 Sft ≈ 82 Sq. m.
Total Floor Area : 2640 Sft. ≈ 246 Sq.m.

The building plan is suitable for a
joint family of 10 to 12 members. In each floor, there
are two standard bed rooms with attached toilets,
kitchen and a hall for dining cum drawing room.
In upper floors, a spacious front verandah is
provided by curtailing the drawing room space. At
ground floor, the space under the stair attached
to the kitchen may be used as a store room.
For privacy, the entrance for upper floor members
should be made separate as shown, by shortening
the length of kitchen by 2'-6".

KITCHEN

5'-0"

3'-6"

DINING

2'-6"

10'-0"

5'-0"

15'-0"

18"

R   O   A   D

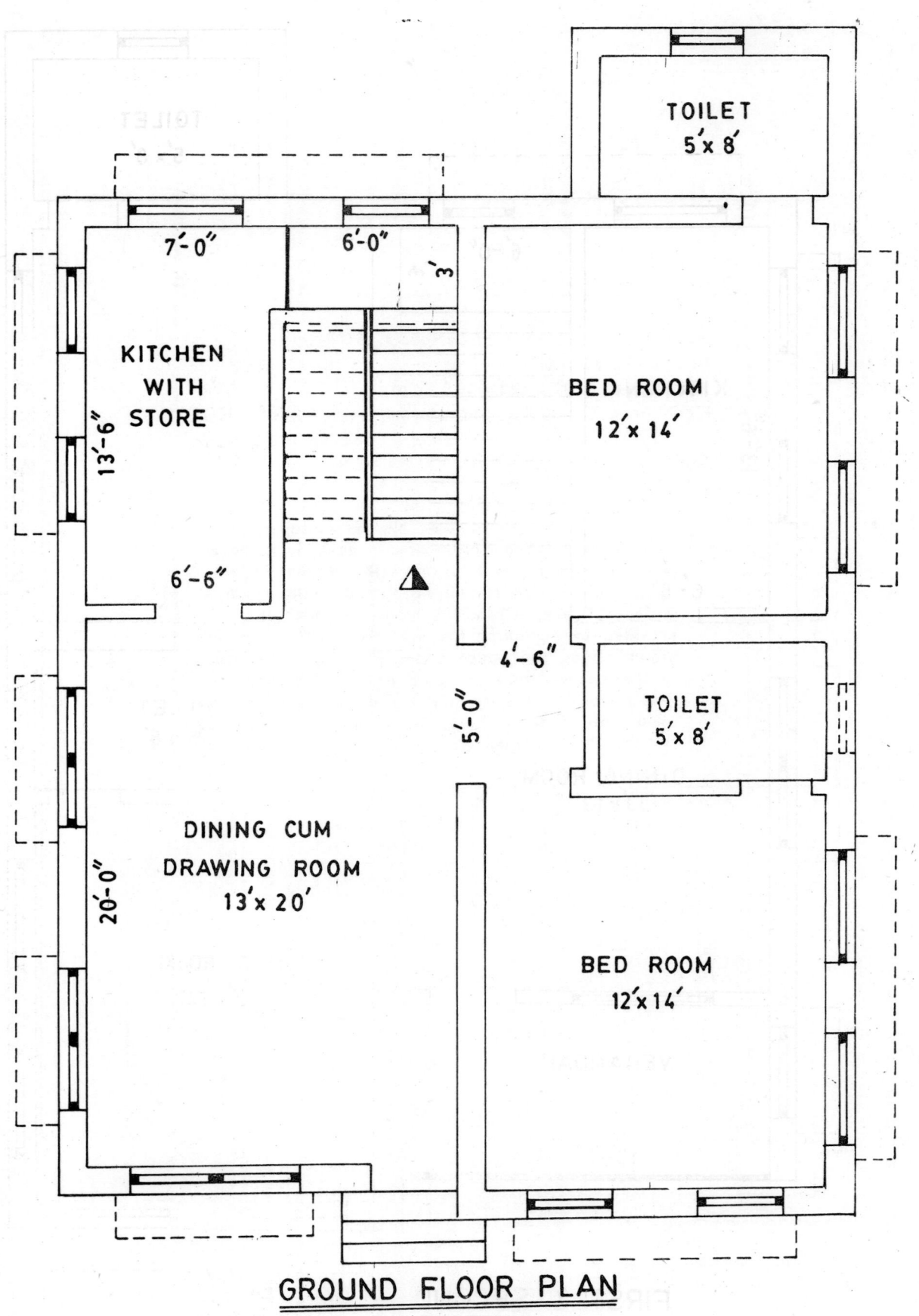

GROUND FLOOR PLAN

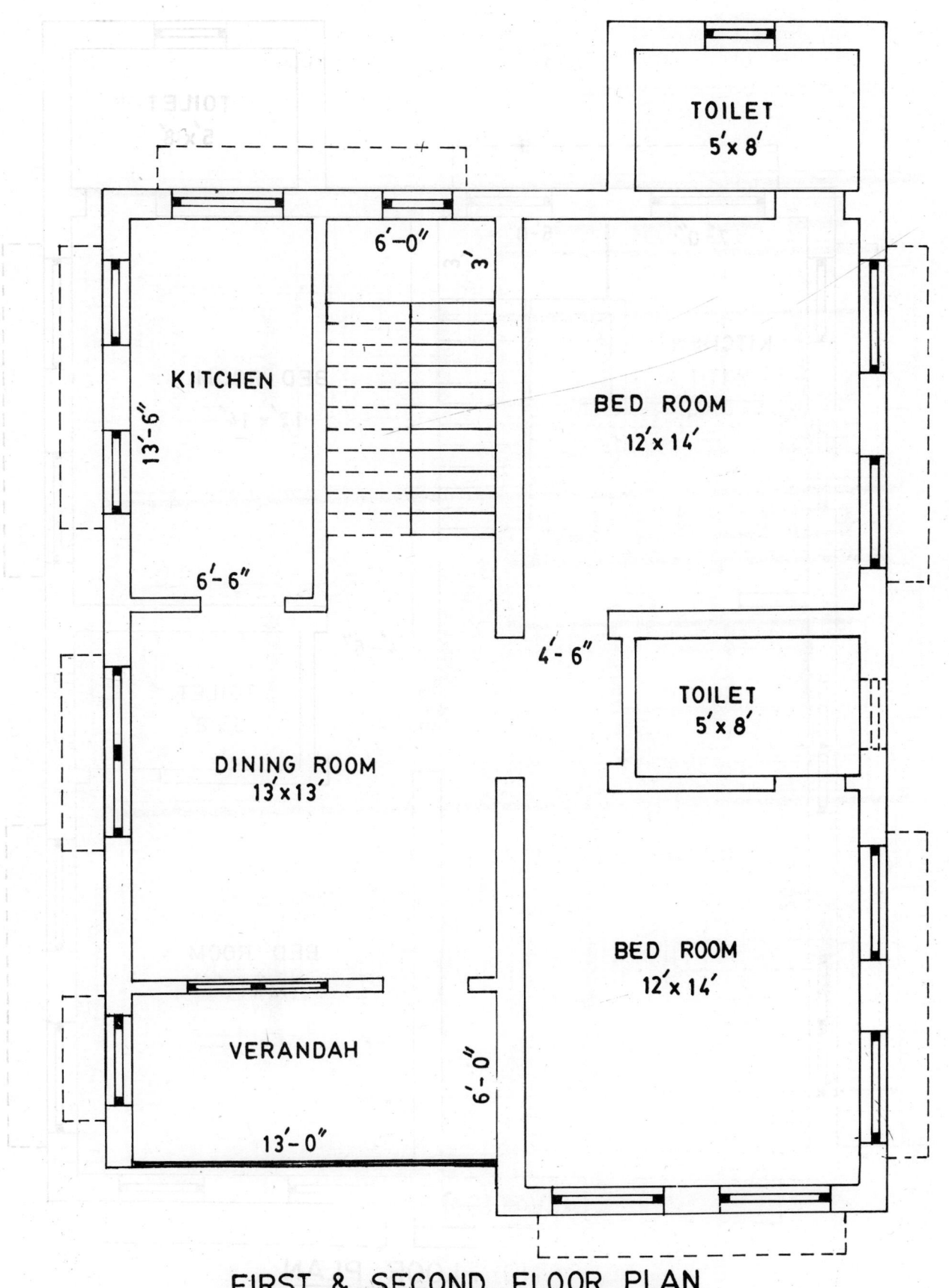

TOILET
5' x 8'

KITCHEN

13'-6"

6'-0"

3'

6'-6"

BED ROOM
12' x 14'

4'-6"

TOILET
5' x 8'

DINING ROOM
13' x 13'

BED ROOM
12' x 14'

VERANDAH

6'-0"

13'-0"

FIRST & SECOND FLOOR PLAN

ELEVATION

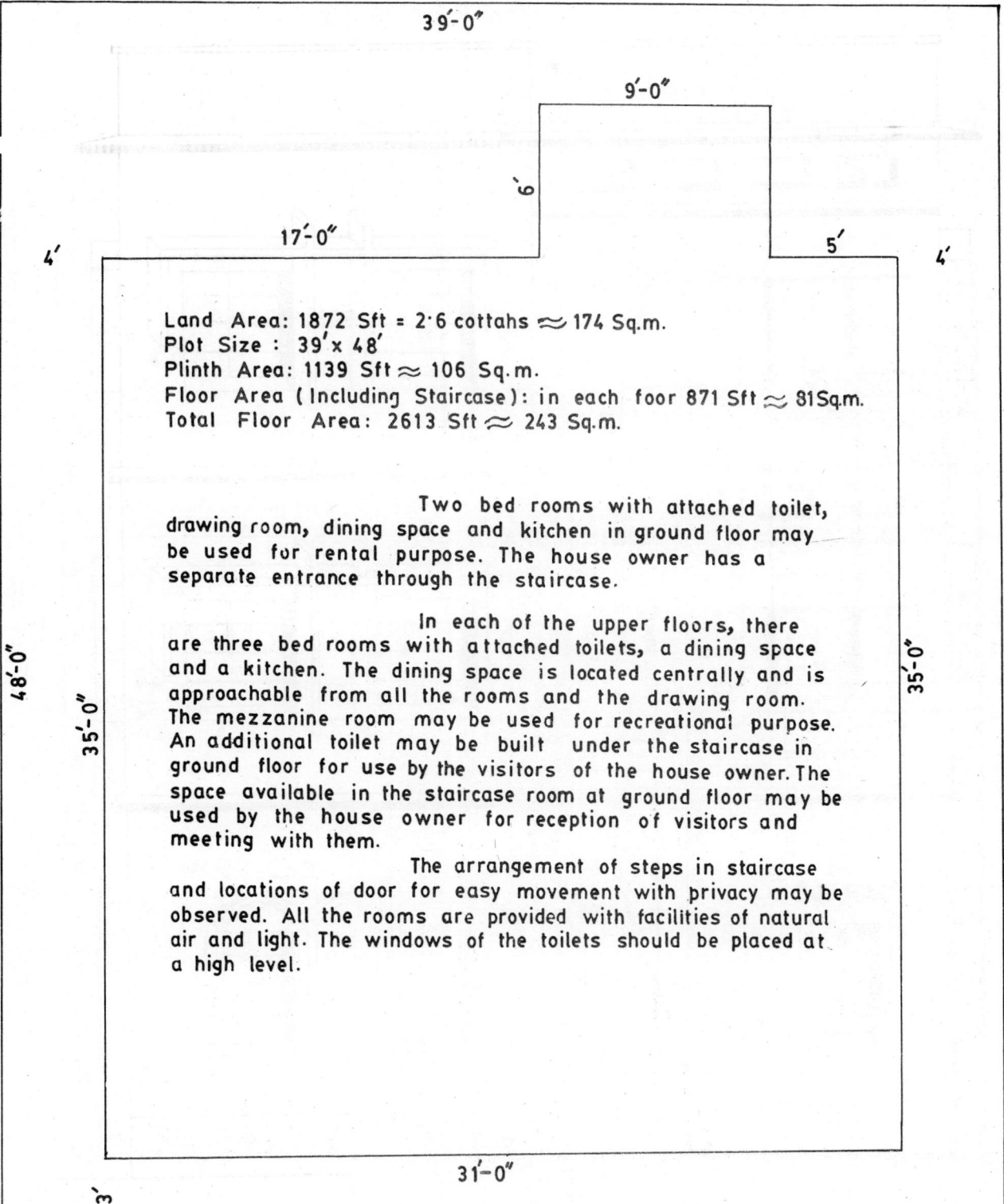

Land Area: 1872 Sft = 2·6 cottahs ≈ 174 Sq.m.
Plot Size : 39' x 48'
Plinth Area: 1139 Sft ≈ 106 Sq.m.
Floor Area (Including Staircase): in each foor 871 Sft ≈ 81Sq.m.
Total Floor Area: 2613 Sft ≈ 243 Sq.m.

Two bed rooms with attached toilet, drawing room, dining space and kitchen in ground floor may be used for rental purpose. The house owner has a separate entrance through the staircase.

In each of the upper floors, there are three bed rooms with attached toilets, a dining space and a kitchen. The dining space is located centrally and is approachable from all the rooms and the drawing room. The mezzanine room may be used for recreational purpose. An additional toilet may be built under the staircase in ground floor for use by the visitors of the house owner. The space available in the staircase room at ground floor may be used by the house owner for reception of visitors and meeting with them.

The arrangement of steps in staircase and locations of door for easy movement with privacy may be observed. All the rooms are provided with facilities of natural air and light. The windows of the toilets should be placed at a high level.

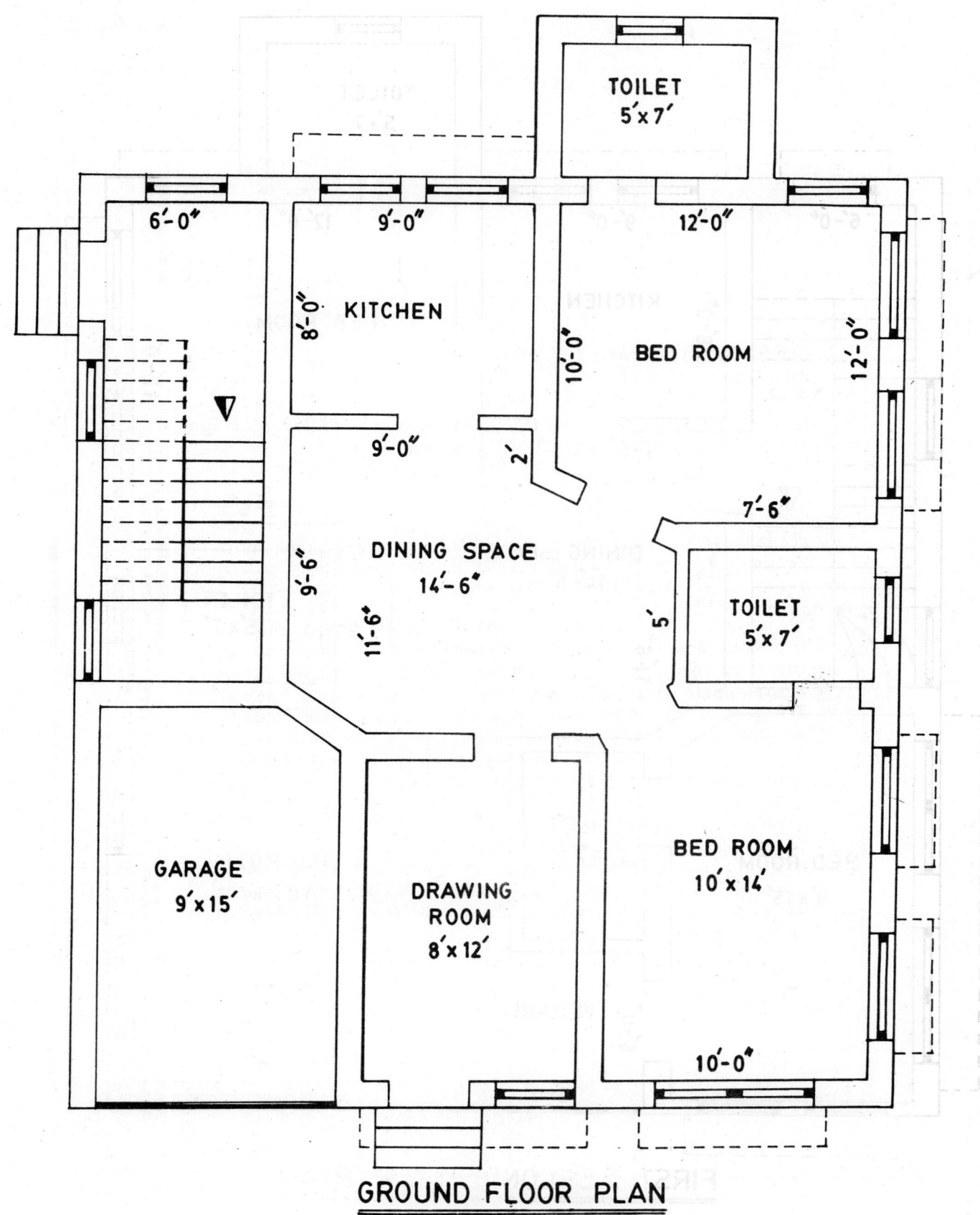

TOILET
5'x 7'

6'-0"

9'-0"

KITCHEN

12'-0"

8'-0"

BED ROOM

12'-0"

9'-0"

10'-0"

2'

7'-6"

DINING SPACE
14'-6"

9'-6"

TOILET
5'x 7'

11'-6"

5'

GARAGE
9'x 15'

DRAWING
ROOM
8'x 12'

BED ROOM
10'x 14'

10'-0"

GROUND FLOOR PLAN

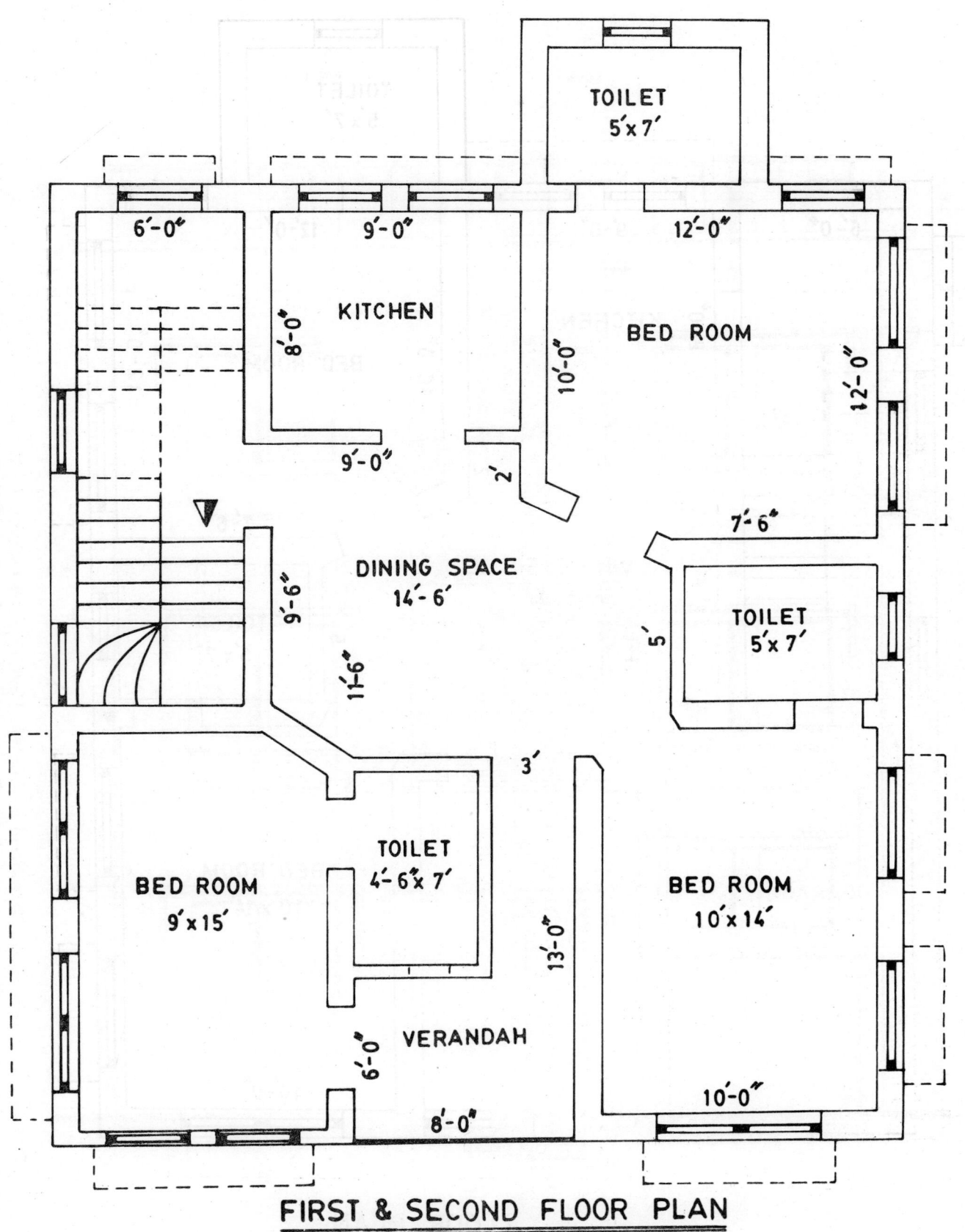

TOILET
5'x 7'

6'-0"

9'-0"

12'-0"

KITCHEN

8'-0"

BED ROOM

10'-0"

12'-0"

9'-0"

2'

7'-6"

DINING SPACE
14'-6'

9'-6"

5'

TOILET
5'x 7'

11'-6"

3'

TOILET
4'-6"x 7'

BED ROOM
9'x 15'

13'-0"

BED ROOM
10'x 14'

6'-0"

VERANDAH

8'-0"

10'-0"

FIRST & SECOND FLOOR PLAN

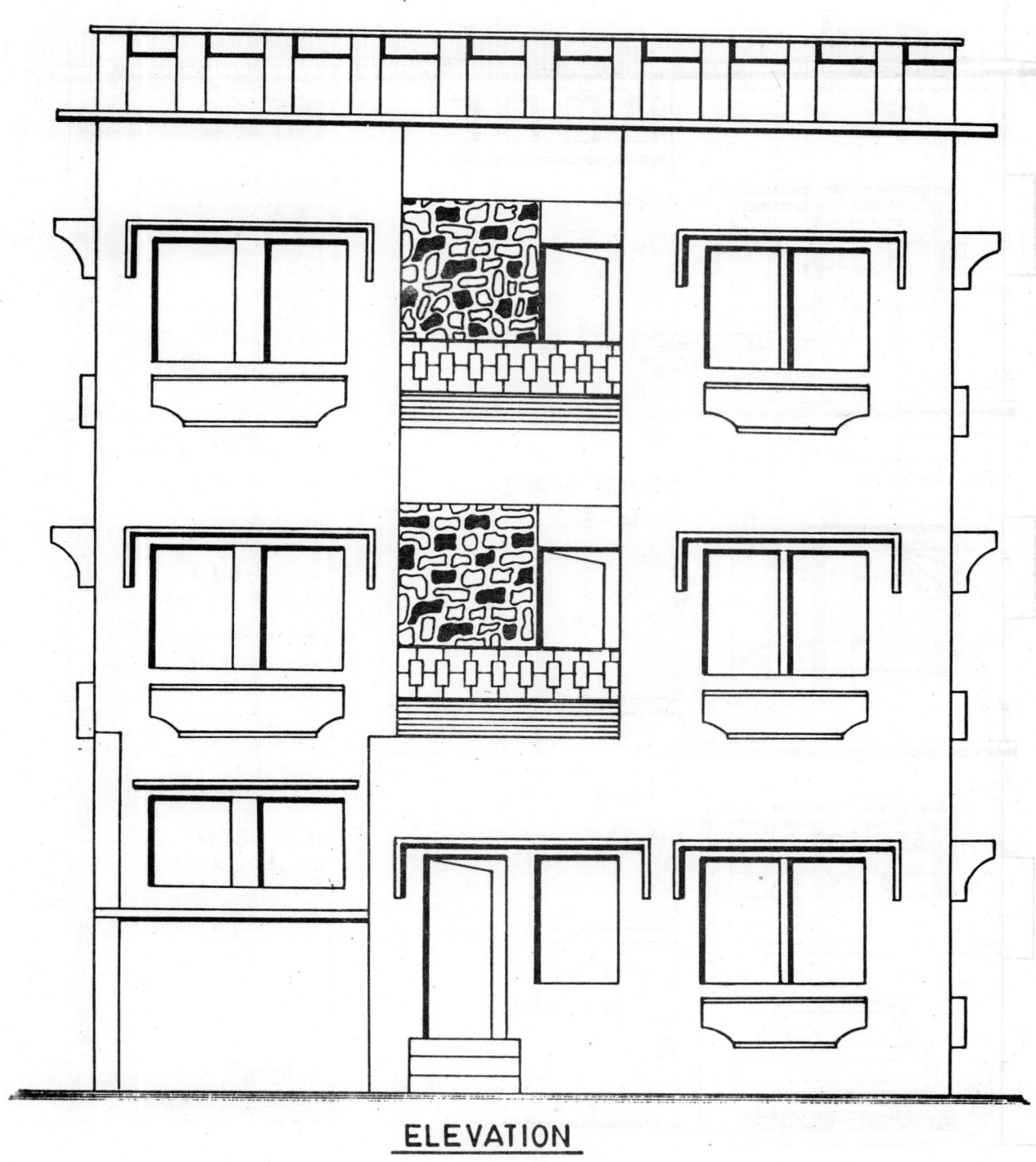

ELEVATION

ELEVATION

Land Area: 1080 Sft = 1'5 cottahs ≈ 100'4 Sq.m.
Plot Size : 30'x 36'
Plinth Area : 582 Sft ≈ 54 Sq.m
Floor Area (Excluding Staircase) in each floor : 345 Sft ≈ 32 Sq.m.

The building plan accomodates drawing room with a front verandah, dining room, kitchen, toilet, staircase and a garage in ground floor. Two bed rooms with toilet and a verandah are provided in the first floor. The drawing room at ground floor may be used as a living room.

The provision of garage space and the staircase layout at ground floor speak of maximum utilisation of space. Although the plan is very compact owing to shortage of land area, it has the facilities of cross ventilation and natural light. Due attention has been given to have safety and security.

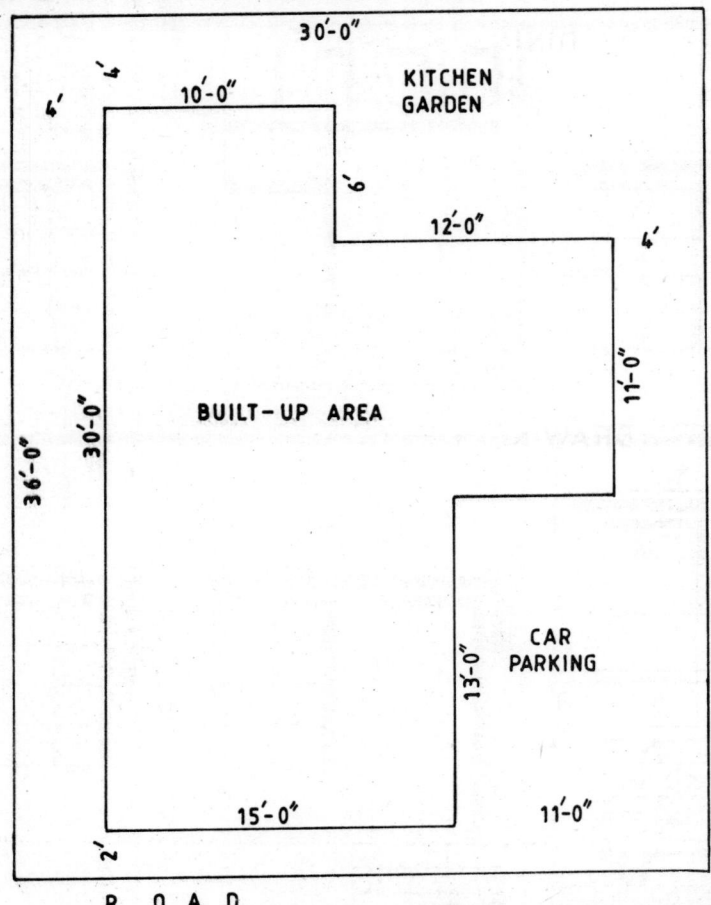

PLAN NO. 22

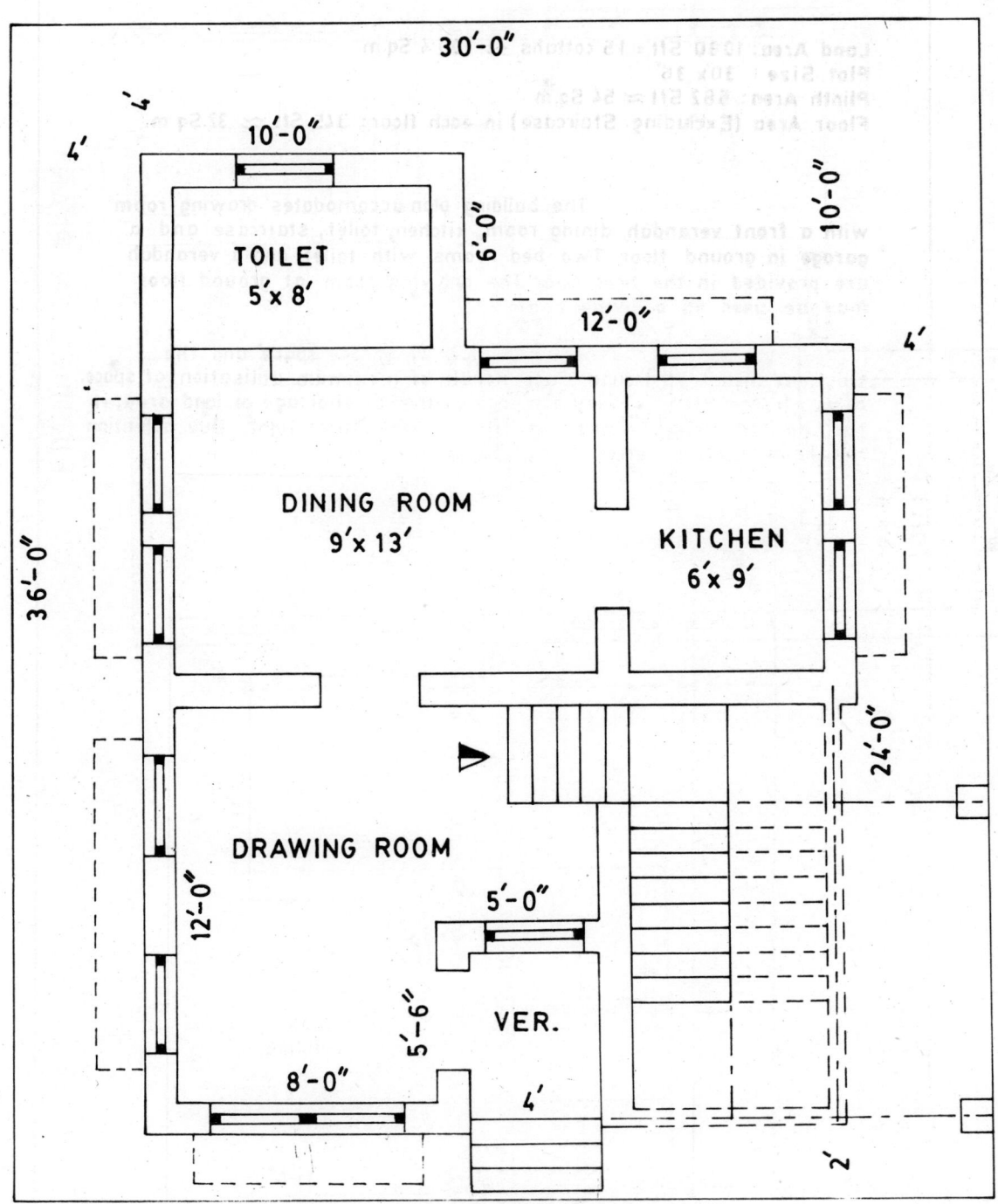

GROUND FLOOR PLAN

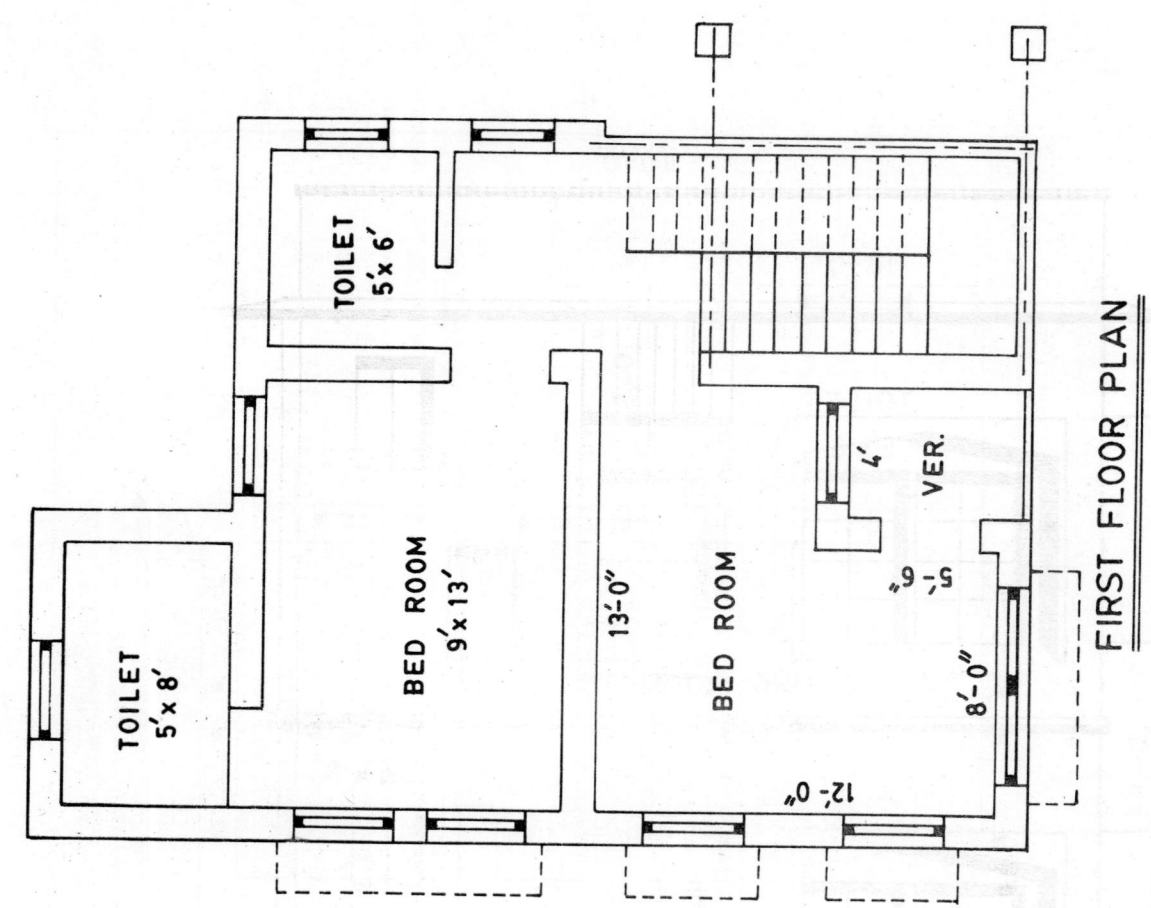

TOILET
5' x 6'

TOILET
5' x 8'

BED ROOM
9' x 13'

13'-0"

BED ROOM

VER.

4'

5'-6"

8'-0"

12'-0"

FIRST FLOOR PLAN

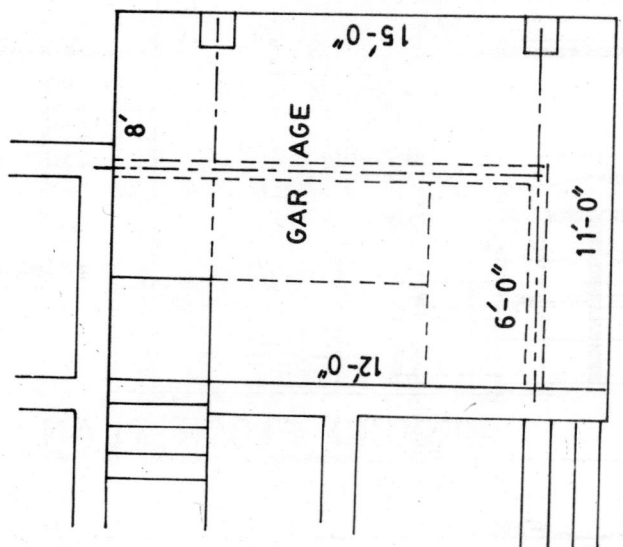

8'

GARAGE

15'-0"

12'-0"

6'-0"

11'-0"

ELEVATION

PLAN NO. 23

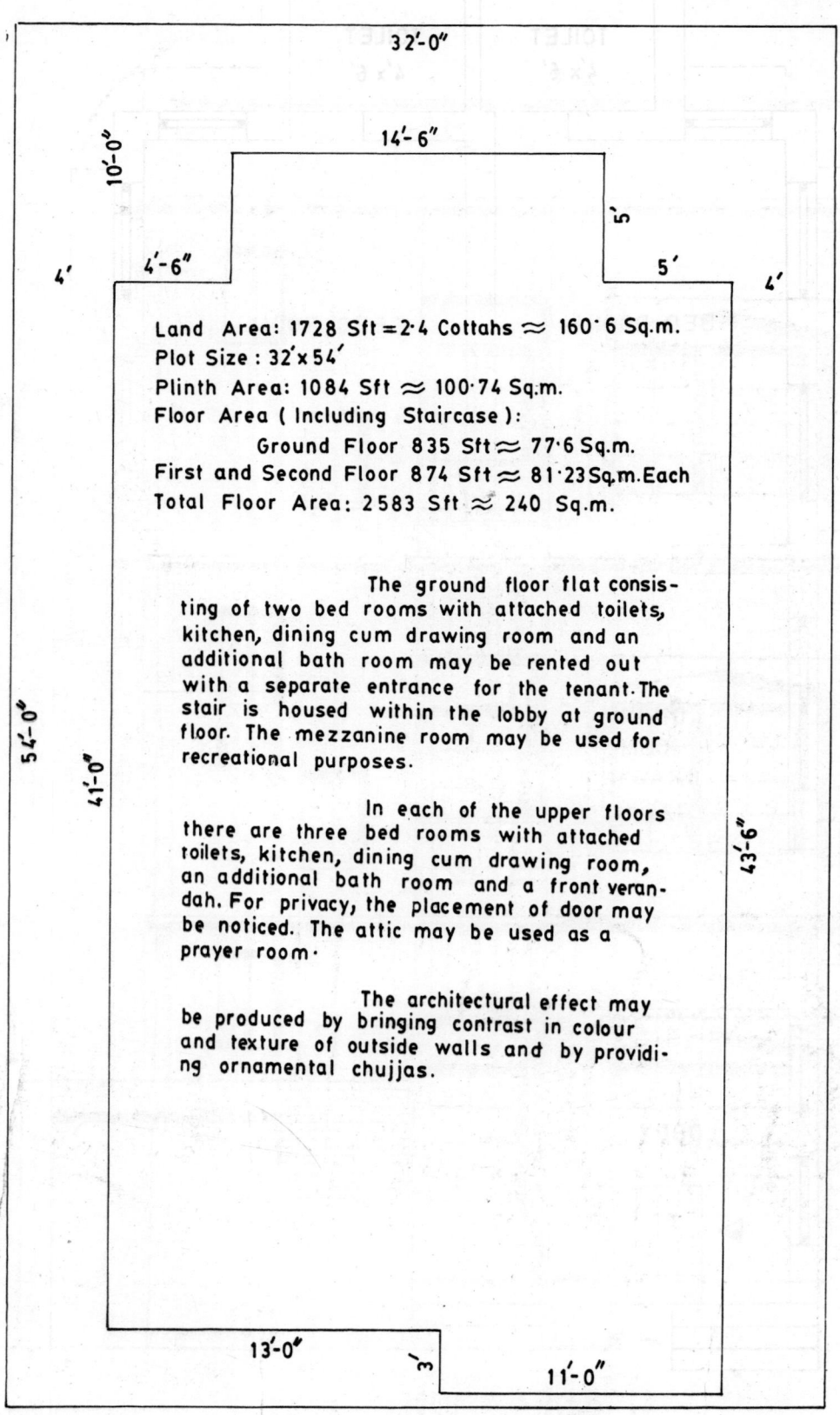

Land Area: 1728 Sft = 2.4 Cottahs ≈ 160.6 Sq.m.
Plot Size : 32' x 54'
Plinth Area: 1084 Sft ≈ 100.74 Sq.m.
Floor Area (Including Staircase):
        Ground Floor 835 Sft ≈ 77.6 Sq.m.
First and Second Floor 874 Sft ≈ 81.23 Sq.m. Each
Total Floor Area: 2583 Sft ≈ 240 Sq.m.

        The ground floor flat consisting of two bed rooms with attached toilets, kitchen, dining cum drawing room and an additional bath room may be rented out with a separate entrance for the tenant. The stair is housed within the lobby at ground floor. The mezzanine room may be used for recreational purposes.

        In each of the upper floors there are three bed rooms with attached toilets, kitchen, dining cum drawing room, an additional bath room and a front verandah. For privacy, the placement of door may be noticed. The attic may be used as a prayer room.

        The architectural effect may be produced by bringing contrast in colour and texture of outside walls and by providing ornamental chujjas.

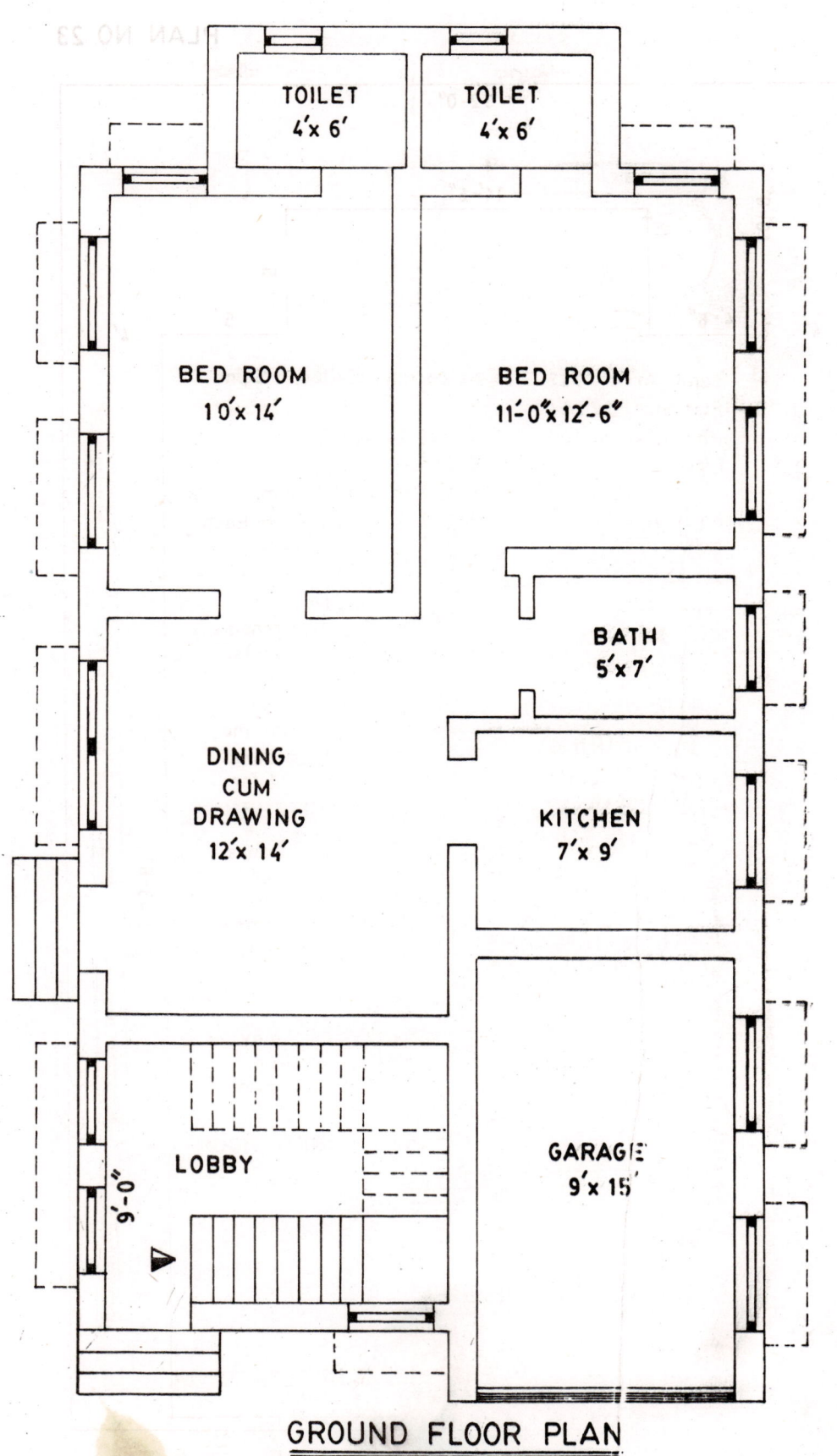

GROUND FLOOR PLAN

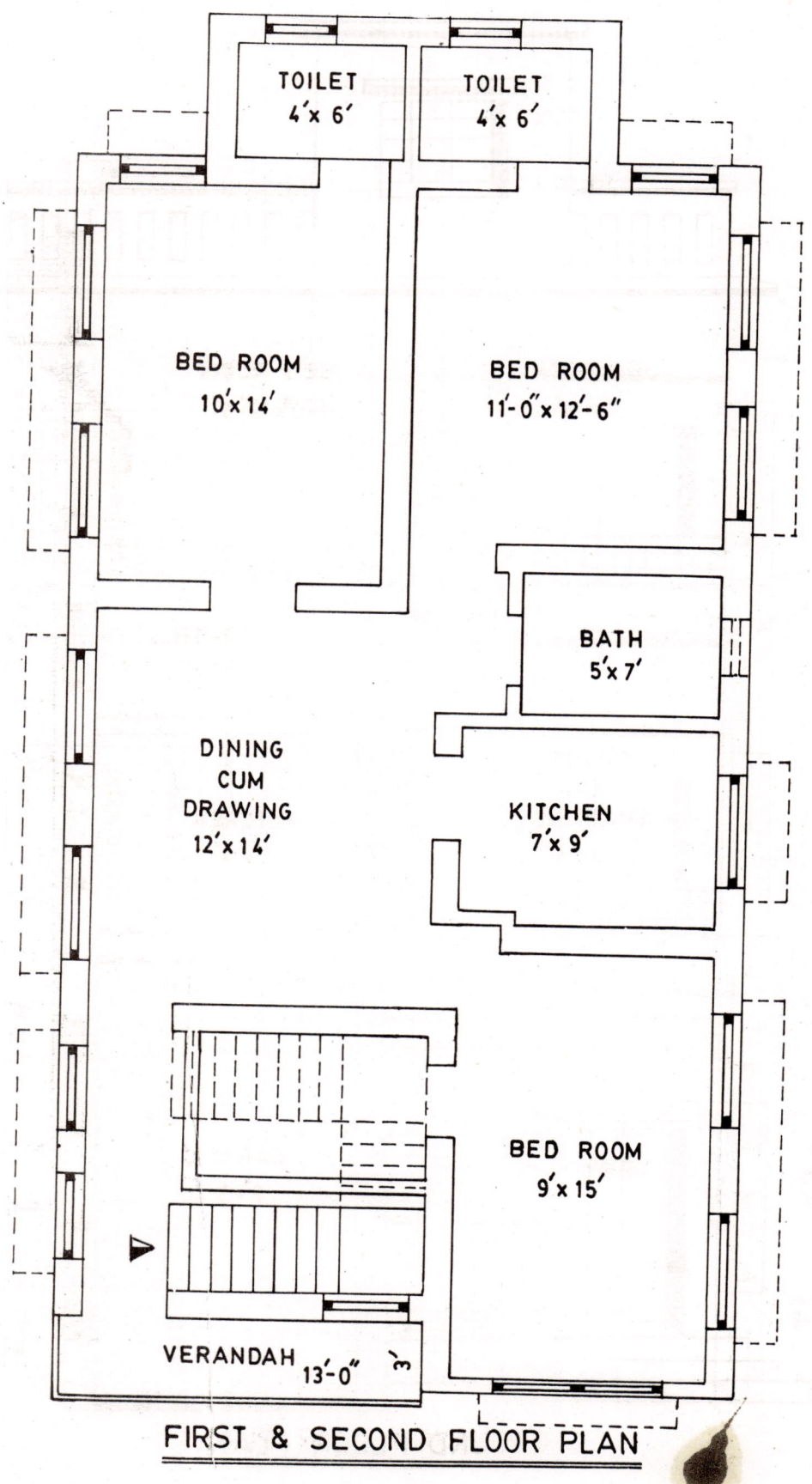

TOILET 4'x 6'

TOILET 4'x 6'

BED ROOM 10'x 14'

BED ROOM 11'-0"x 12'-6"

BATH 5'x 7'

DINING CUM DRAWING 12'x 14'

KITCHEN 7'x 9'

BED ROOM 9'x 15'

VERANDAH 13'-0" x 3'

FIRST & SECOND FLOOR PLAN

ELEVATION

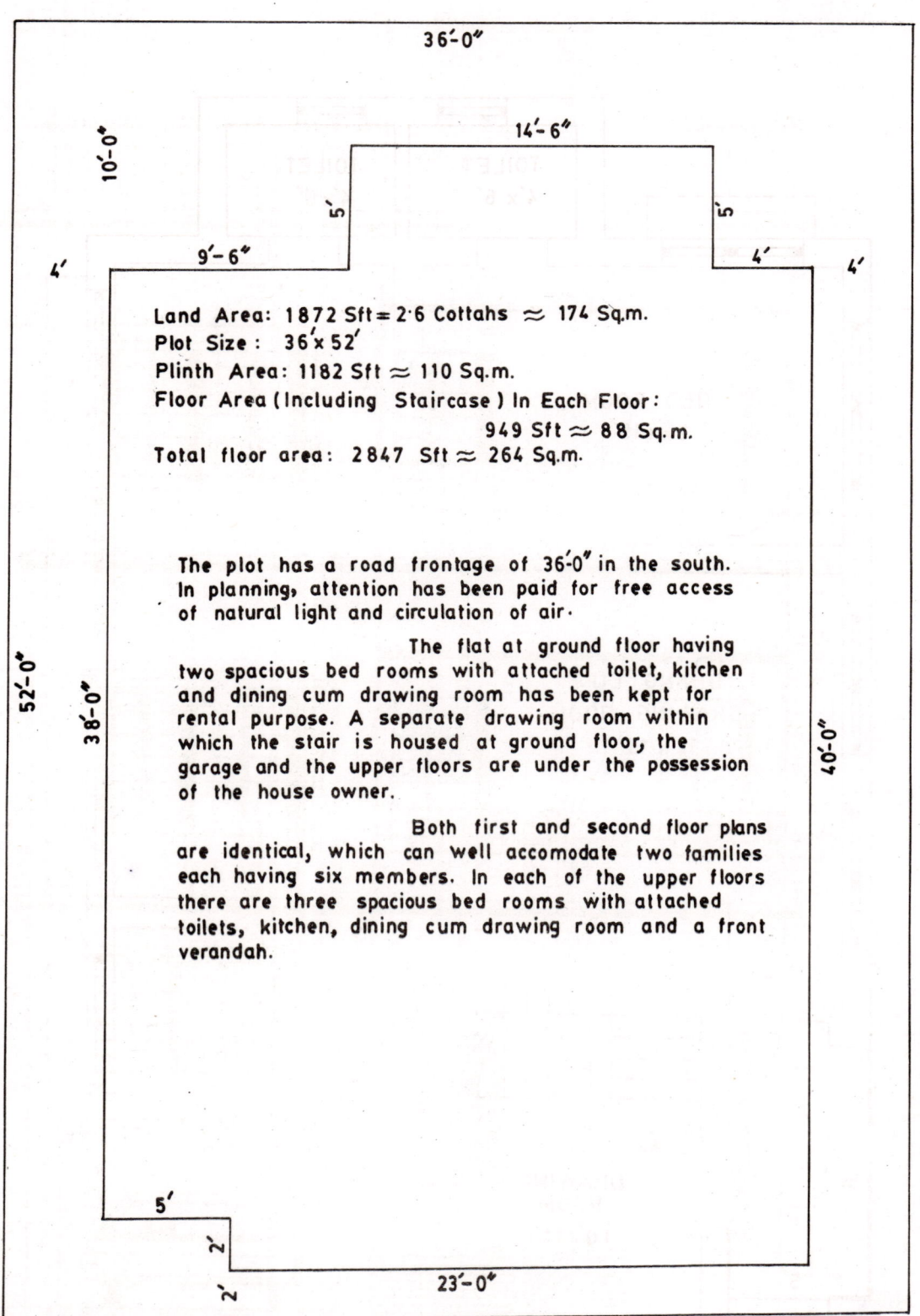

36'-0"

14'-6"

10'-0"

5'

5'

9'-6"

4'

4'

4'

Land Area: 1872 Sft = 2·6 Cottahs ≈ 174 Sq.m.
Plot Size : 36' x 52'
Plinth Area: 1182 Sft ≈ 110 Sq.m.
Floor Area (Including Staircase) In Each Floor:
                    949 Sft ≈ 88 Sq.m.
Total floor area: 2847 Sft ≈ 264 Sq.m.

The plot has a road frontage of 36'-0" in the south. In planning, attention has been paid for free access of natural light and circulation of air·

The flat at ground floor having two spacious bed rooms with attached toilet, kitchen and dining cum drawing room has been kept for rental purpose. A separate drawing room within which the stair is housed at ground floor, the garage and the upper floors are under the possession of the house owner.

Both first and second floor plans are identical, which can well accomodate two families each having six members. In each of the upper floors there are three spacious bed rooms with attached toilets, kitchen, dining cum drawing room and a front verandah.

52'-0"

36'-0"

40'-0"

5'

2'

2'

23'-0"

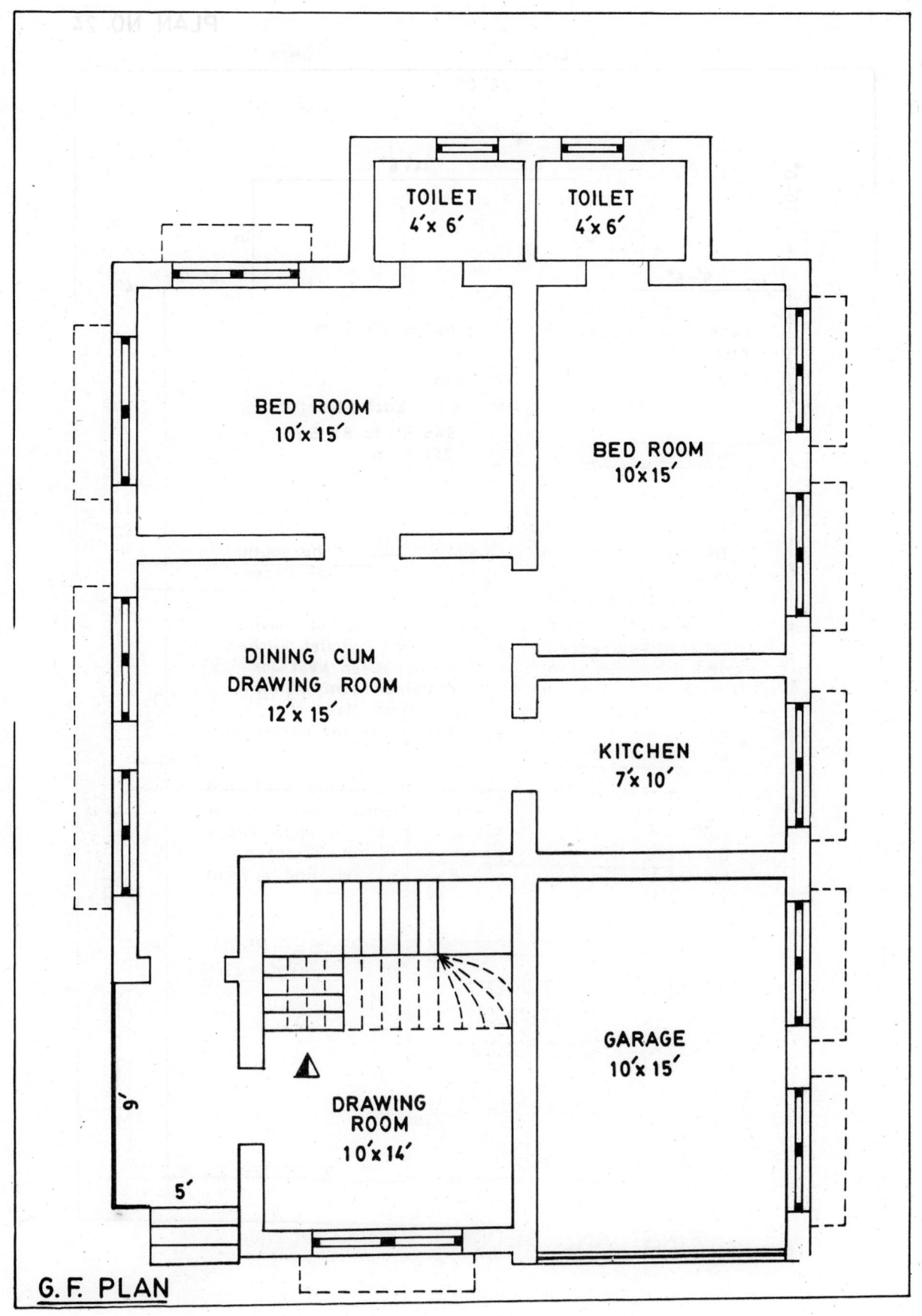

TOILET
4'x 6'

TOILET
4'x 6'

BED ROOM
10'x 15'

BED ROOM
10'x 15'

DINING CUM
DRAWING ROOM
12'x 15'

KITCHEN
7'x 10'

GARAGE
10'x 15'

DRAWING
ROOM
10'x 14'

9'

5'

G.F. PLAN

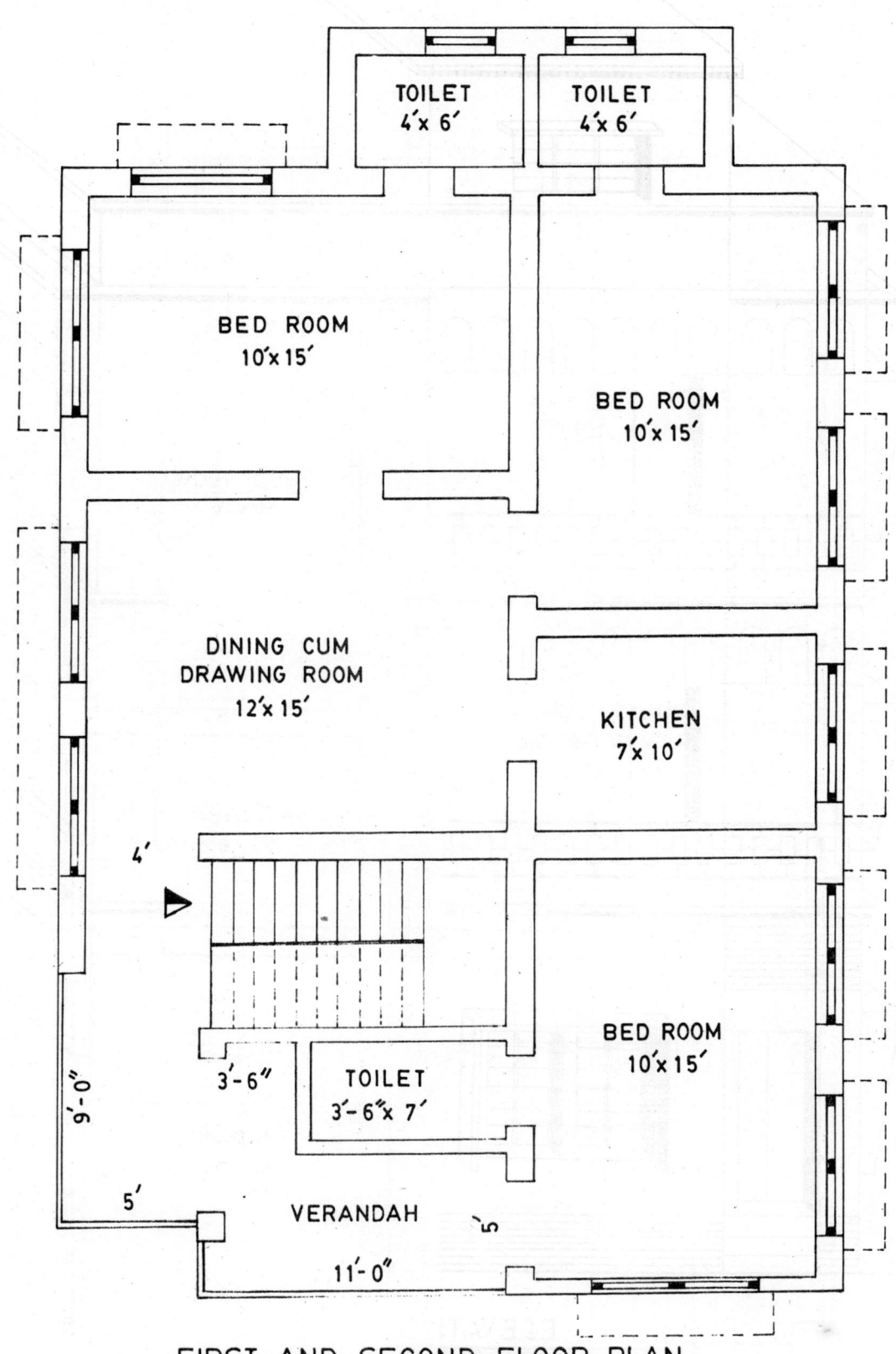

FIRST AND SECOND FLOOR PLAN

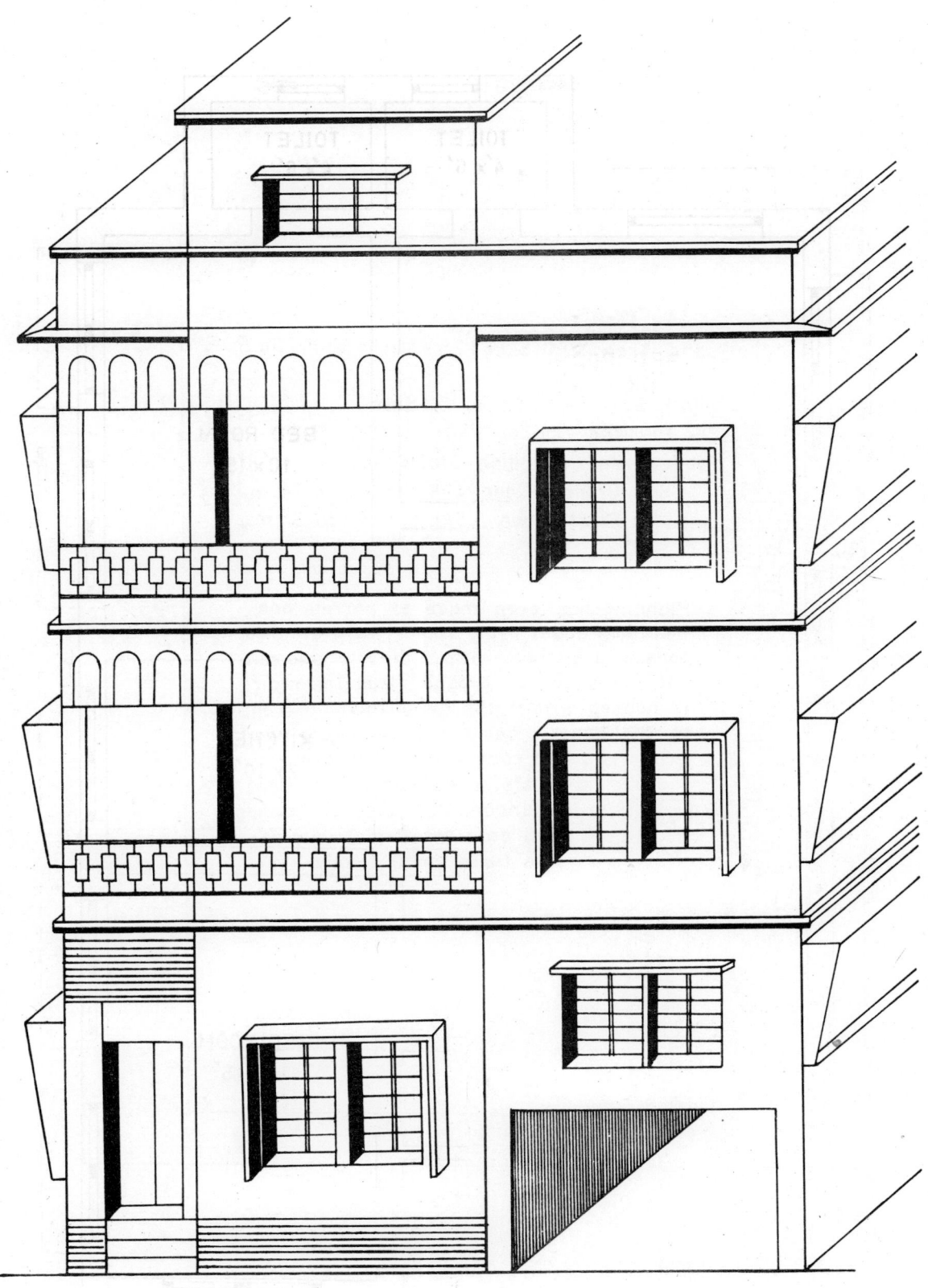

ELEVATION

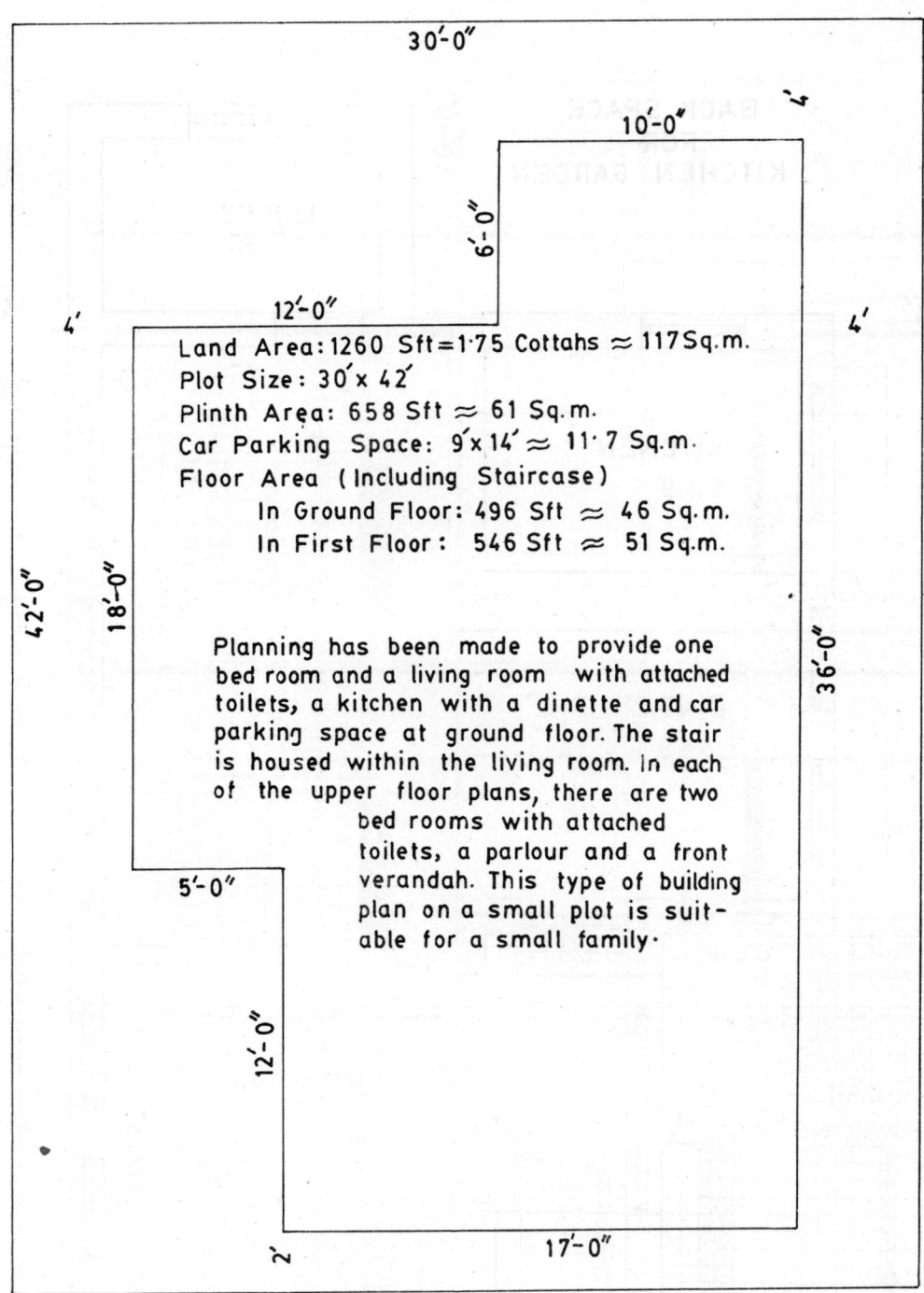

30'-0"

10'-0"

6'-0"

12'-0"

4'

4'

42'-0"

18'-0"

36'-0"

Land Area: 1260 Sft = 1·75 Cottahs ≈ 117 Sq.m.
Plot Size: 30' x 42'
Plinth Area: 658 Sft ≈ 61 Sq.m.
Car Parking Space: 9' x 14' ≈ 11·7 Sq.m.
Floor Area (Including Staircase)
    In Ground Floor: 496 Sft ≈ 46 Sq.m.
    In First Floor: 546 Sft ≈ 51 Sq.m.

Planning has been made to provide one
bed room and a living room with attached
toilets, a kitchen with a dinette and car
parking space at ground floor. The stair
is housed within the living room. In each
of the upper floor plans, there are two
bed rooms with attached
toilets, a parlour and a front
verandah. This type of building
plan on a small plot is suitable for a small family·

5'-0"

12'-0"

2'

17'-0"

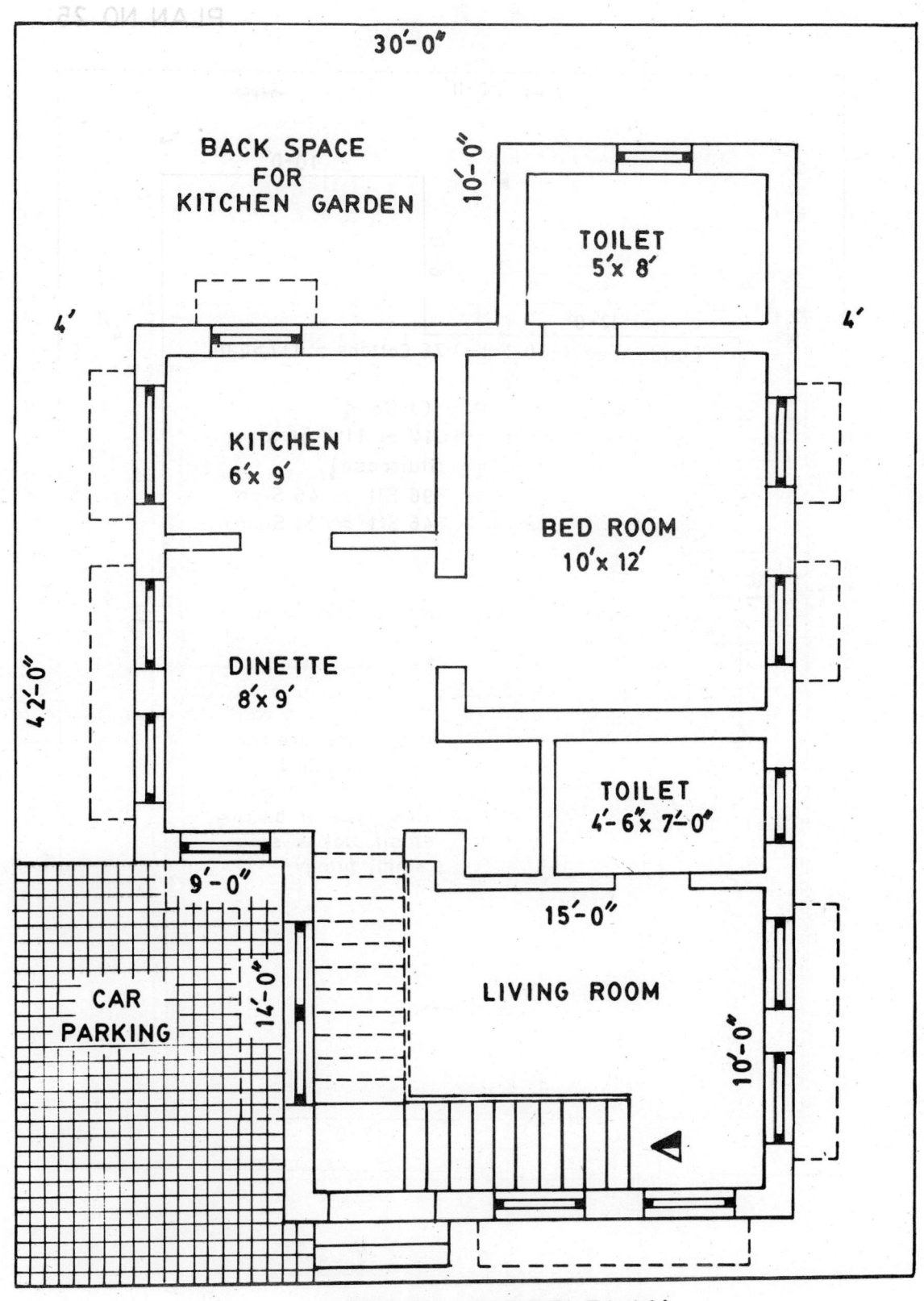

GROUND FLOOR PLAN

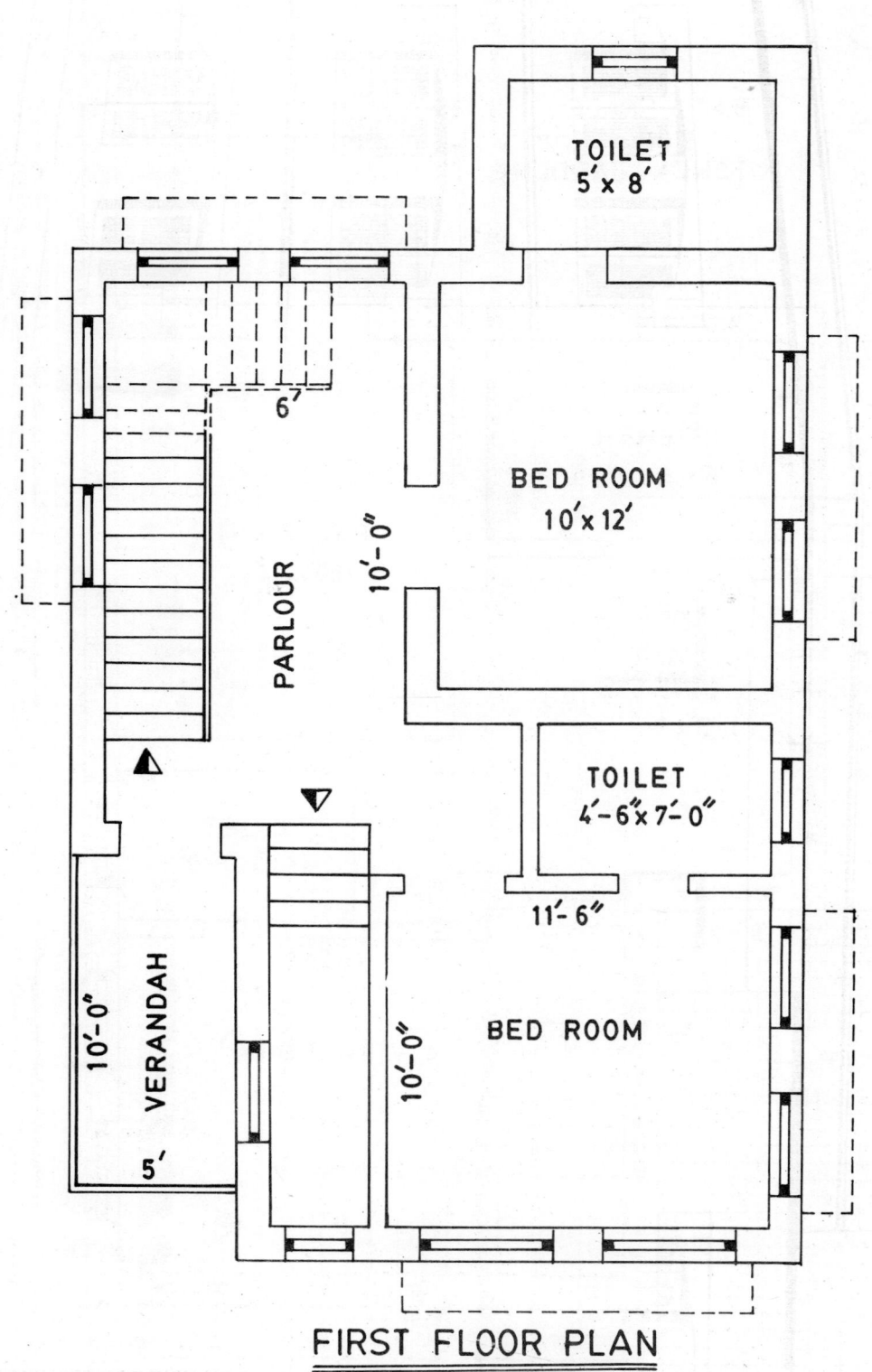

TOILET
5' x 8'

BED ROOM
10'x 12'

10'-0"

PARLOUR

6'

TOILET
4'-6"x 7'-0"

11'-6"

10'-0"

VERANDAH

10'-0"

BED ROOM

10'-0"

5'

FIRST FLOOR PLAN

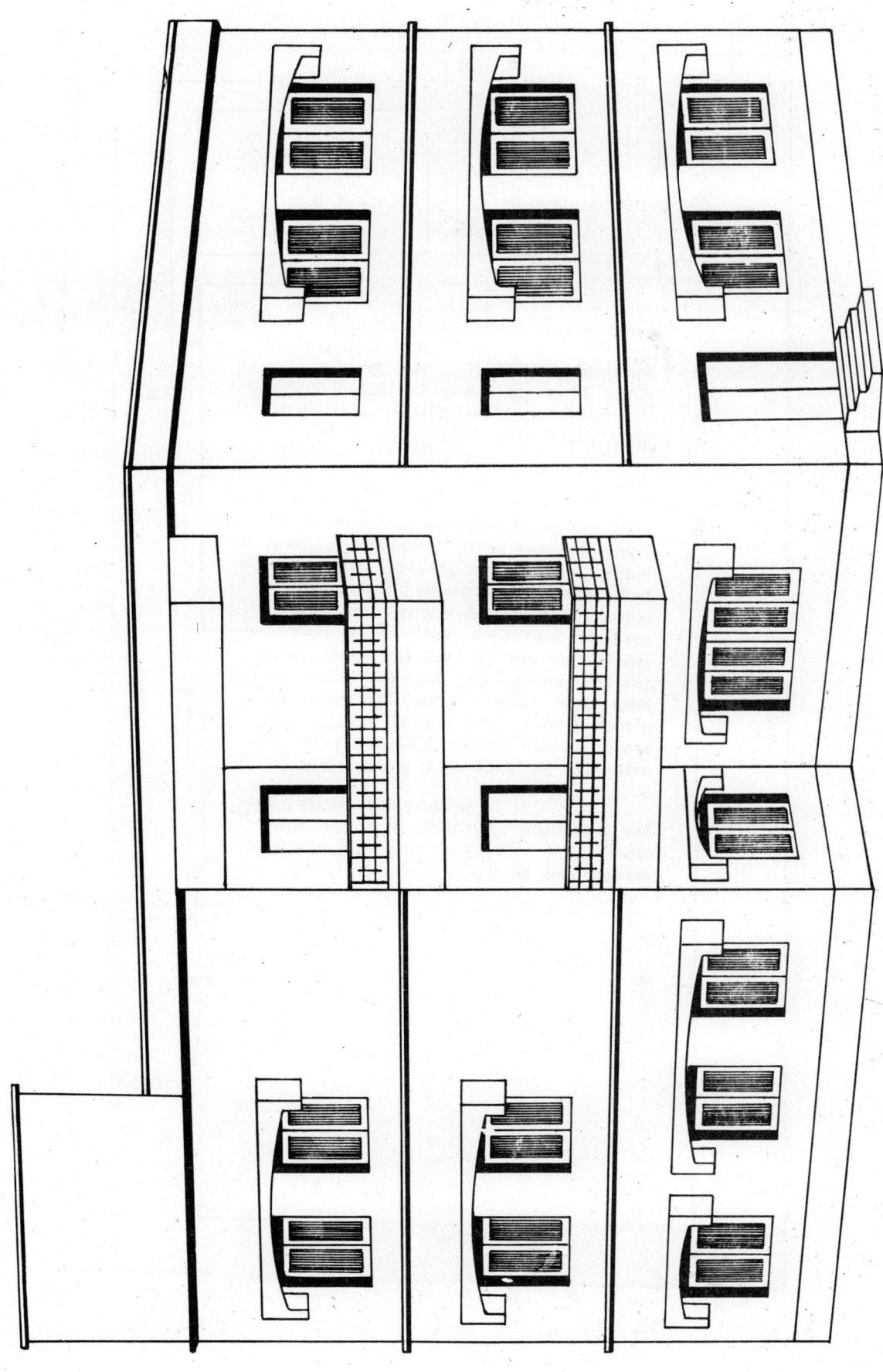

PLAN NO. 26

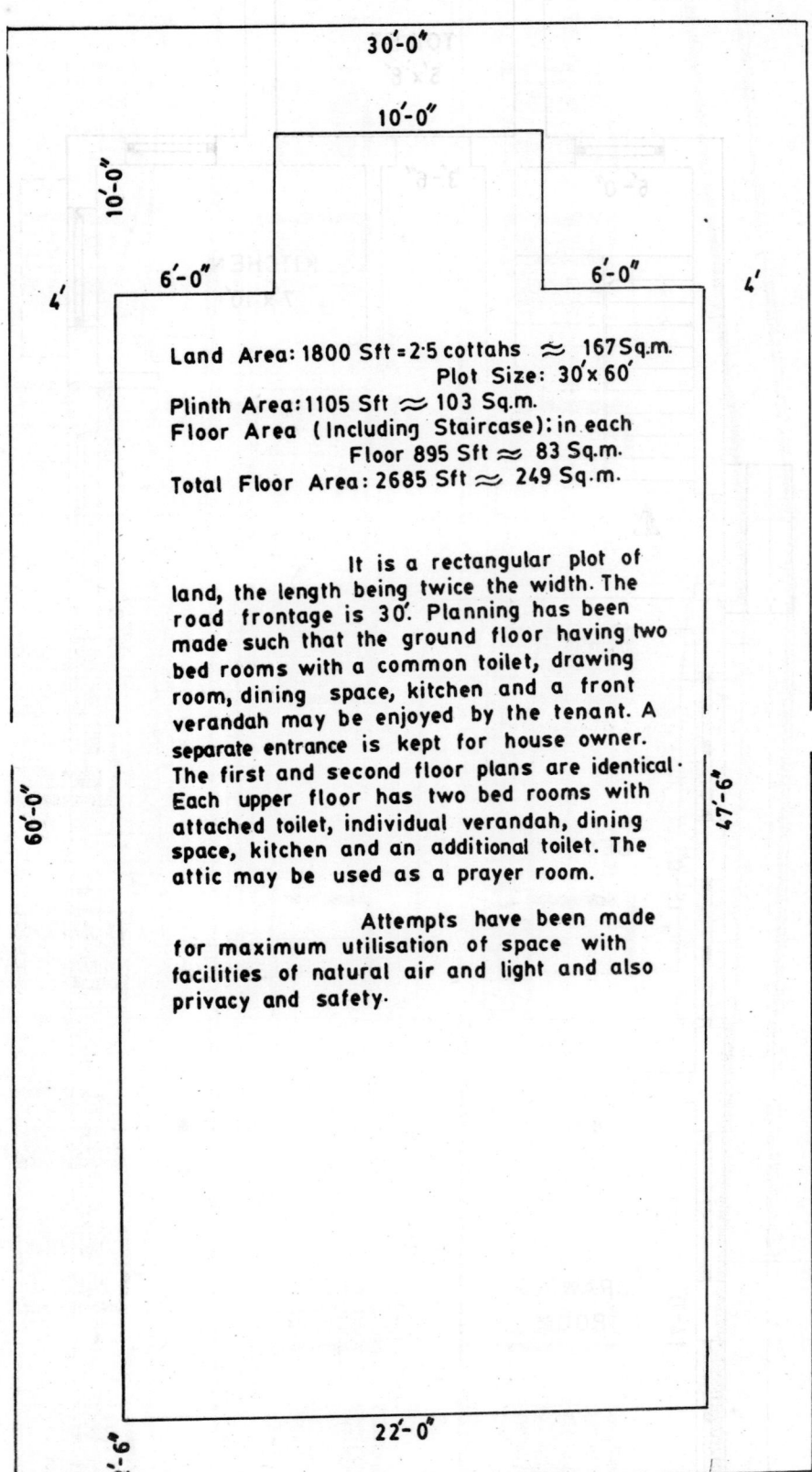

Land Area: 1800 Sft = 2·5 cottahs ≈ 167 Sq.m.
Plot Size: 30'x 60'
Plinth Area: 1105 Sft ≈ 103 Sq.m.
Floor Area (Including Staircase): in each Floor 895 Sft ≈ 83 Sq.m.
Total Floor Area: 2685 Sft ≈ 249 Sq.m.

It is a rectangular plot of land, the length being twice the width. The road frontage is 30'. Planning has been made such that the ground floor having two bed rooms with a common toilet, drawing room, dining space, kitchen and a front verandah may be enjoyed by the tenant. A separate entrance is kept for house owner. The first and second floor plans are identical. Each upper floor has two bed rooms with attached toilet, individual verandah, dining space, kitchen and an additional toilet. The attic may be used as a prayer room.

Attempts have been made for maximum utilisation of space with facilities of natural air and light and also privacy and safety.

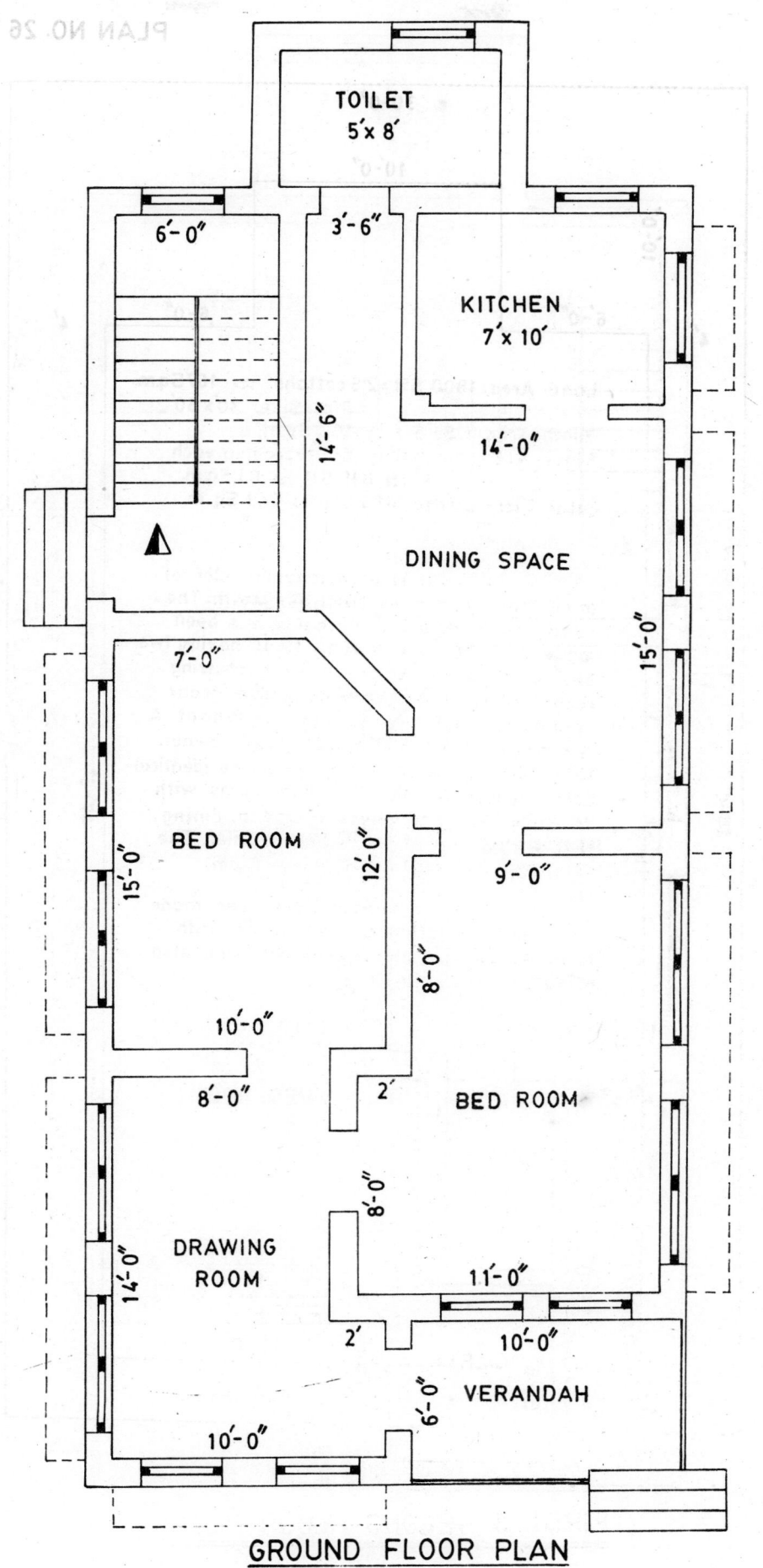

GROUND FLOOR PLAN

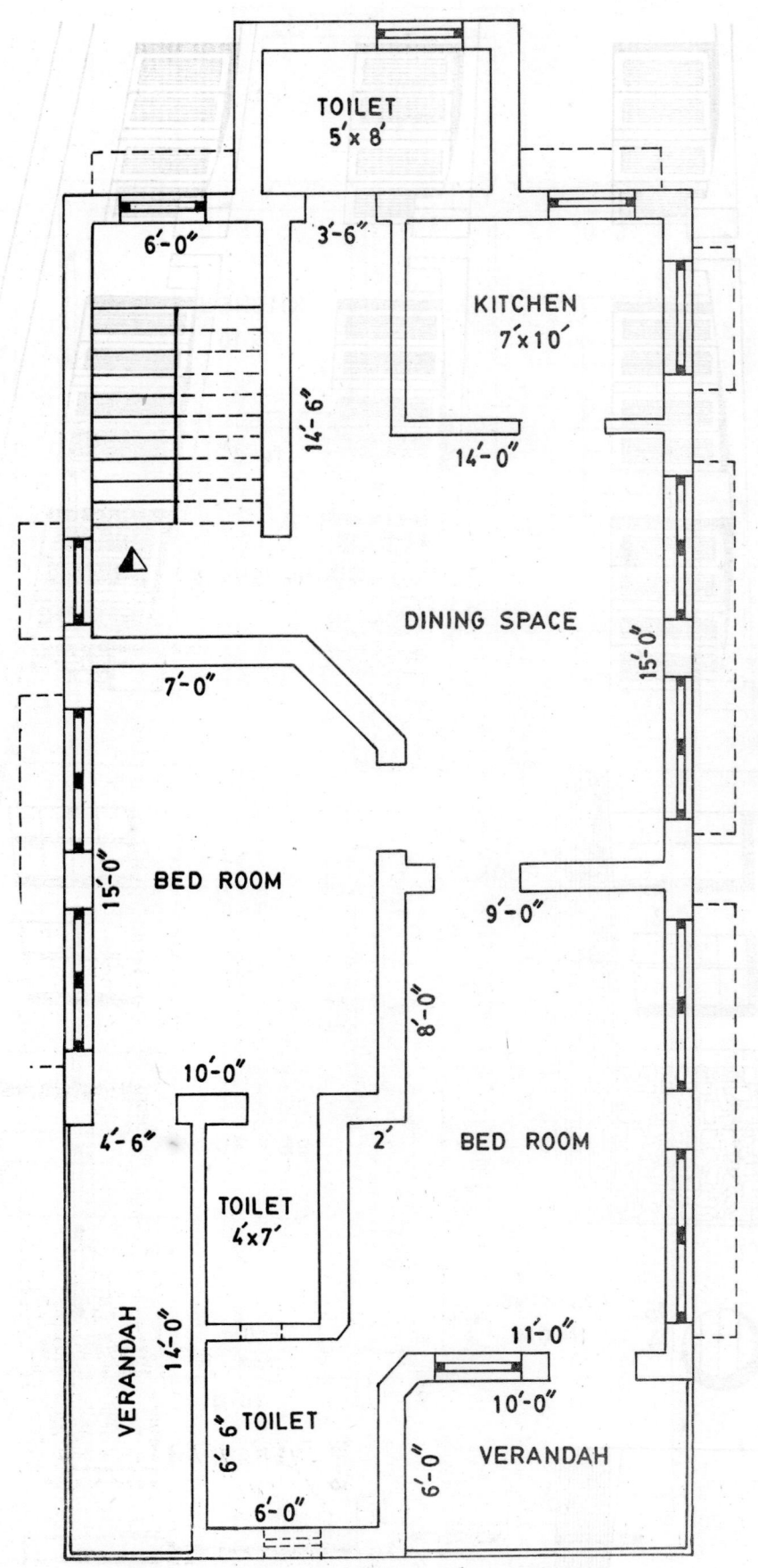

TOILET
5' x 8'

6'-0"   3'-6"

KITCHEN
7' x 10'

14'-6"

14'-0"

DINING SPACE

15'-0"

7'-0"

BED ROOM

15'-0"

9'-0"

8'-0"

10'-0"

2'

BED ROOM

4'-6"

TOILET
4' x 7'

11'-0"

VERANDAH

14'-0"

6'-6"   TOILET

10'-0"

6'-0"

VERANDAH

6'-0"

FIRST & SECOND FLOOR PLAN

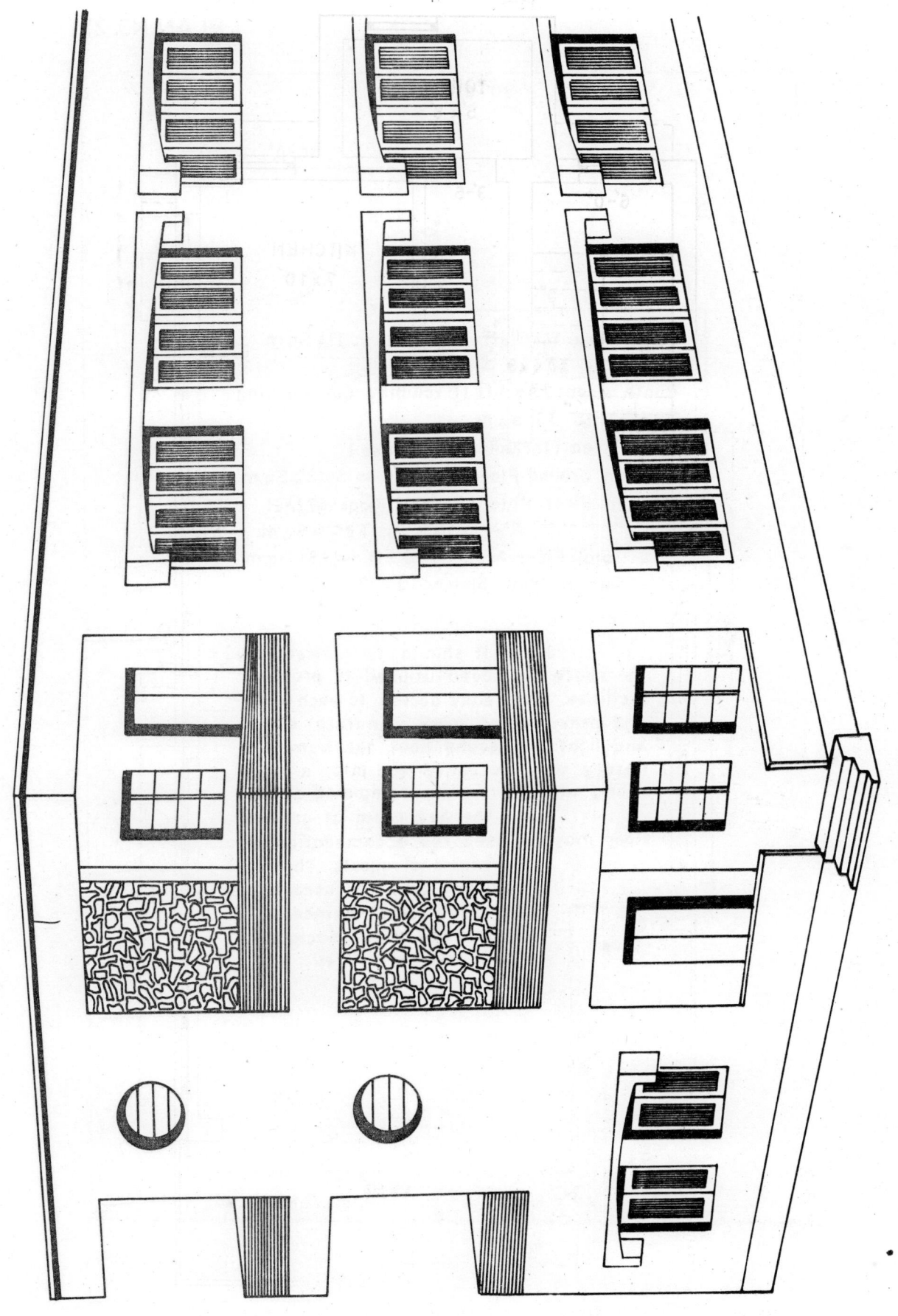

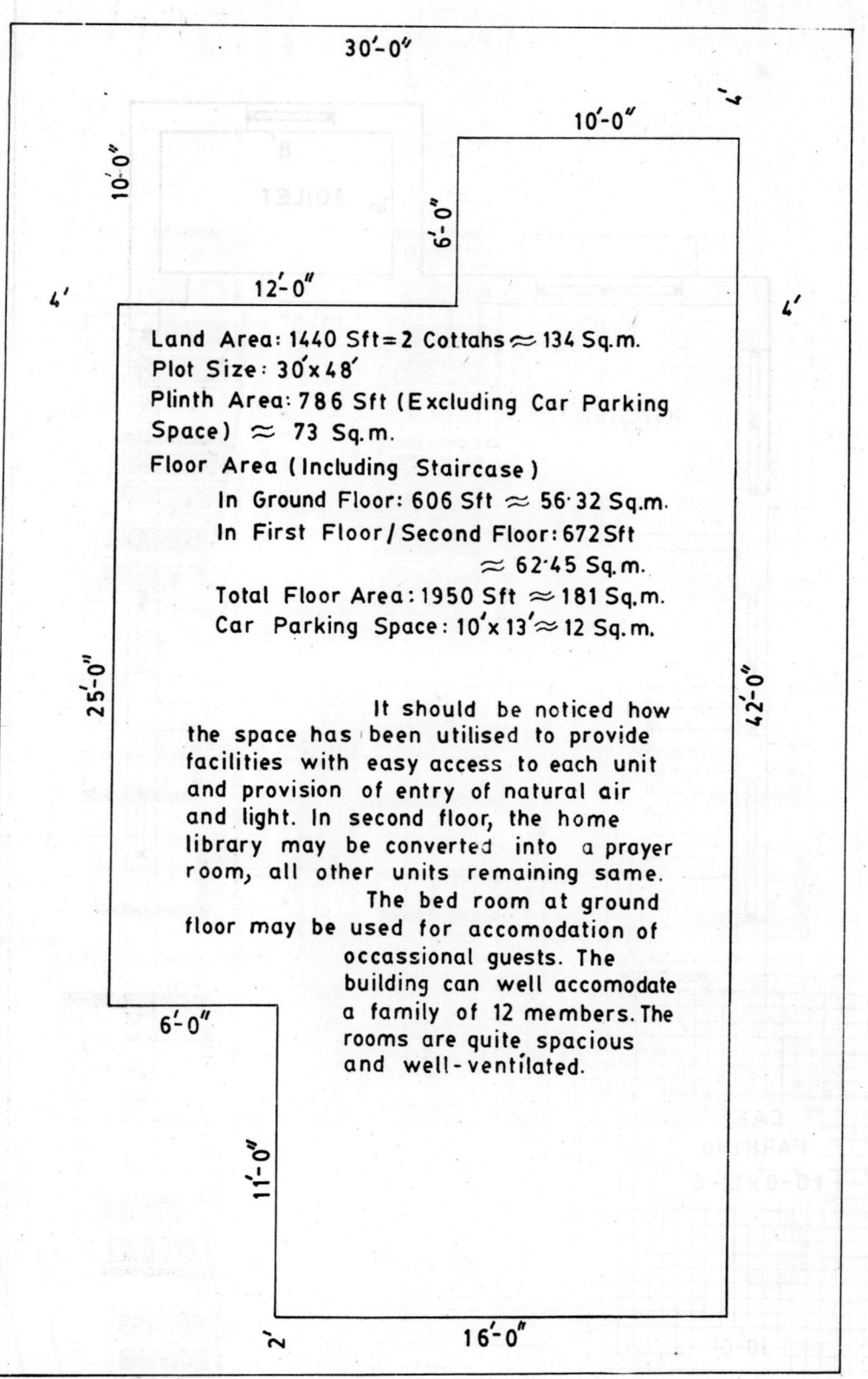

Land Area: 1440 Sft = 2 Cottahs ≈ 134 Sq.m.
Plot Size: 30'x48'
Plinth Area: 786 Sft (Excluding Car Parking Space) ≈ 73 Sq.m.
Floor Area (Including Staircase)
In Ground Floor: 606 Sft ≈ 56.32 Sq.m.
In First Floor/Second Floor: 672 Sft ≈ 62.45 Sq.m.
Total Floor Area: 1950 Sft ≈ 181 Sq.m.
Car Parking Space: 10'x13' ≈ 12 Sq.m.

It should be noticed how the space has been utilised to provide facilities with easy access to each unit and provision of entry of natural air and light. In second floor, the home library may be converted into a prayer room, all other units remaining same.

The bed room at ground floor may be used for accomodation of occassional guests. The building can well accomodate a family of 12 members. The rooms are quite spacious and well-ventilated.

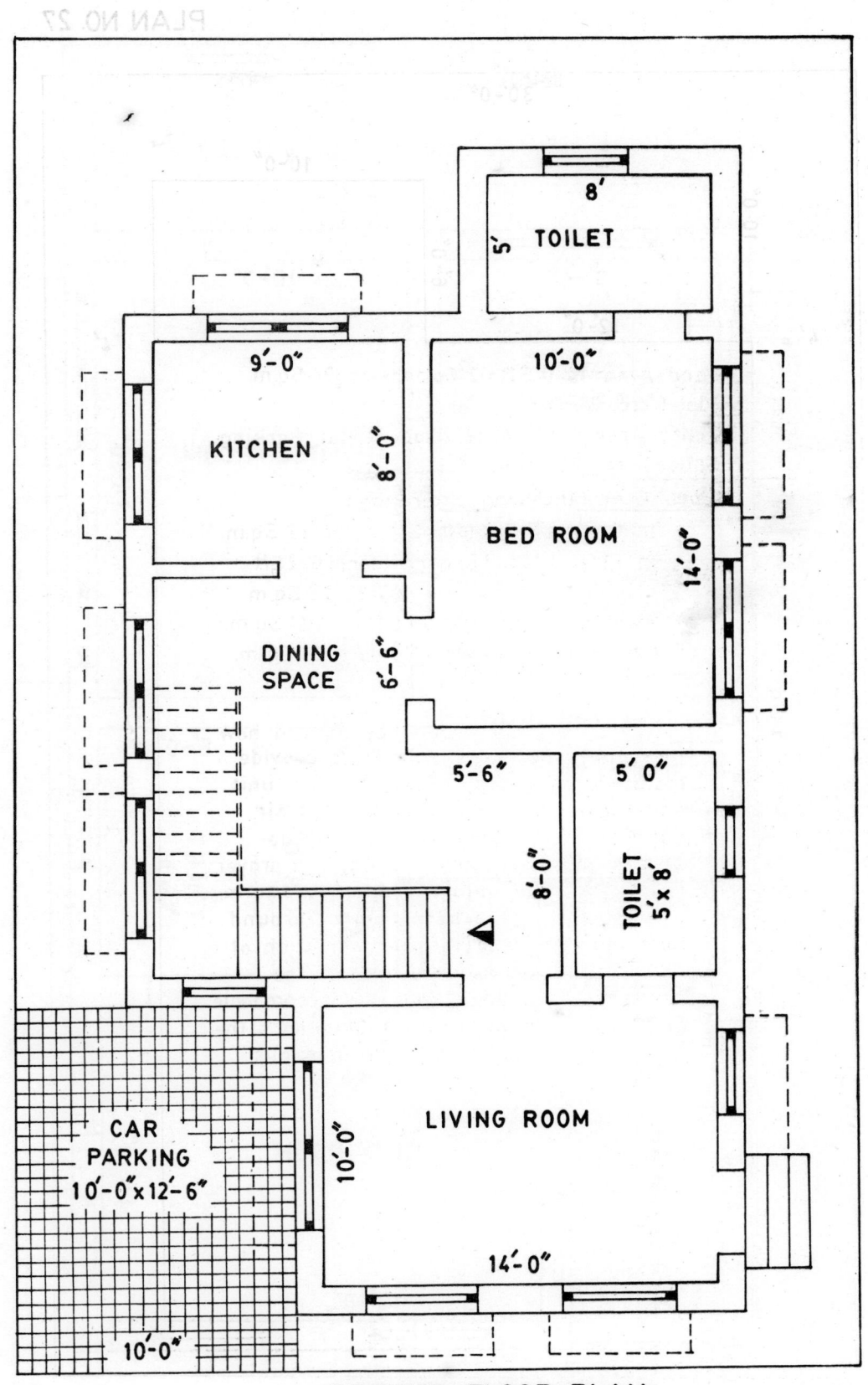

GROUND FLOOR PLAN

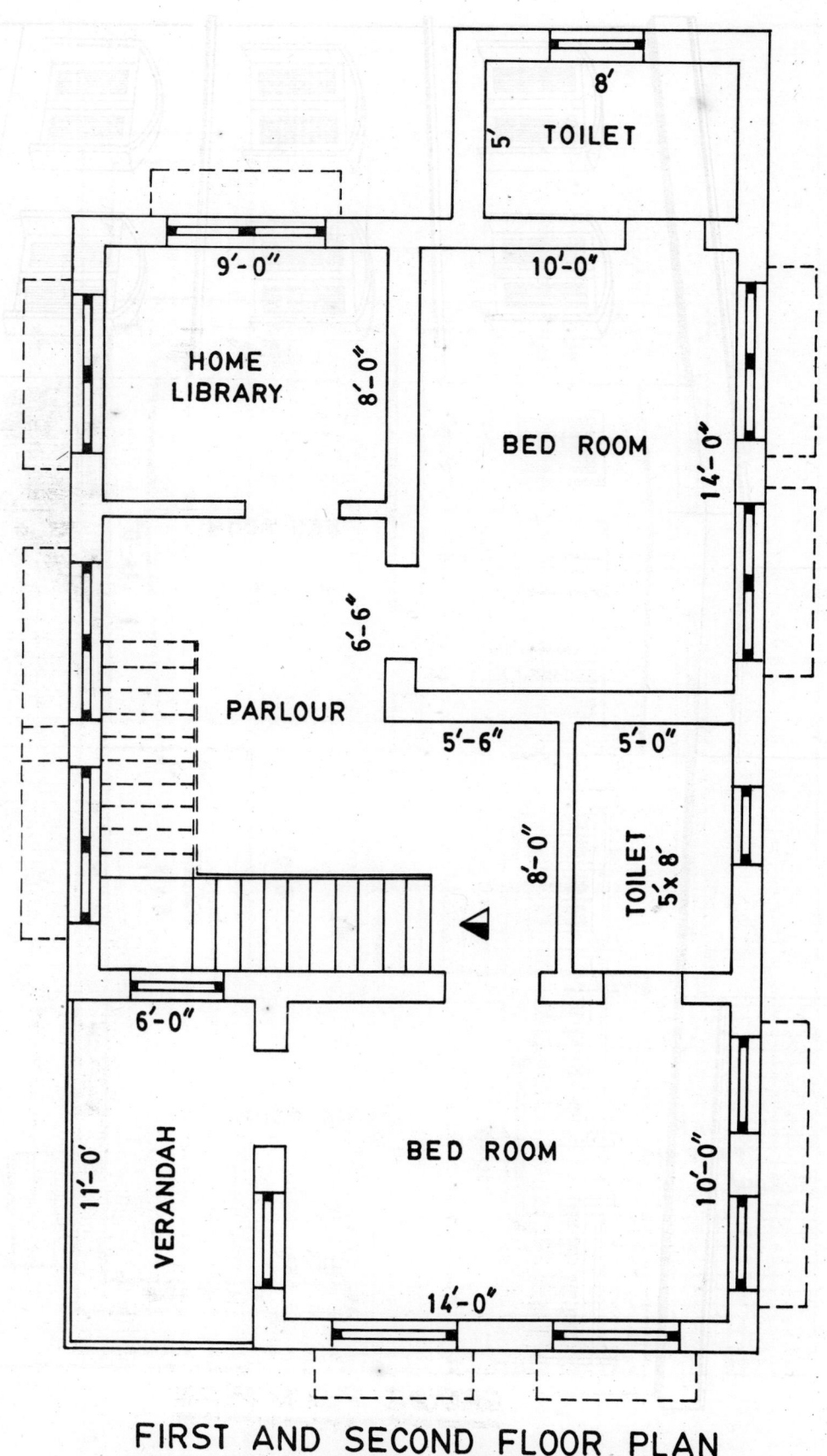

FIRST AND SECOND FLOOR PLAN

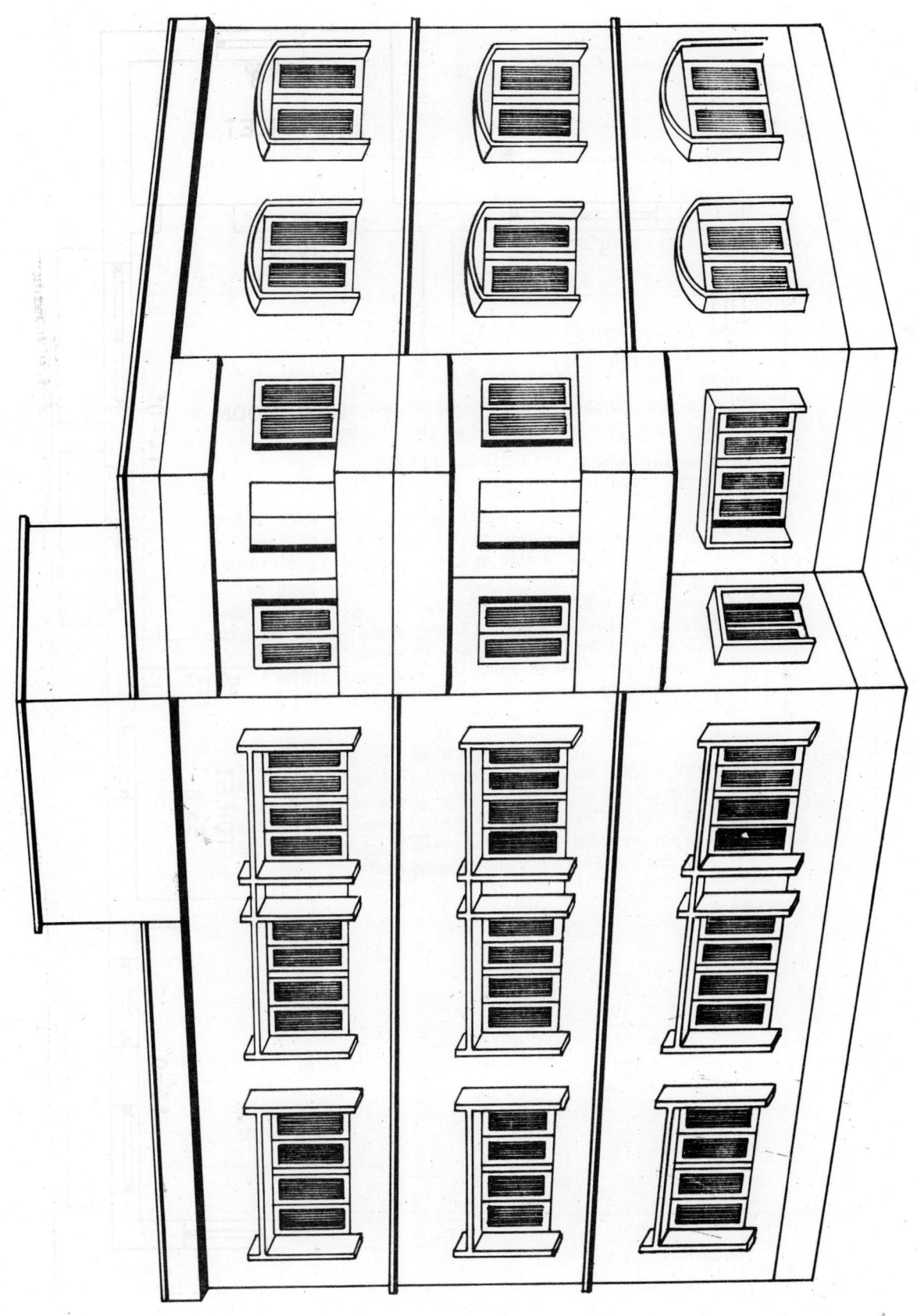

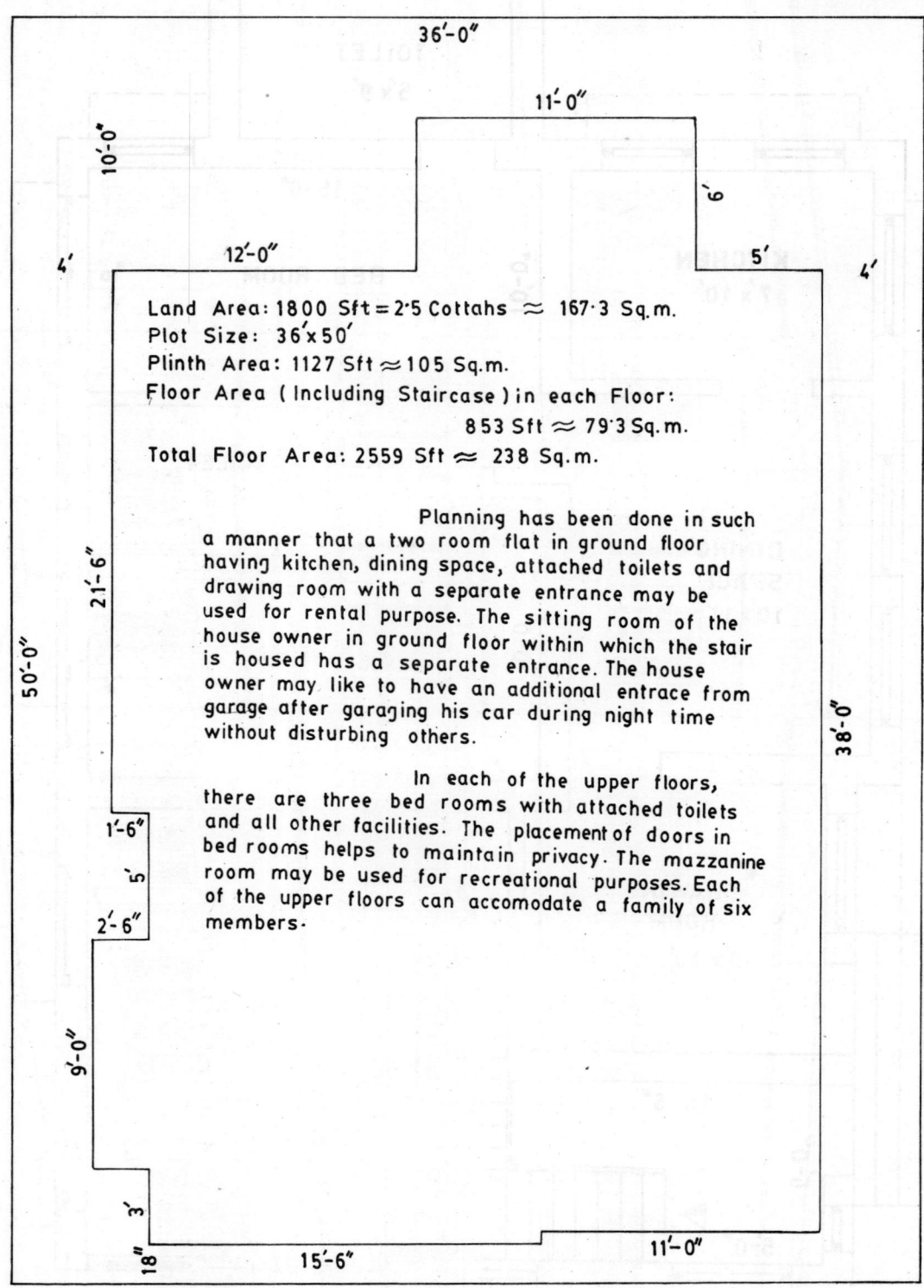

Land Area: 1800 Sft = 2·5 Cottahs ≈ 167·3 Sq.m.
Plot Size: 36'x50'
Plinth Area: 1127 Sft ≈ 105 Sq.m.
Floor Area (Including Staircase) in each Floor:
853 Sft ≈ 79·3 Sq.m.
Total Floor Area: 2559 Sft ≈ 238 Sq.m.

Planning has been done in such a manner that a two room flat in ground floor having kitchen, dining space, attached toilets and drawing room with a separate entrance may be used for rental purpose. The sitting room of the house owner in ground floor within which the stair is housed has a separate entrance. The house owner may like to have an additional entrace from garage after garaging his car during night time without disturbing others.

In each of the upper floors, there are three bed rooms with attached toilets and all other facilities. The placement of doors in bed rooms helps to maintain privacy. The mazzanine room may be used for recreational purposes. Each of the upper floors can accomodate a family of six members.

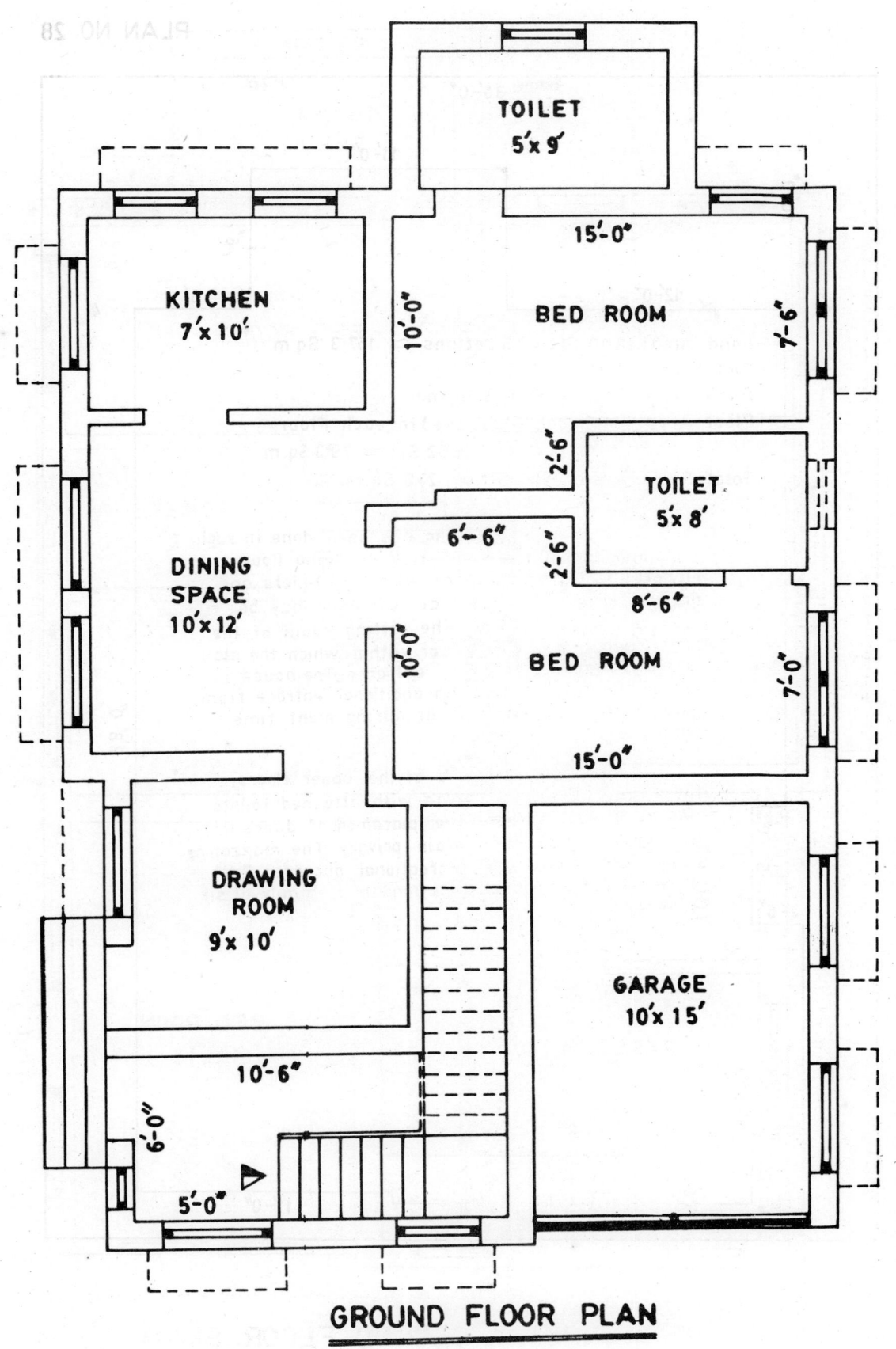

PLAN NO. 28

GROUND FLOOR PLAN

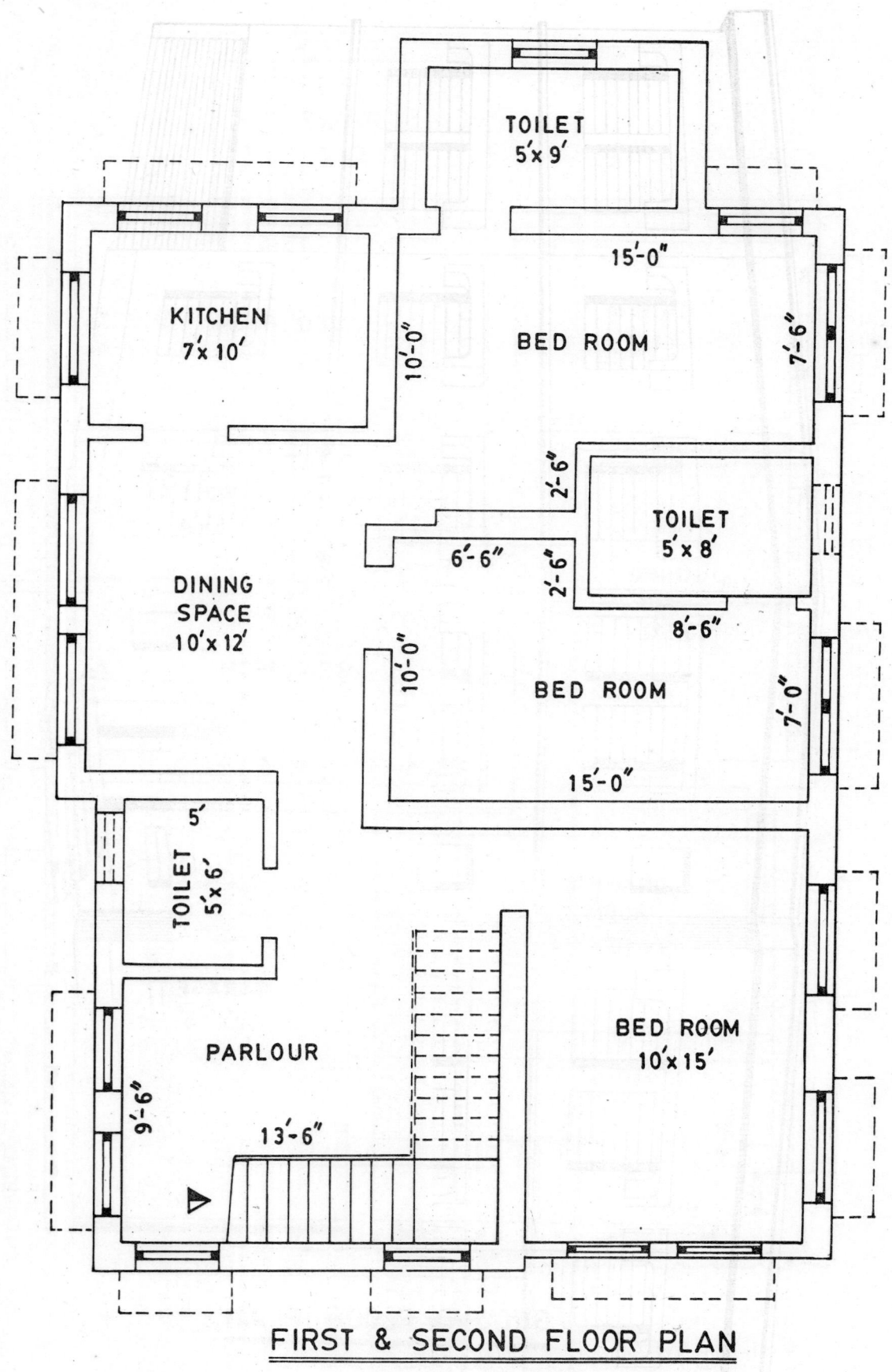

FIRST & SECOND FLOOR PLAN

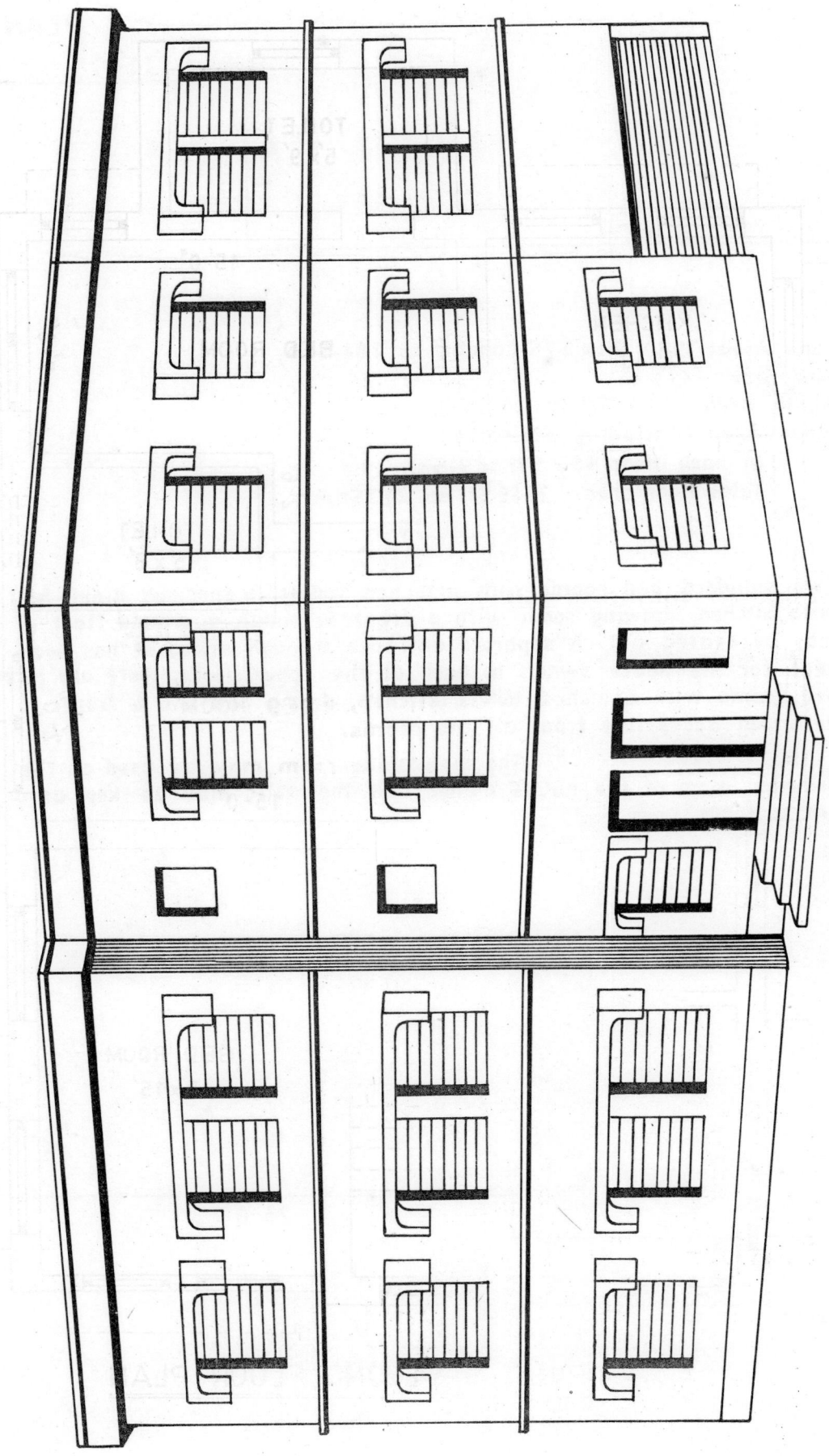

PLAN NO. 29

10'-0"

6'-0"

2'

24'-0"

Land Area: 1980 Sft = 2·75 Cottahs ≈ 184 Sq.n.
Plot Size : 44'x 45'
Plinth Area : 1282 Sft ≈ 119 Sq.m.
Floor Area (Including Staircase):
    In each floor 982 Sft ≈ 91Sq.m.
    Total floor area : 2946 Sft ≈ 273 Sq.m.

Two standard bed rooms with attached toilets, a spacious dining hall
with kitchen, drawing room with a front verandah in ground floor
may be rented out. A separate entrance through staircase has been
kept for the house owner. In each of the upper floors, there are three
bed rooms with attached toilets, kitchen, dining hall and a front
verandah accessible from all the rooms.

            The mezzanine room may be used as the
drawing room of the house owner and the attic may be kept as a
prayer room.

35'-0"

33'-6"

16'-0"

25'-0"

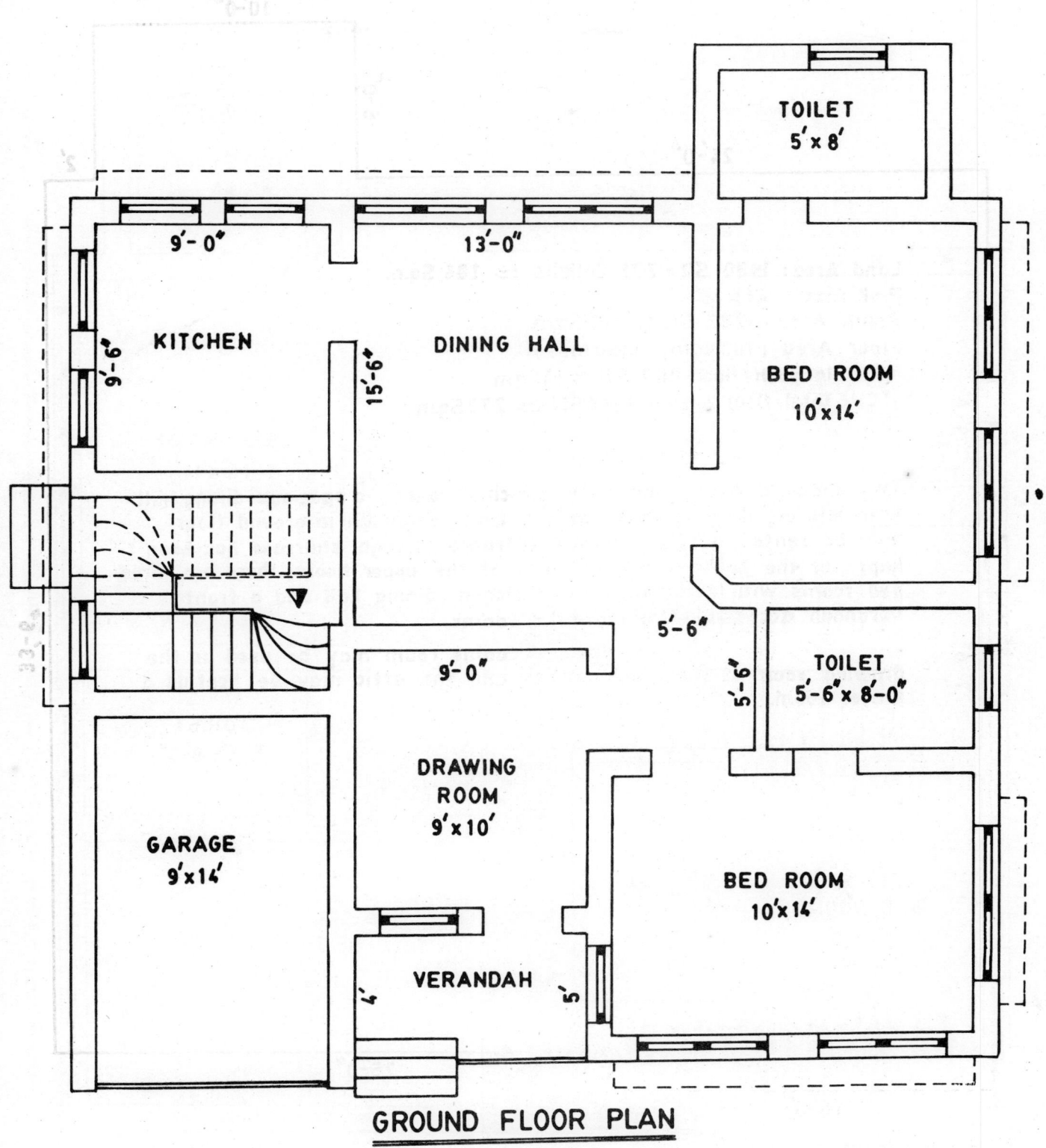

GROUND FLOOR PLAN

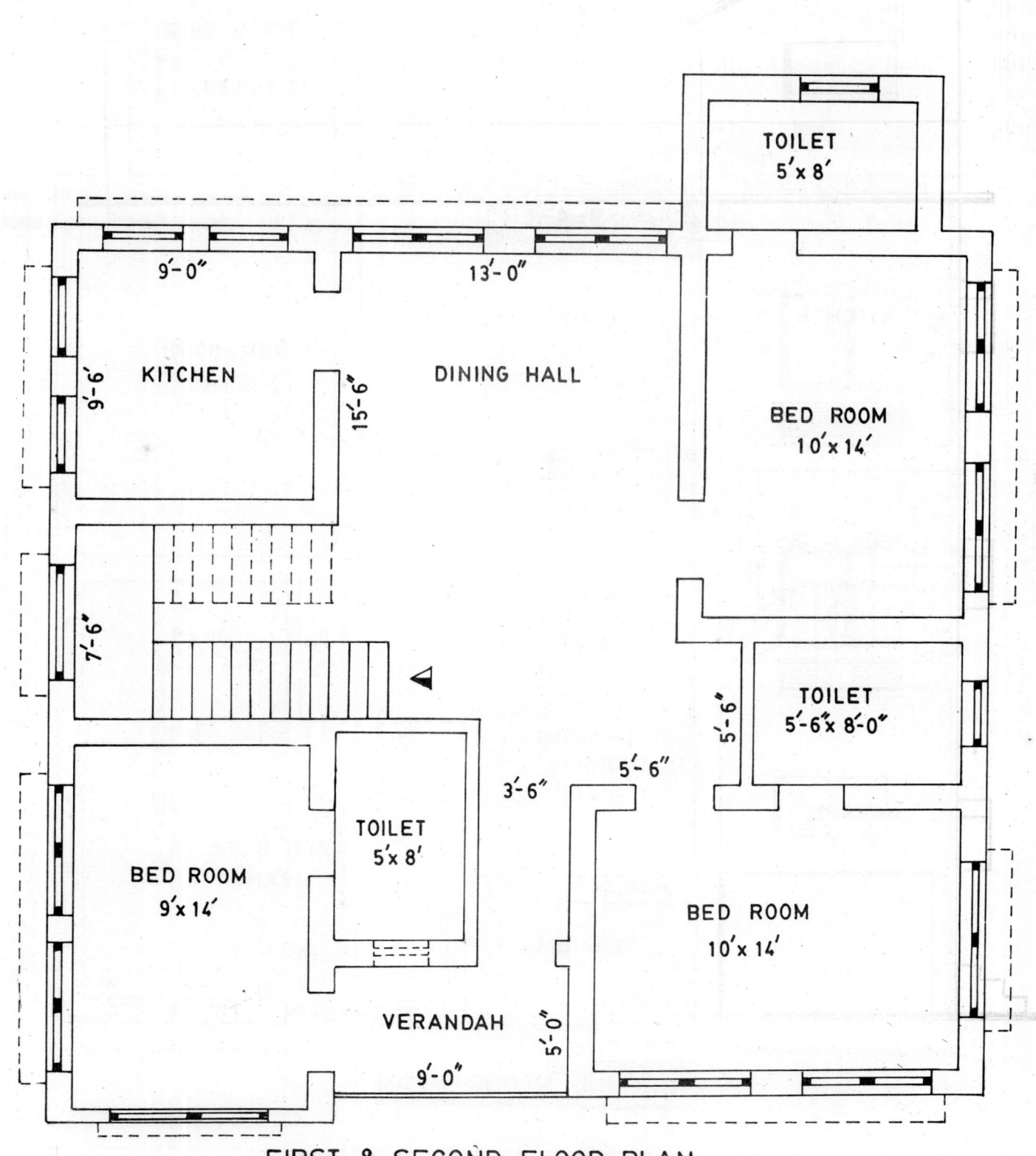

FIRST & SECOND FLOOR PLAN

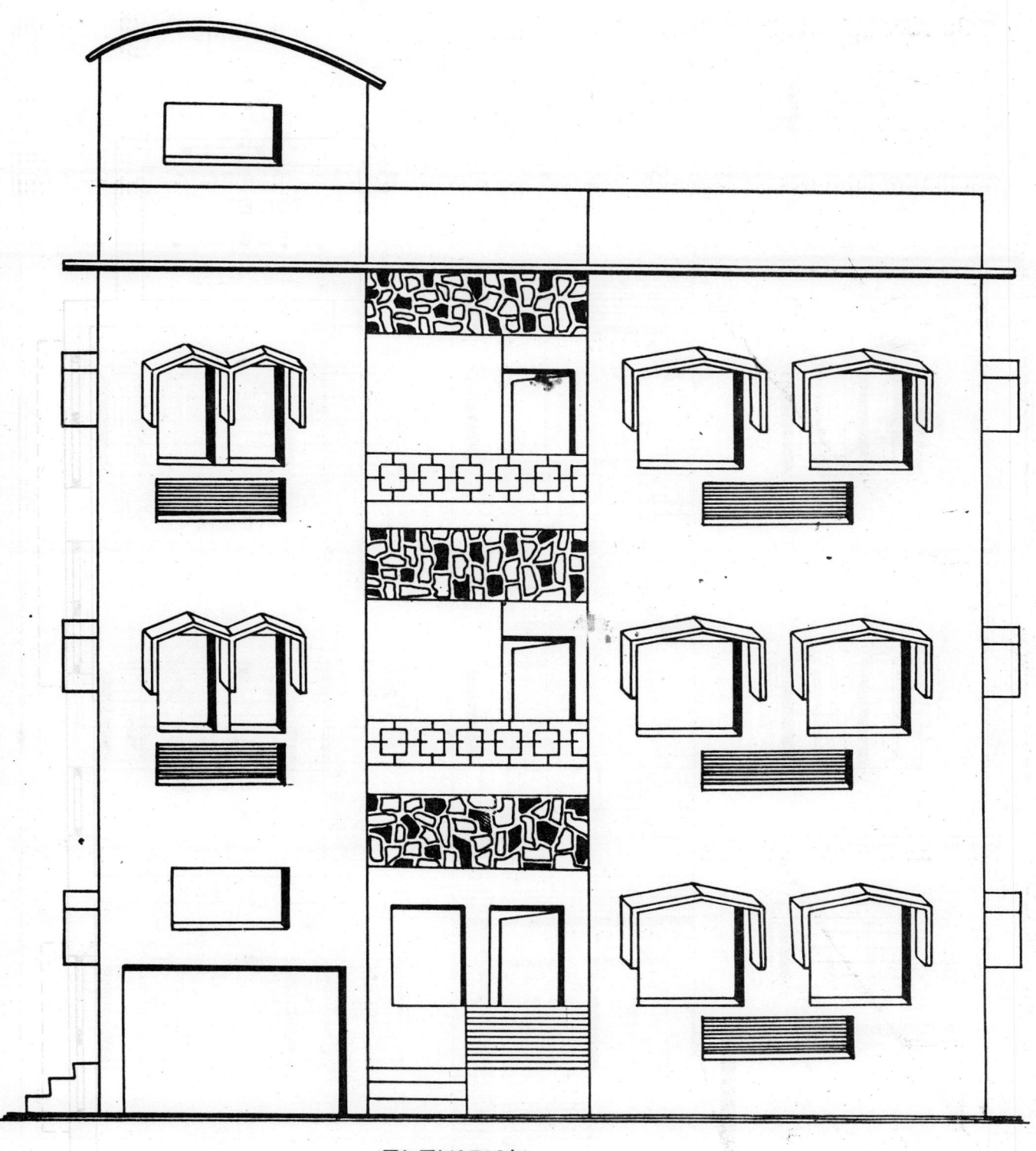

ELEVATION

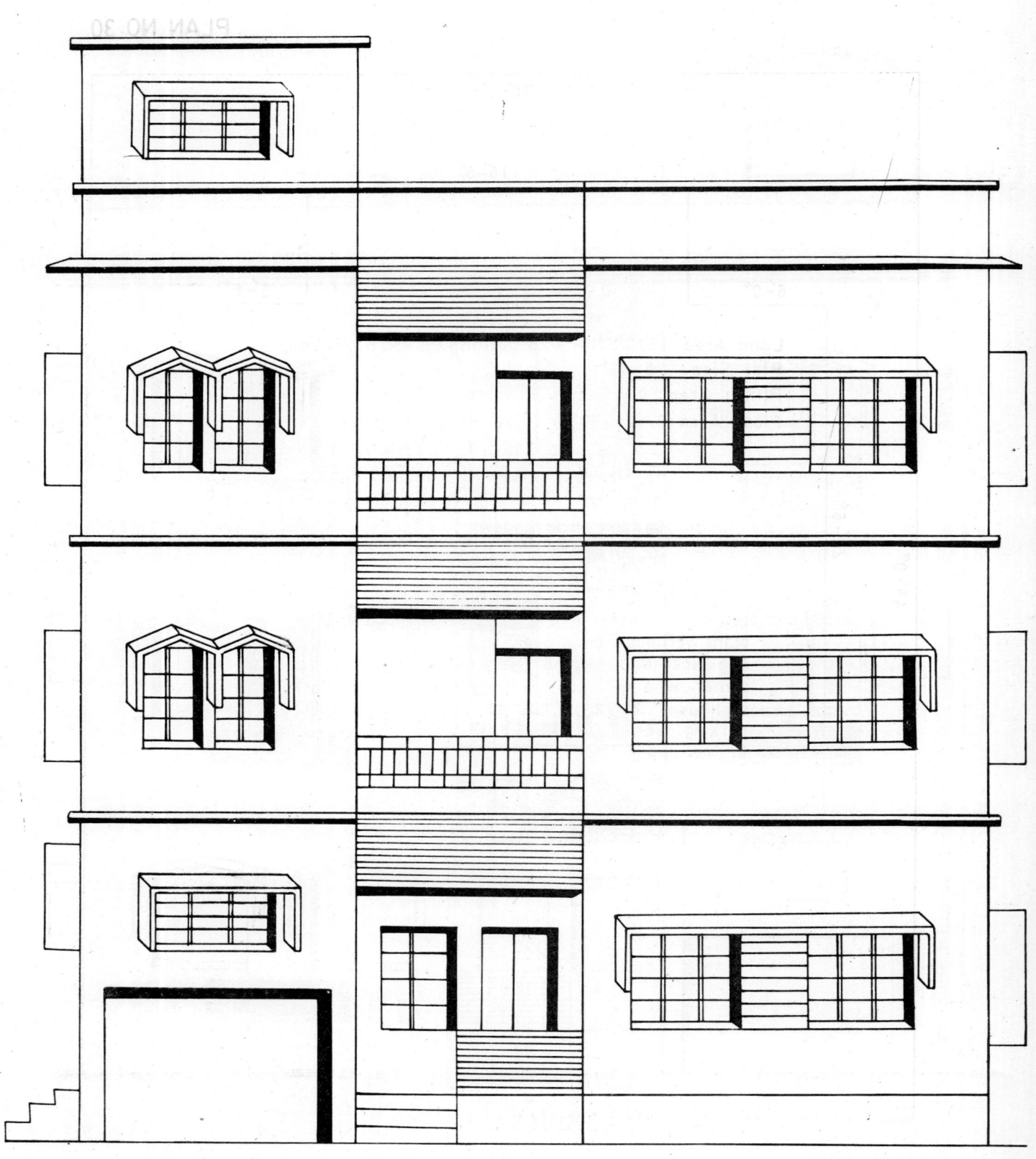

ELEVATION

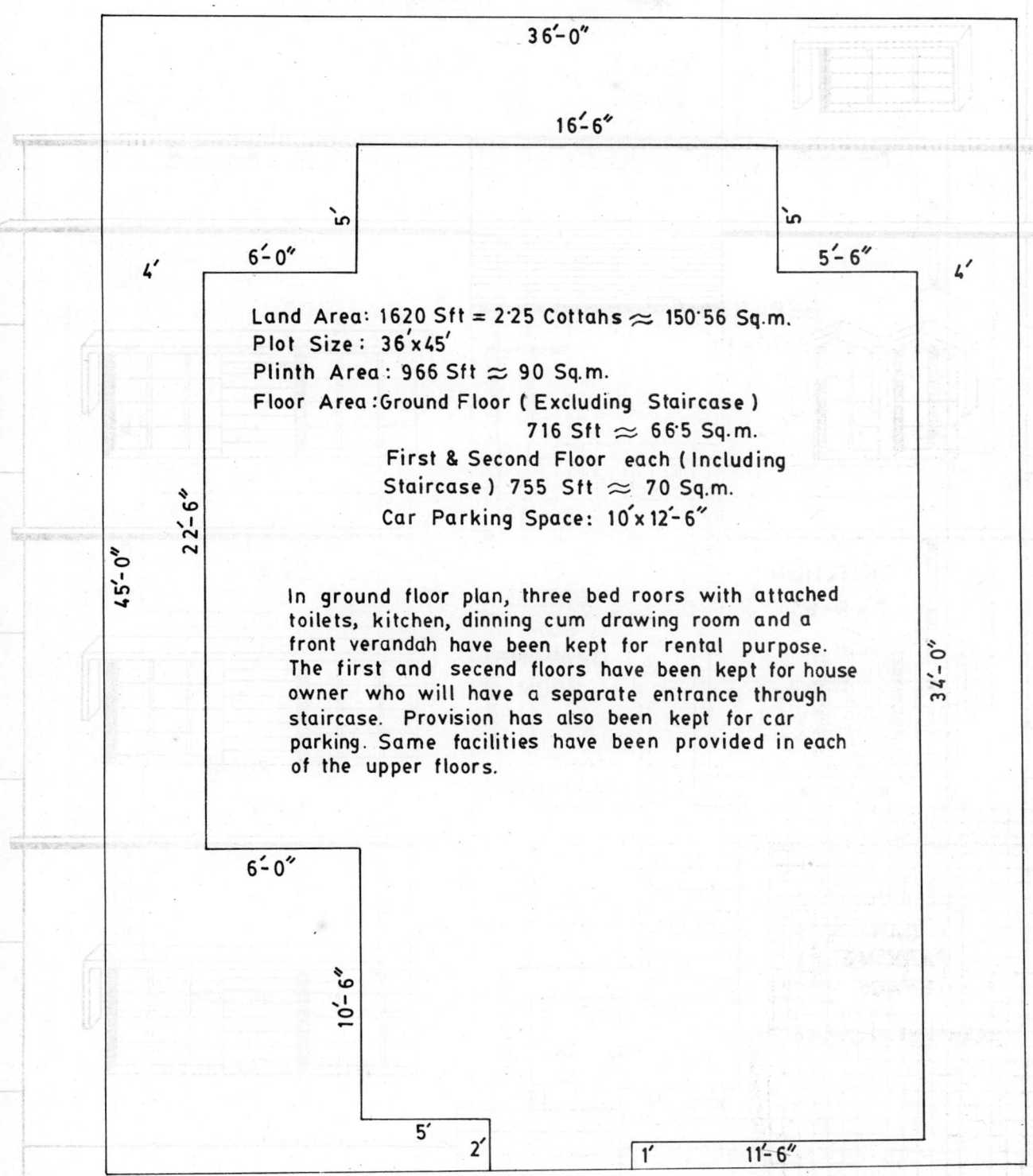

Land Area: 1620 Sft = 2·25 Cottahs ≈ 150·56 Sq.m.
Plot Size: 36'x45'
Plinth Area: 966 Sft ≈ 90 Sq.m.
Floor Area: Ground Floor (Excluding Staircase)
716 Sft ≈ 66·5 Sq.m.
First & Second Floor each (Including
Staircase) 755 Sft ≈ 70 Sq.m.
Car Parking Space: 10'x 12'–6"

In ground floor plan, three bed roors with attached toilets, kitchen, dinning cum drawing room and a front verandah have been kept for rental purpose. The first and second floors have been kept for house owner who will have a separate entrance through staircase. Provision has also been kept for car parking. Same facilities have been provided in each of the upper floors.

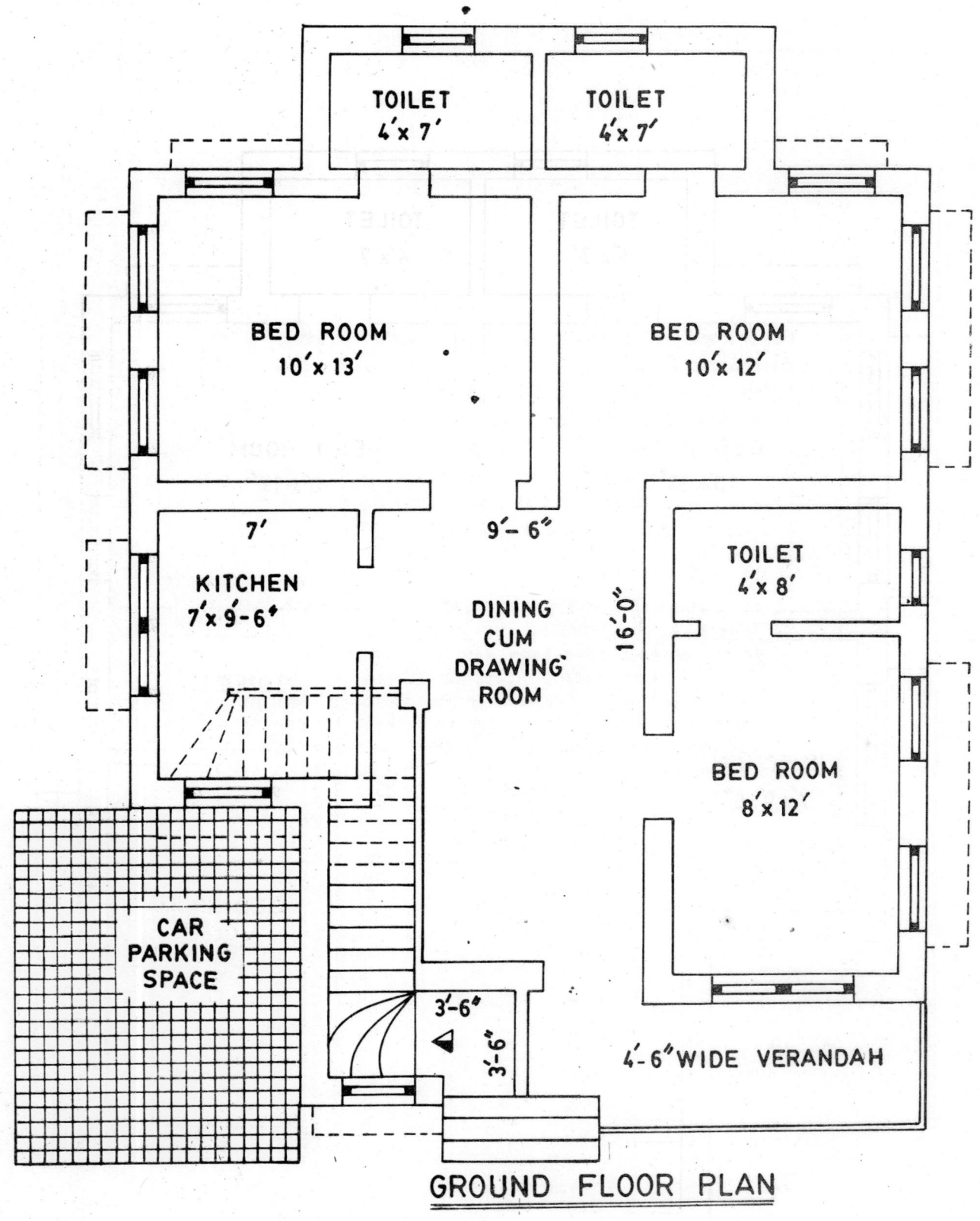

TOILET
4'x 7'

TOILET
4'x 7'

BED ROOM
10'x 13'

BED ROOM
10'x 12'

7'

KITCHEN
7'x 9'-6"

9'-6"

DINING
CUM
DRAWING
ROOM

16'-0"

TOILET
4'x 8'

BED ROOM
8'x 12'

CAR
PARKING
SPACE

3'-6"

3'-6"

4'-6" WIDE VERANDAH

GROUND FLOOR PLAN

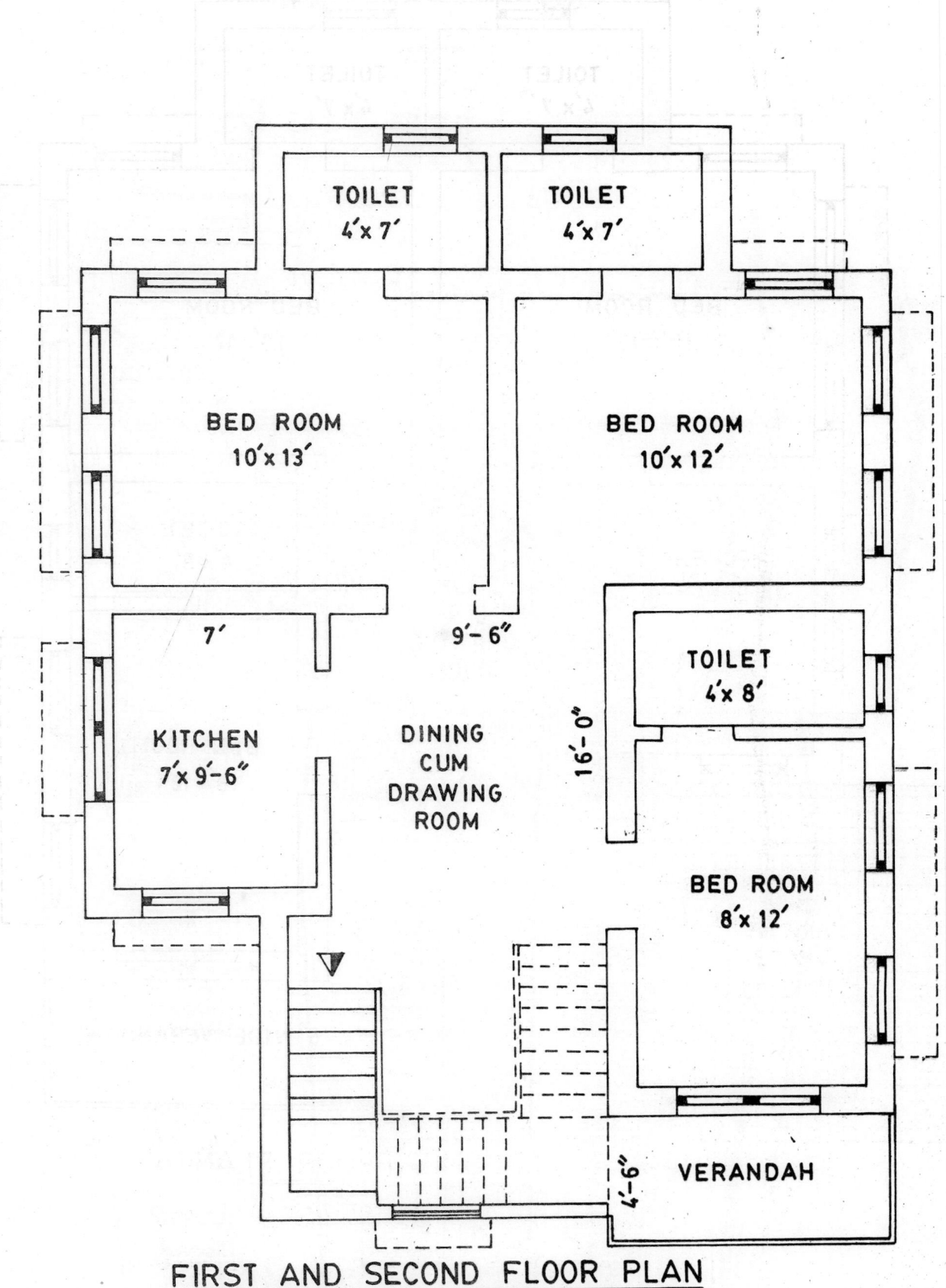

TOILET
4'x 7'

TOILET
4'x 7'

BED ROOM
10'x 13'

BED ROOM
10'x 12'

7'

9'- 6"

TOILET
4'x 8'

KITCHEN
7'x 9'-6"

DINING
CUM
DRAWING
ROOM

16'- 0"

BED ROOM
8'x 12'

VERANDAH

7'-6"

FIRST AND SECOND FLOOR PLAN

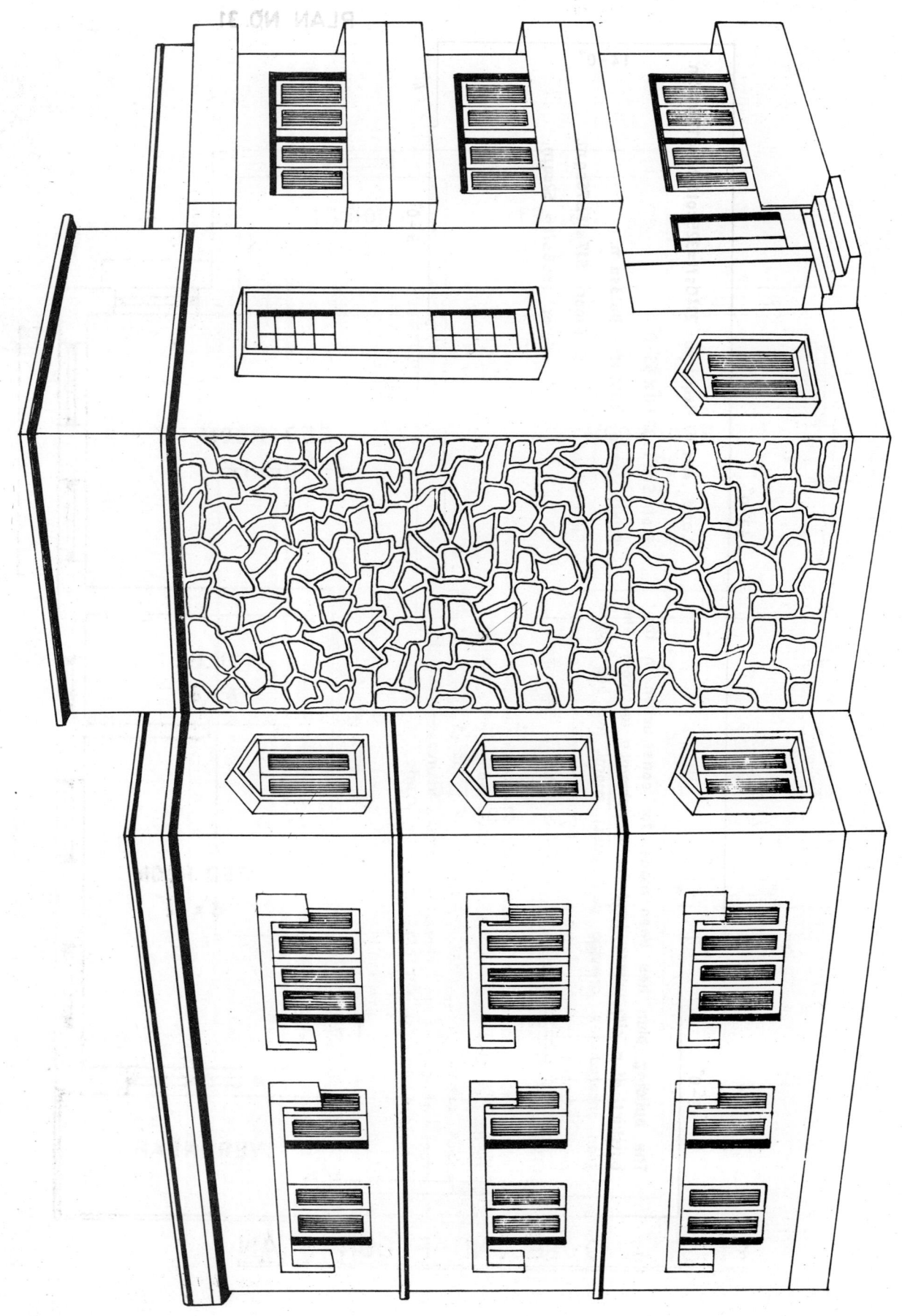

PLAN NO. 31

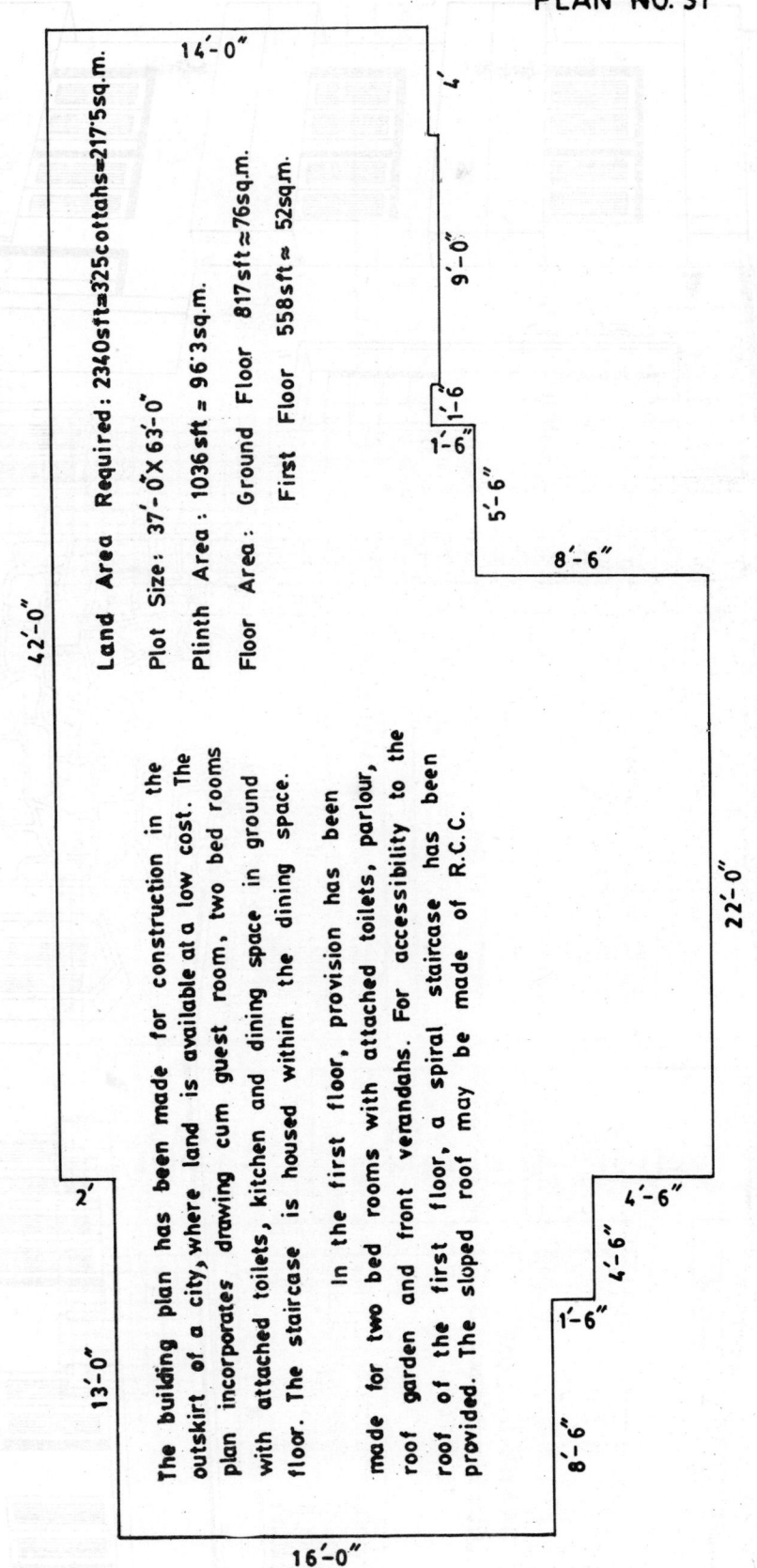

14'-0"

42'-0"

2'

13'-0"

16'-0"

9'-0"

1'-6"
1'-6"

5'-6"

8'-6"

22'-0"

4'-6"

4'-6"

1'-6"

8'-6"

Land Area Required: 2340sft=325cottahs=217·5sq.m.

Plot Size: 37'-0"X 63'-0"

Plinth Area: 1036sft = 96·3sq.m.

Floor Area: Ground Floor 817sft≈76sq.m.

First Floor 558sft≈ 52sq.m.

The building plan has been made for construction in the outskirt of a city, where land is available at a low cost. The plan incorporates drawing cum guest room, two bed rooms with attached toilets, kitchen and dining space in ground floor. The staircase is housed within the dining space.

In the first floor, provision has been made for two bed rooms with attached toilets, parlour, roof garden and front verandahs. For accessibility to the roof of the first floor, a spiral staircase has been provided. The sloped roof may be made of R.C.C.

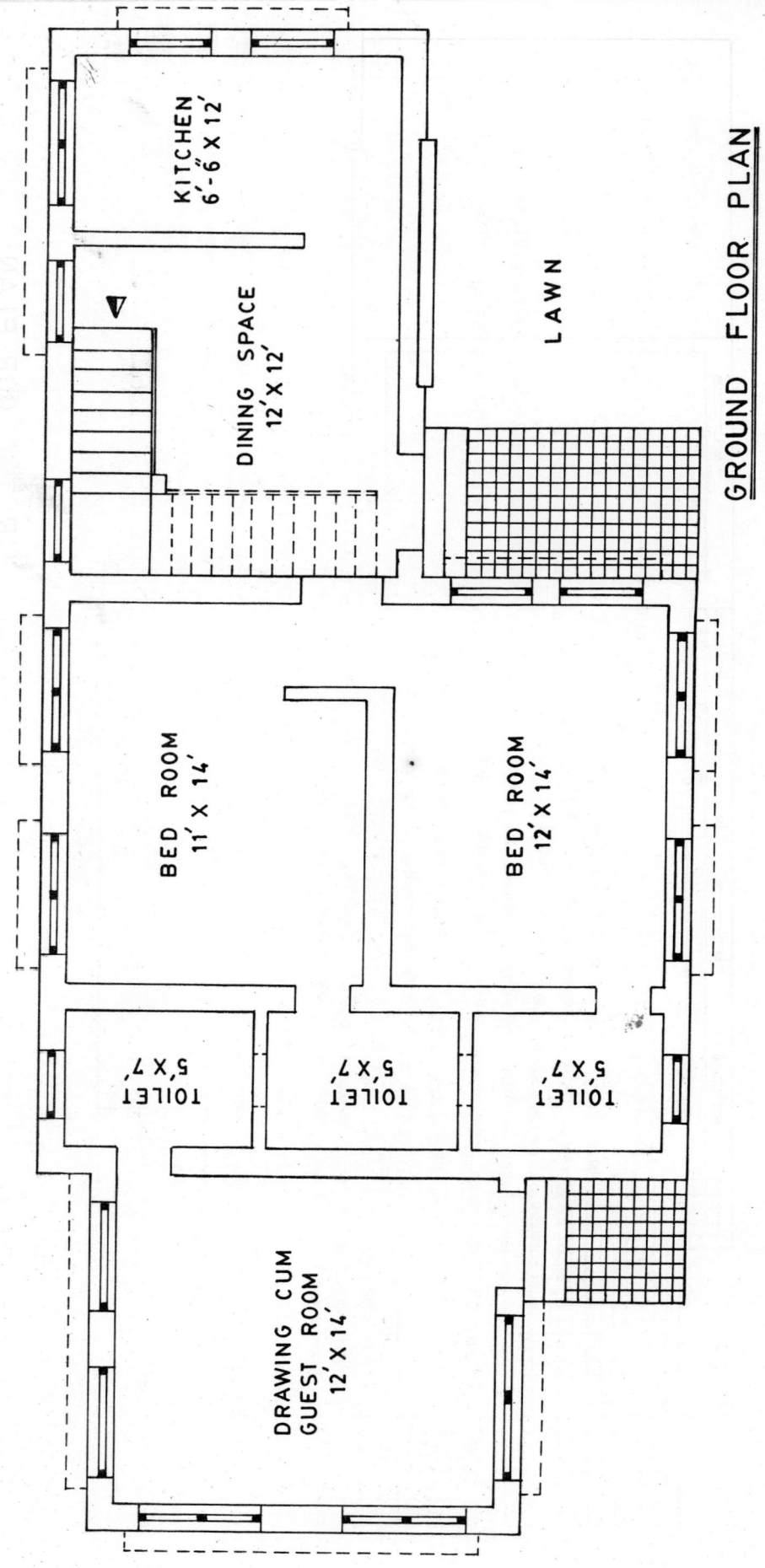

KITCHEN
6'-6" X 12'

DINING SPACE
12' X 12'

LAWN

BED ROOM
11' X 14'

BED ROOM
12' X 14'

TOILET
5' X 7'

TOILET
5' X 7'

TOILET
5' X 7'

DRAWING CUM
GUEST ROOM
12' X 14'

GROUND FLOOR PLAN

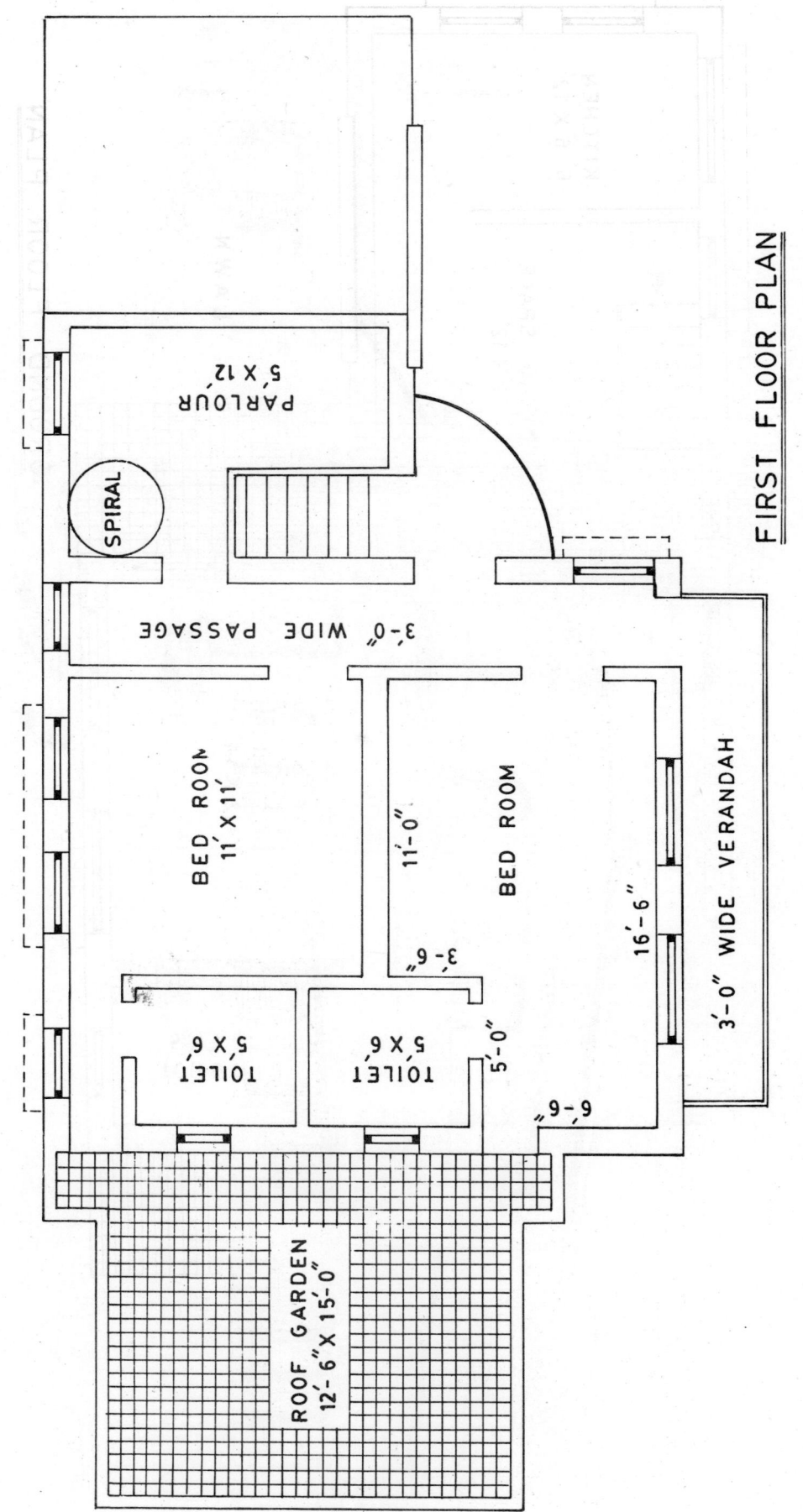

FIRST FLOOR PLAN

PLAN NO.32

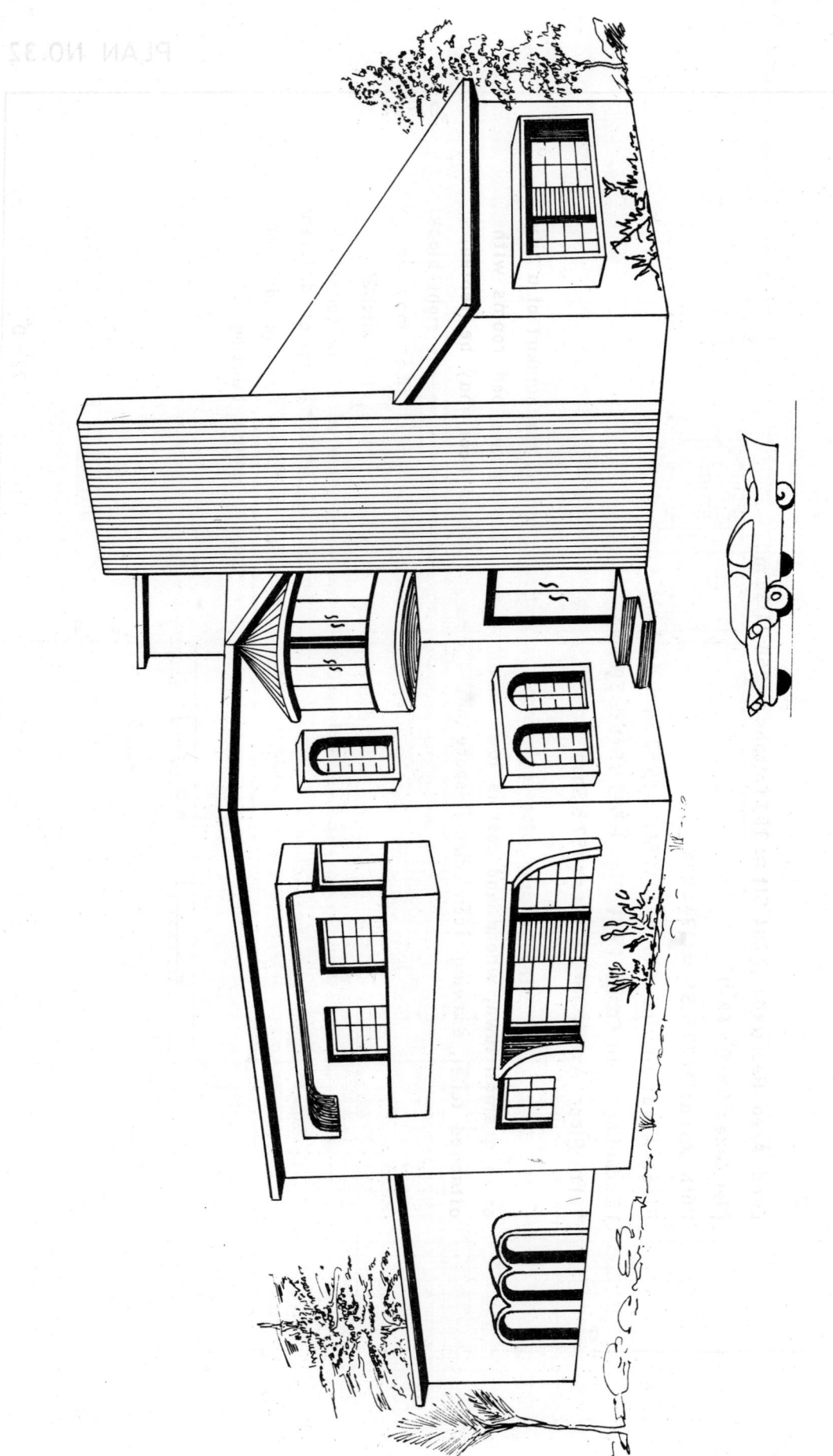

PLAN NO.32

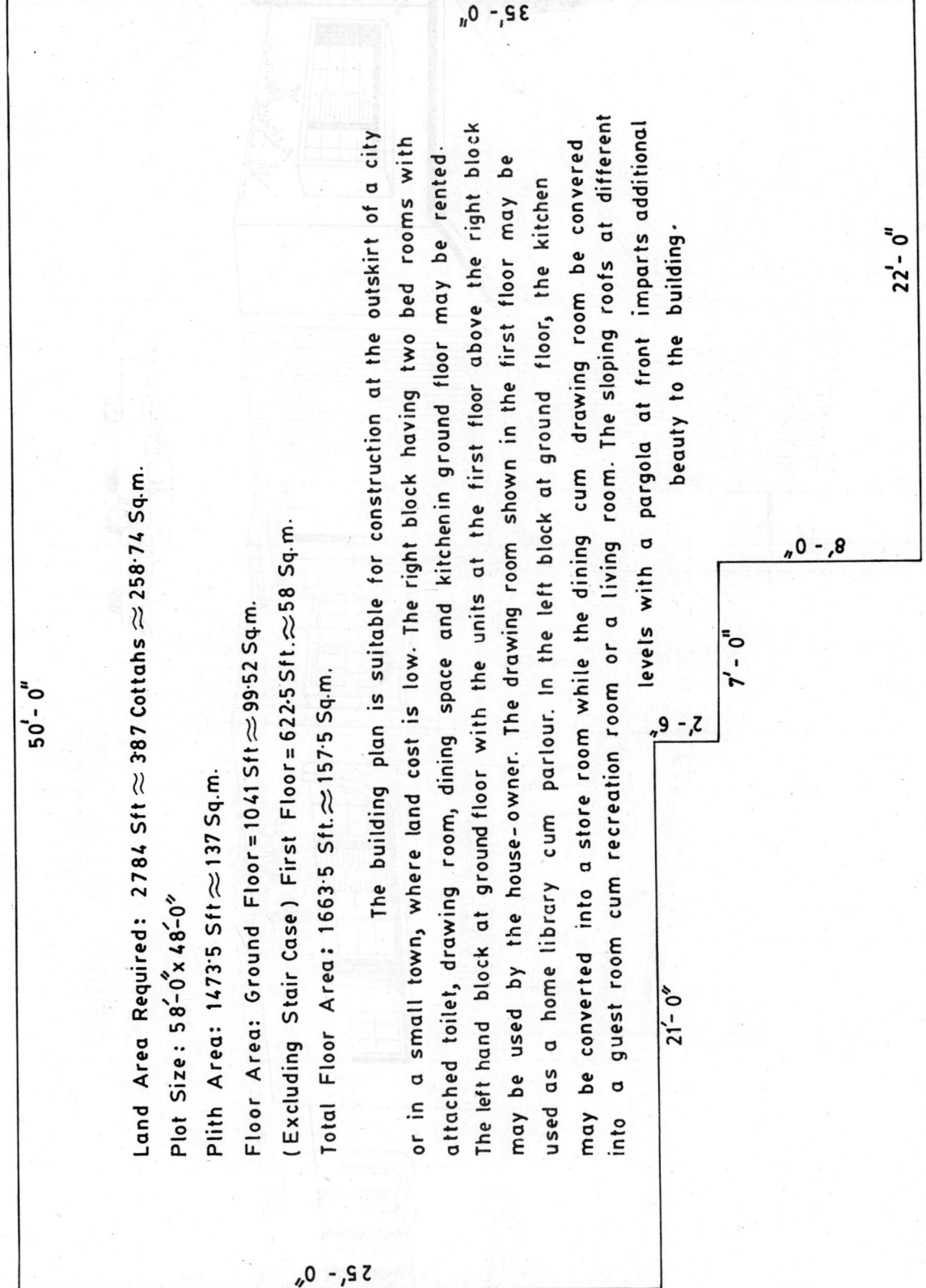

Land Area Required: 2784 Sft ≈ 387 Cottahs ≈ 258·74 Sq.m.

Plot Size: 58'-0"x 48'-0"

Plith Area: 1473·5 Sft ≈ 137 Sq.m.

Floor Area: Ground Floor=1041 Sft ≈ 99·52 Sq.m.

(Excluding Stair Case) First Floor = 622·5 Sft. ≈ 58 Sq.m.

Total Floor Area: 1663·5 Sft. ≈ 157·5 Sq.m.

The building plan is suitable for construction at the outskirt of a city or in a small town, where land cost is low. The right block having two bed rooms with attached toilet, drawing room, dining space and kitchen in ground floor may be rented. The left hand block at ground floor with the units at the first floor above the right block may be used by the house-owner. The drawing room shown in the first floor may be used as a home library cum parlour. In the left block at ground floor, the kitchen may be converted into a store room while the dining cum drawing room be converted into a guest room cum recreation room or a living room. The sloping roofs at different levels with a pargola at front imparts additional beauty to the building.

35'-0"

22'-0"

8'-0"

7'-0"

2'-6"

21'-0"

50'-0"

25'-0"

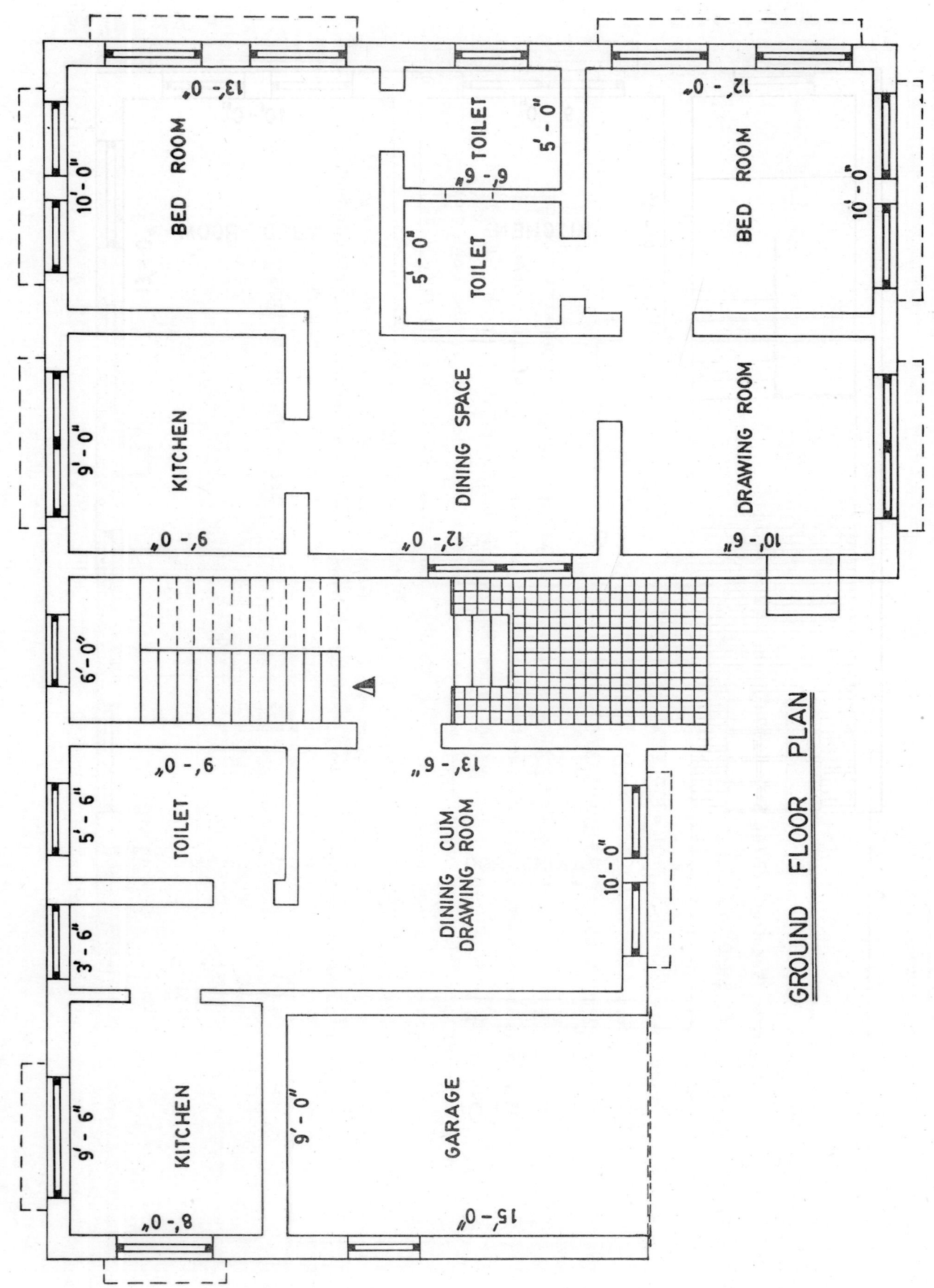

GROUND FLOOR PLAN

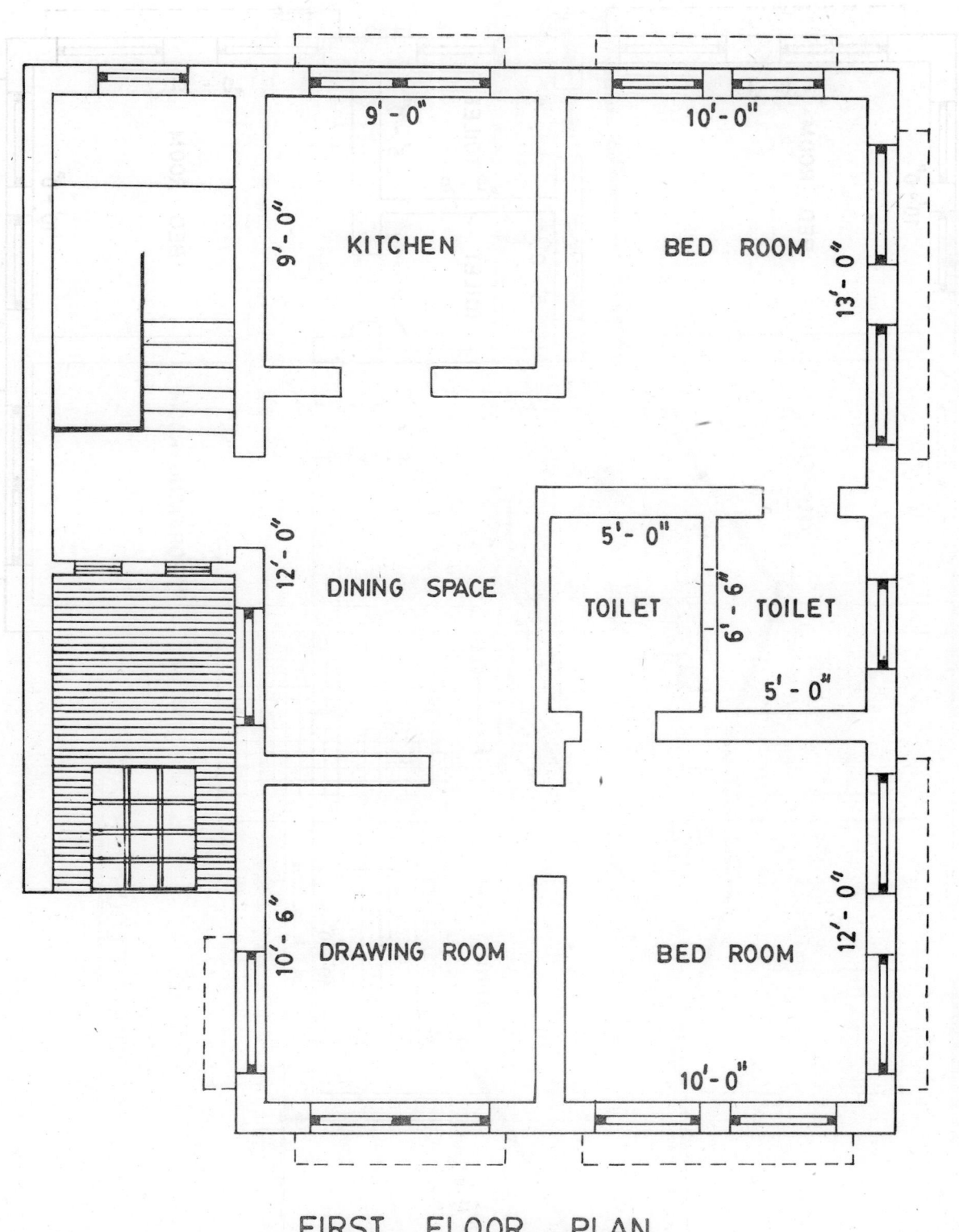

FIRST FLOOR PLAN

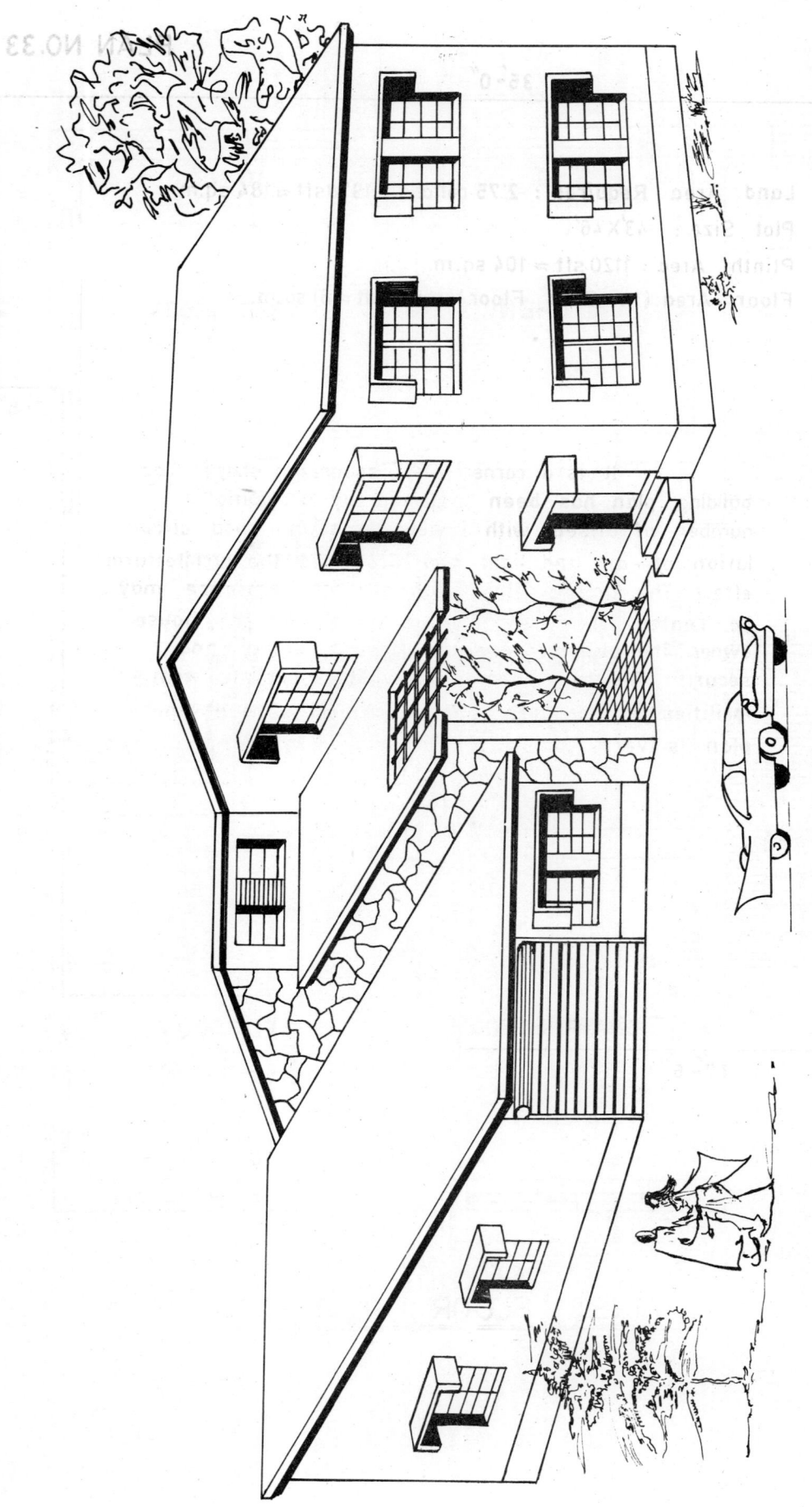

PLAN NO. 33

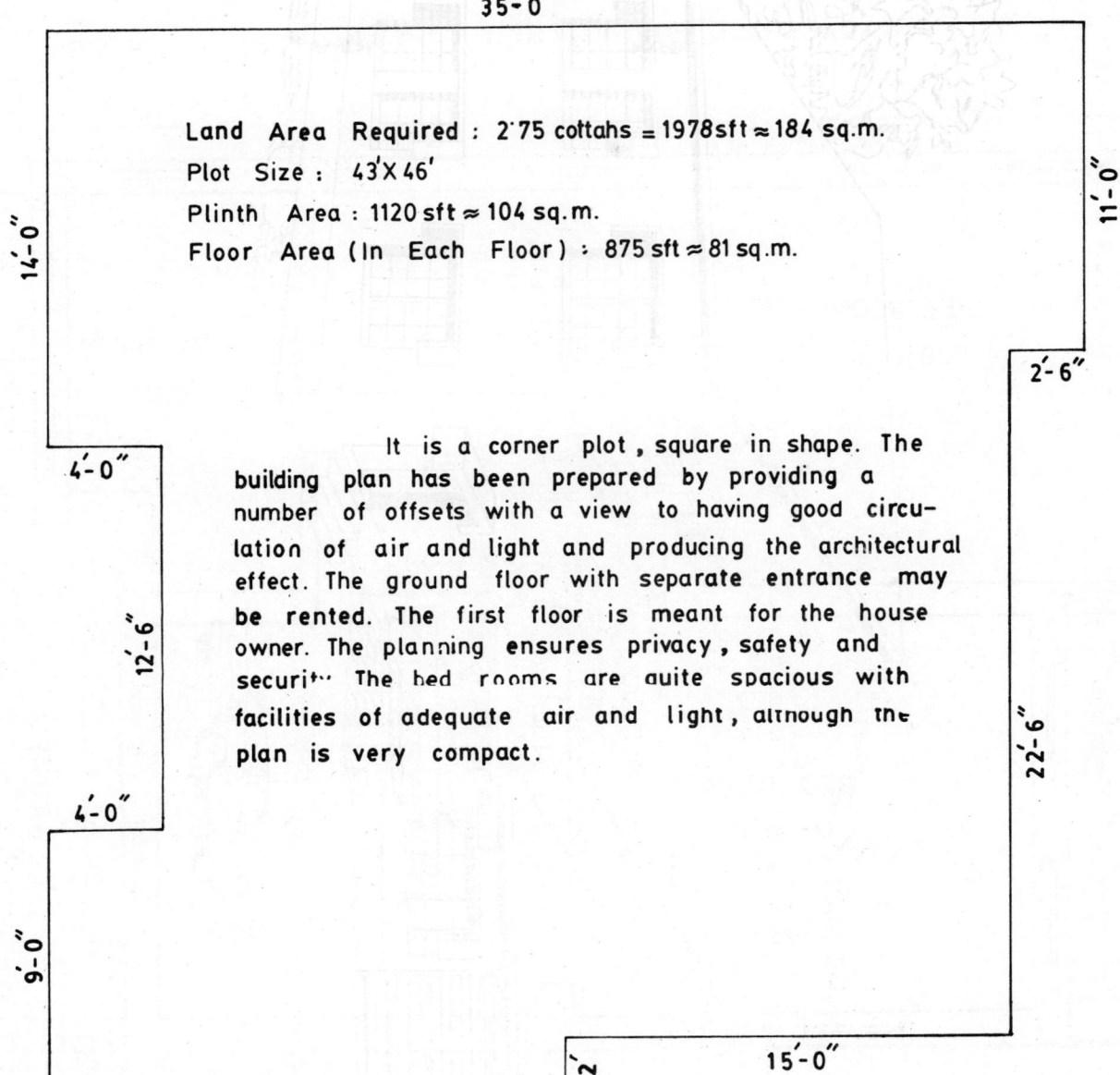

Land Area Required : 2:75 cottahs = 1978 sft ≈ 184 sq.m.

Plot Size : 43'X 46'

Plinth Area : 1120 sft ≈ 104 sq.m.

Floor Area (In Each Floor) : 875 sft ≈ 81 sq.m.

It is a corner plot, square in shape. The building plan has been prepared by providing a number of offsets with a view to having good circulation of air and light and producing the architectural effect. The ground floor with separate entrance may be rented. The first floor is meant for the house owner. The planning ensures privacy, safety and security. The bed rooms are quite spacious with facilities of adequate air and light, although the plan is very compact.

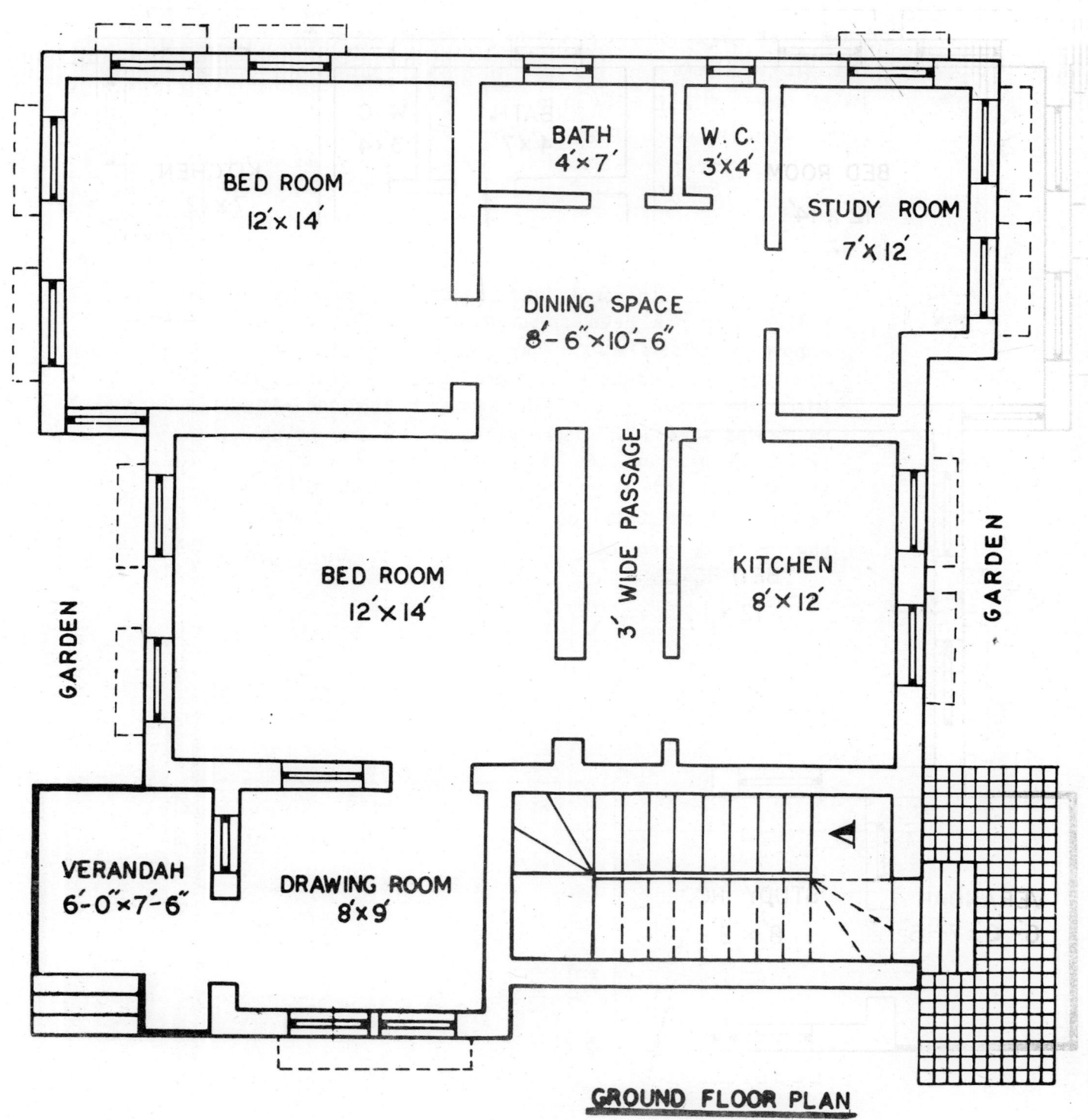

BACK SPACE FOR KITCHEN GARDEN

BED ROOM
12' × 14'

BATH
4' × 7'

W. C.
3' × 4'

STUDY ROOM
7' × 12'

DINING SPACE
8'-6" × 10'-6"

GARDEN

BED ROOM
12' × 14'

3' WIDE PASSAGE

KITCHEN
8' × 12'

GARDEN

VERANDAH
6'-0" × 7'-6"

DRAWING ROOM
8' × 9'

GROUND FLOOR PLAN

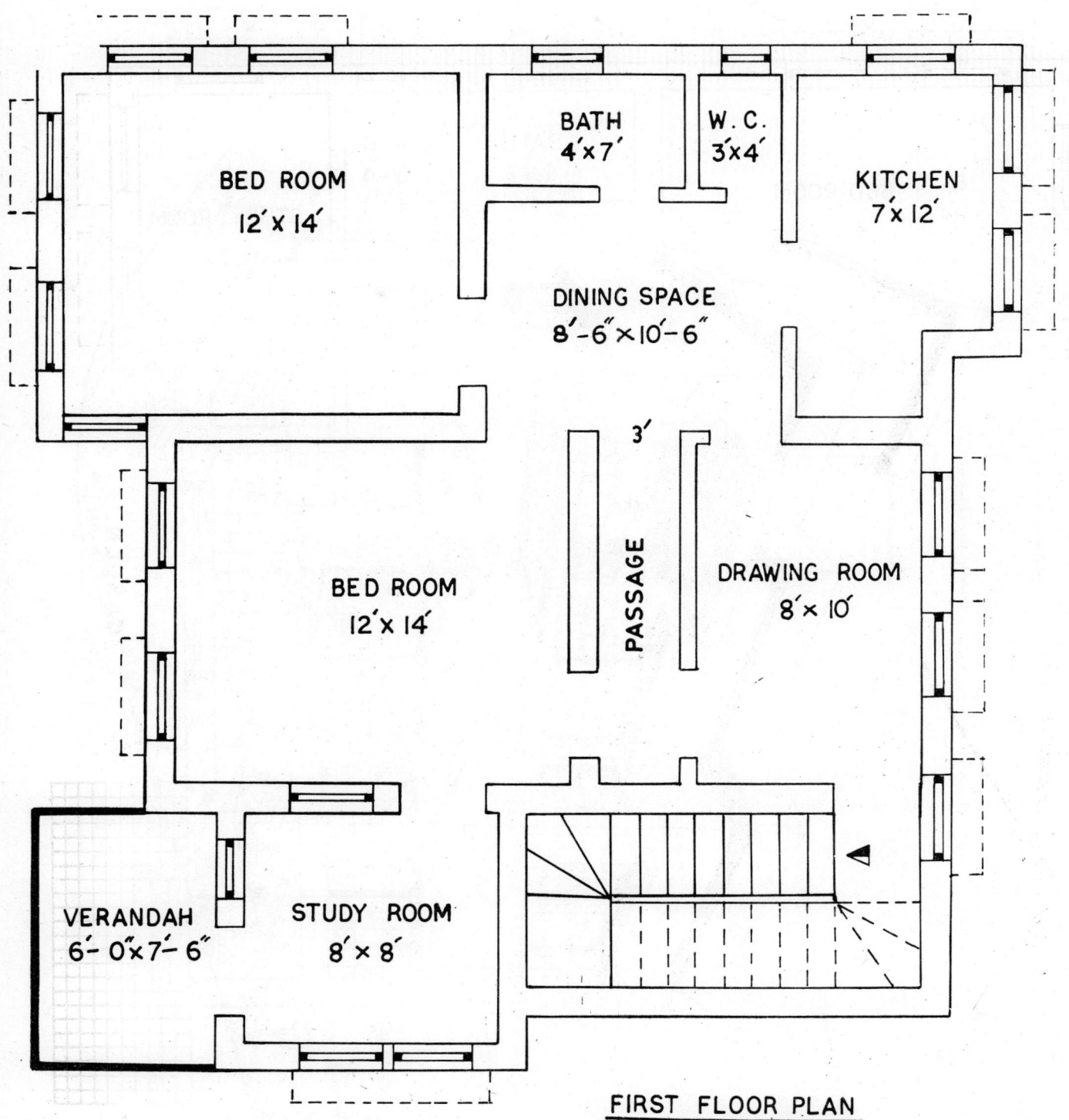

BED ROOM
12' x 14'

BATH
4' x 7'

W. C.
3' x 4'

KITCHEN
7' x 12'

DINING SPACE
8'-6" x 10'-6"

BED ROOM
12' x 14'

3'

PASSAGE

DRAWING ROOM
8' x 10'

VERANDAH
6'-0" x 7'-6"

STUDY ROOM
8' x 8'

FIRST FLOOR PLAN

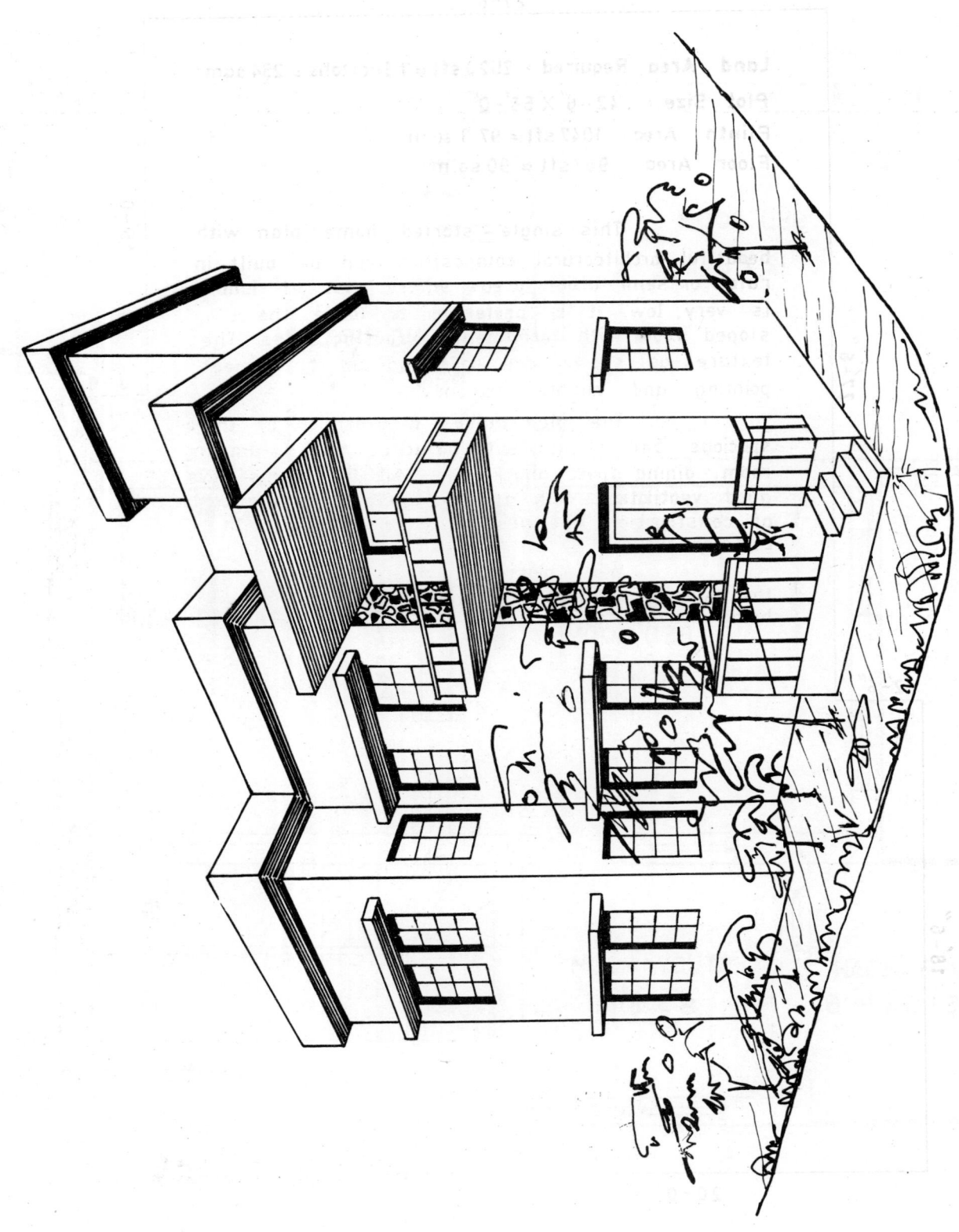

PLAN NO. 34

27'-6"

15'-0"

Land Area Required : 2520 sft = 3.5 cottahs ≈ 234 sqm.

Plot Size : 42'-6" X 58'-0"

Plinth Area : 1047 sft ≈ 97.3 sq.m.

Floor Area : 967 sft ≈ 90 sq.m.

This single – storied home plan with beautiful architectural composition can be built in rural or semi – urban areas where cost of land is very low. It is preferred to make the sloped roof with ornamental roofing tiles. The texture as shown can be produced by cement pointing and rubble masonry.

The plan makes a provision of three spacious bed rooms with attached toilets, drawing room, dining space and kitchen. All the rooms have good ventilation. With a view to reducing the cost of construction, the plan is made compact with common walls.

With plantation of trees and shrubs, the building will have a pleasant look. This will bring harmonious effect.

3'-6"

2'-6"

2'-6"

2'

2'-6"

27'-6"

10'-0"

11'-0"

2'-6"

18'-6"

13'-0"

20'-0"

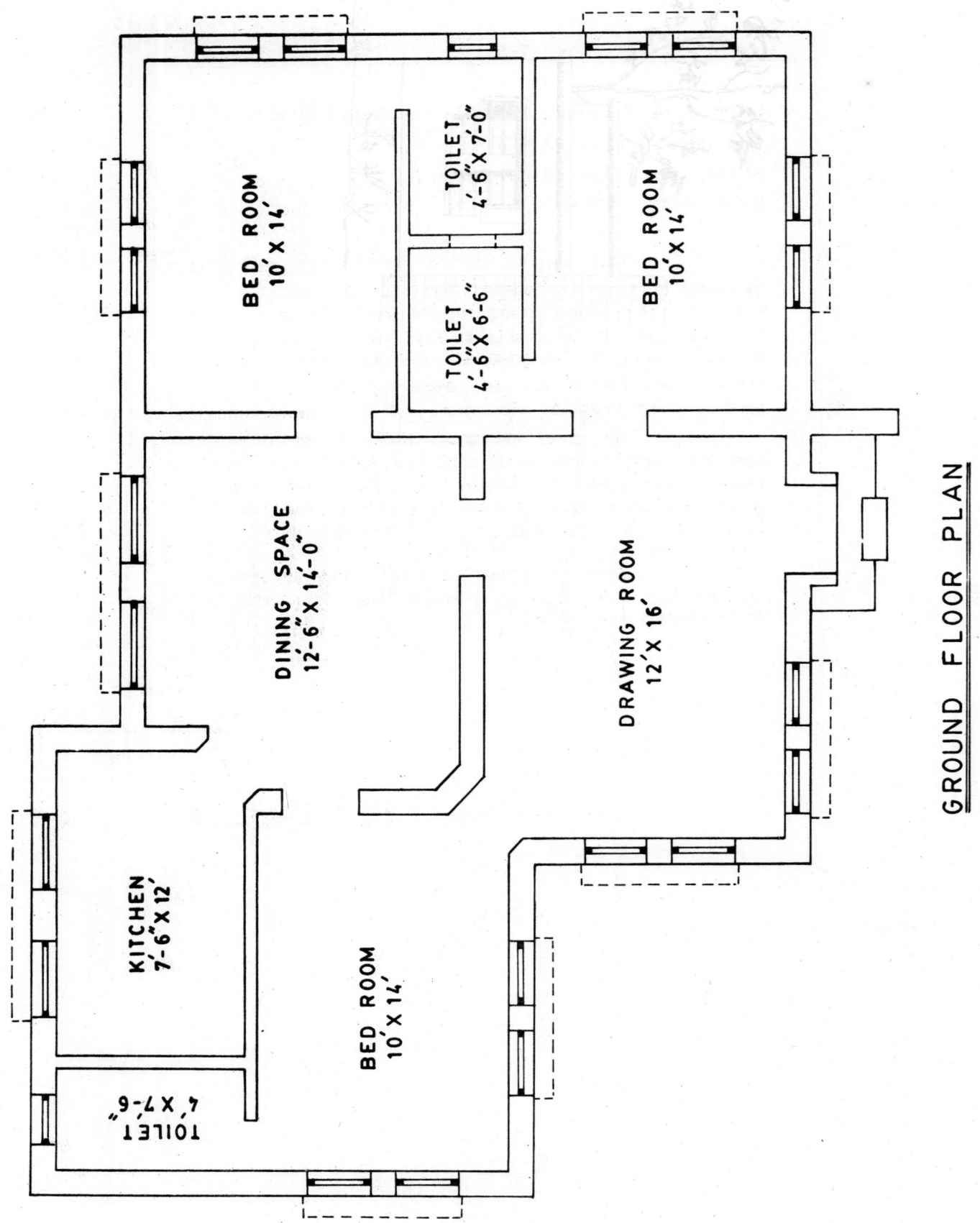

GROUND FLOOR PLAN

BED ROOM
10' X 14'

TOILET
4'-6" X 7'-0"

TOILET
4'-6" X 6'-6"

BED ROOM
10' X 14'

DINING SPACE
12'-6" X 14'-0"

DRAWING ROOM
12' X 16'

KITCHEN
7'-6" X 12'

BED ROOM
10' X 14'

TOILET
4' X 7'-6"

PLAN NO. 35

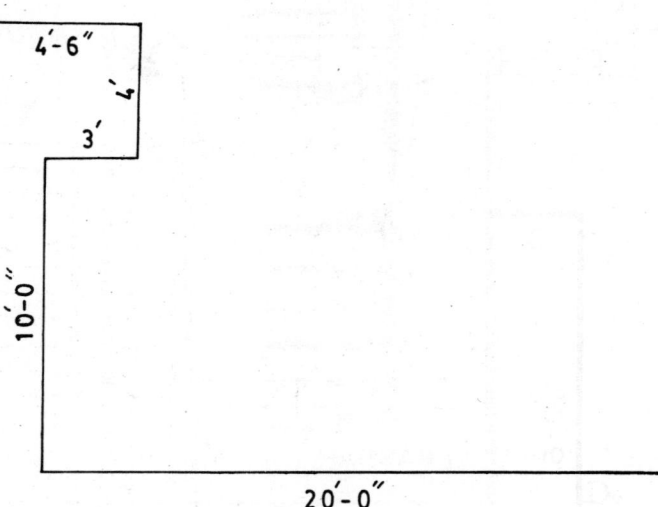

22'-6"

18'-0"

Land Area Required : 2.5 cottahs = 1800 sft. ≈ 167.3 sq. m.

Plot Size : 30'-0" x 60'-0"

Floor Area In Each Floor : 716 sft ≈ 66.5 sq. m.
 (excluding staircase area)

Plinth Area : 939 sft. ≈ 87.2 sq.m.

      The planning has been done in a rectangular strip of land with 30'-0" road frontage. There are two bed rooms with attached toilet, bath, kitchen, dining room, drawing room and a front verandah in each floor. This type of building can well accomodate a large joint family of Middle Income Group.

Alternatively, the drawing room with front verandah and staircase in ground floor can be separated from the rest part by building a wall in place of curtain shown in plan and a two-room flat with attached bath and toilet, kitchen and dining space may be rented. But, a separate entry is to be made for the tenant through the dining space, which will then be used as drawing cum dining.

1'

12'-6"

44'-9"

4'-6"

4'

3'

10'-0"

20'-0"

R   O   A   D

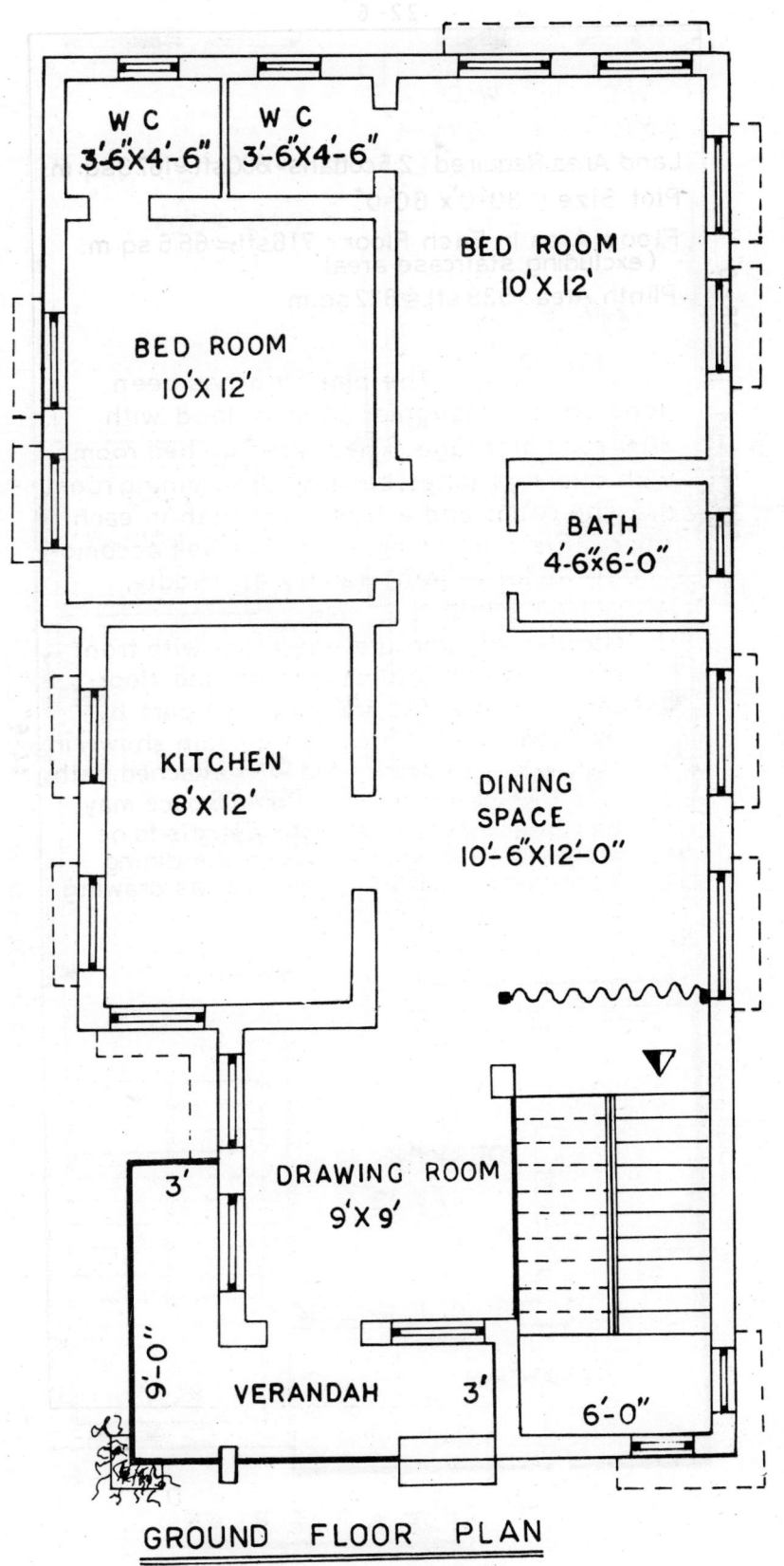

GROUND FLOOR PLAN

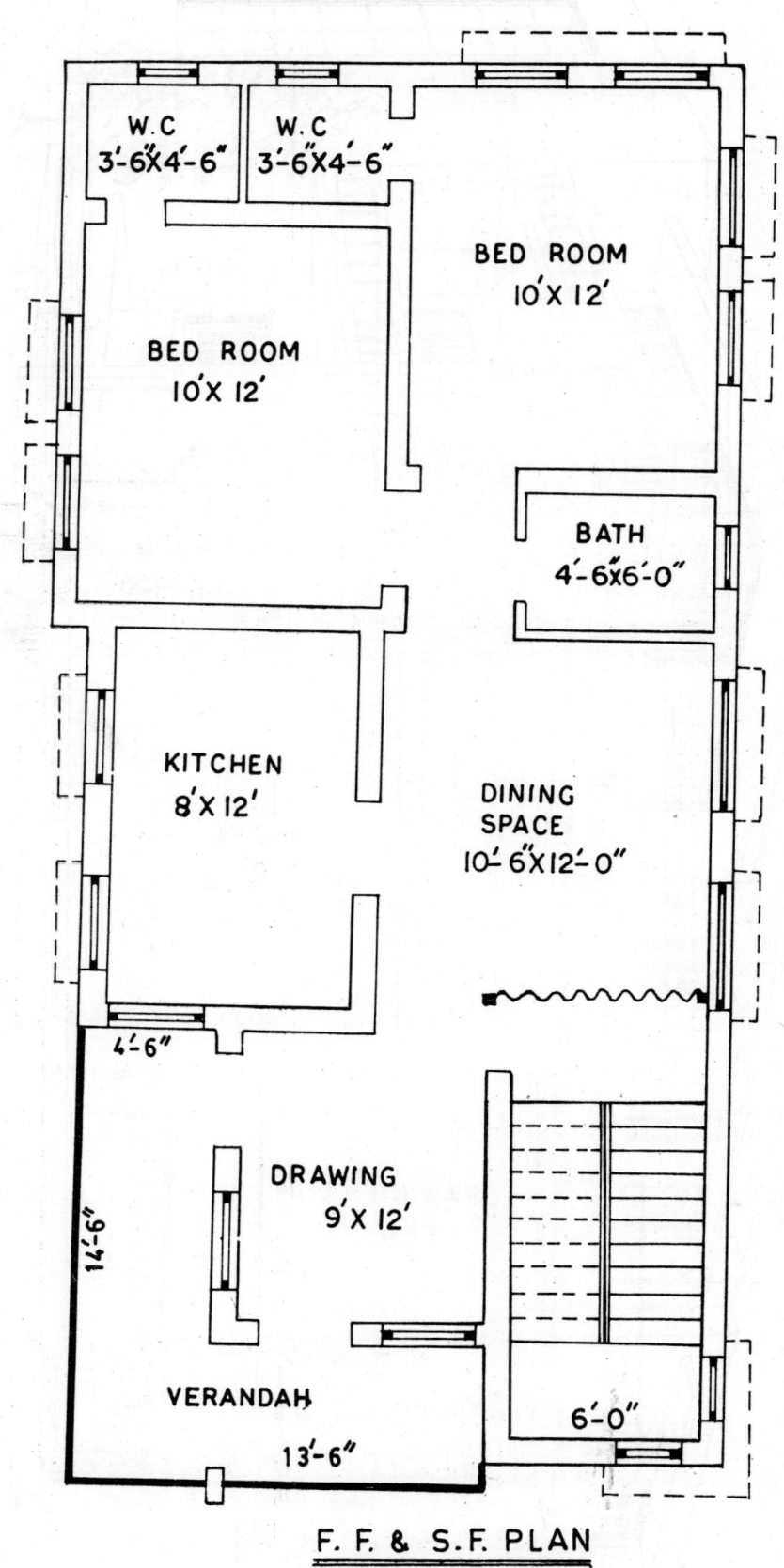

W.C
3'-6"X4'-6"

W.C
3'-6"X4'-6"

BED ROOM
10'X 12'

BED ROOM
10'X 12'

BATH
4'-6"X6'-0"

KITCHEN
8'X 12'

DINING SPACE
10'-6"X12'-0"

4'-6"

DRAWING
9'X 12'

14'-6"

VERANDAH

13'-6"

6'-0"

F. F. & S.F. PLAN

PLAN NO. 36

15'-0"

5'

13'-0"

31'-0"

28'-6"

Land Area Required : 1980 sft = 2'75 cottahs ≈ 184 sq.m.

Plot Size : 36'X 55'

Plinth Area : 965 sft ≈ 90 sq.m.

Floor Area In Each Floor : 768 sft = 71'4 sq.m.

In planning this building , it was thought of providing more number of spacious rooms with maximum utilisation of area. The stair-case having three flights and quarter landings is housed within the lobby. All bed rooms have attached toilets. A study room is provided in first floor. This two-storied building can well accomodate a family of 12 members.

In case if it is needed, the building may be made three storied with a total foor area of 2304 sft (214 Sq.m.) to accomodate 3 families in 3 floors. In such a case, each floor should be made as per first floor plan.

10'-0"

2'

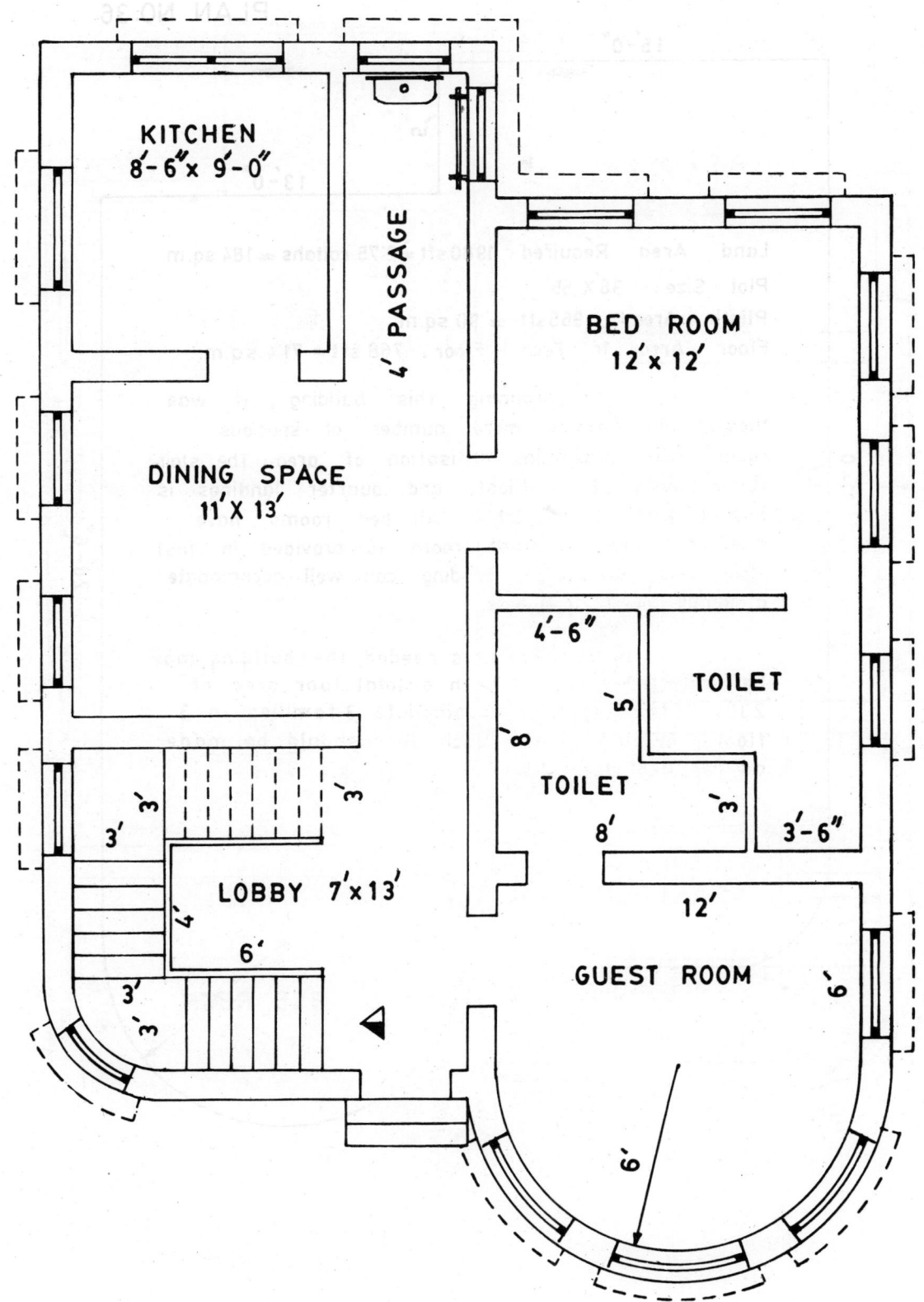

GROUND FLOOR PLAN

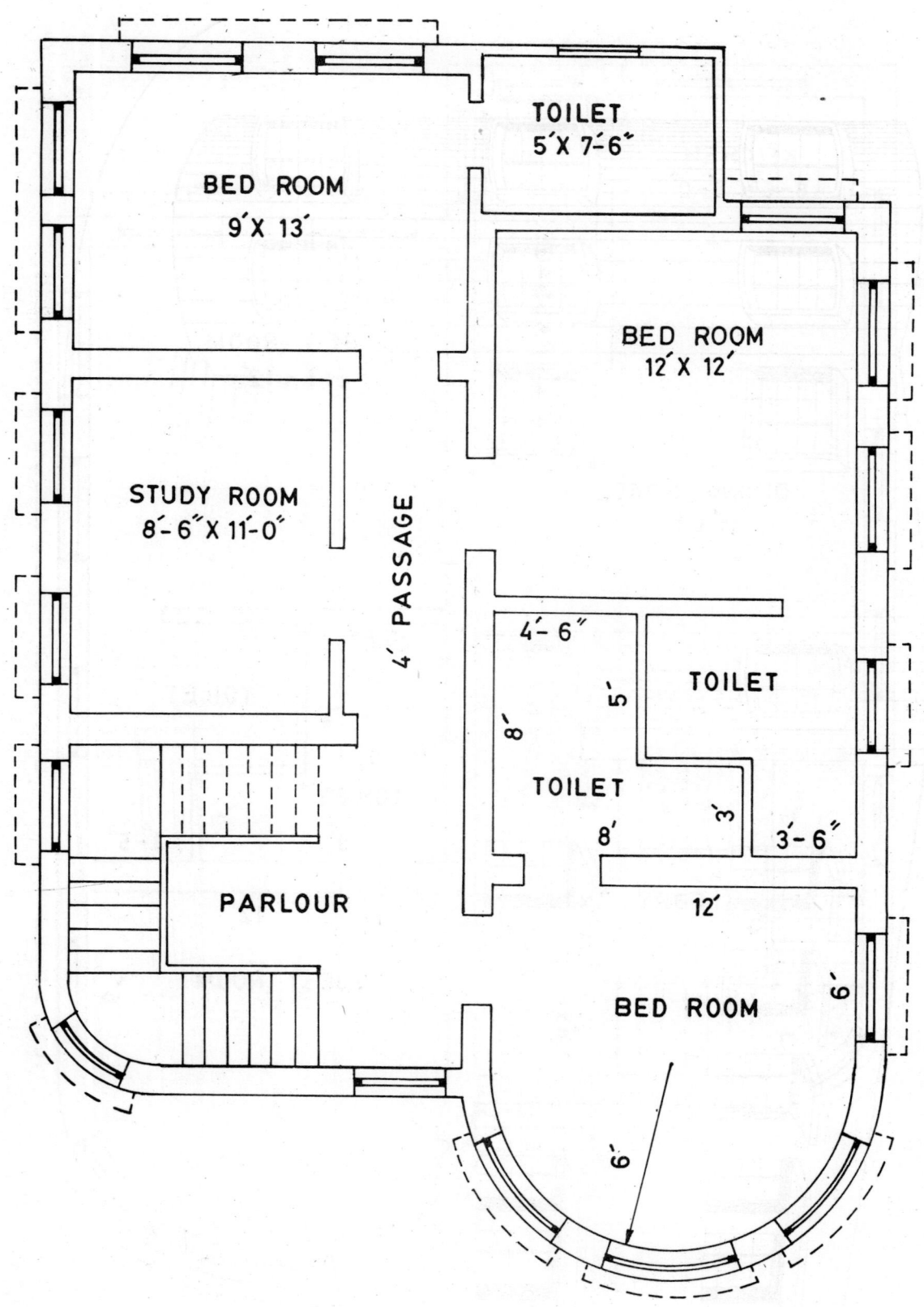

FIRST FLOOR PLAN

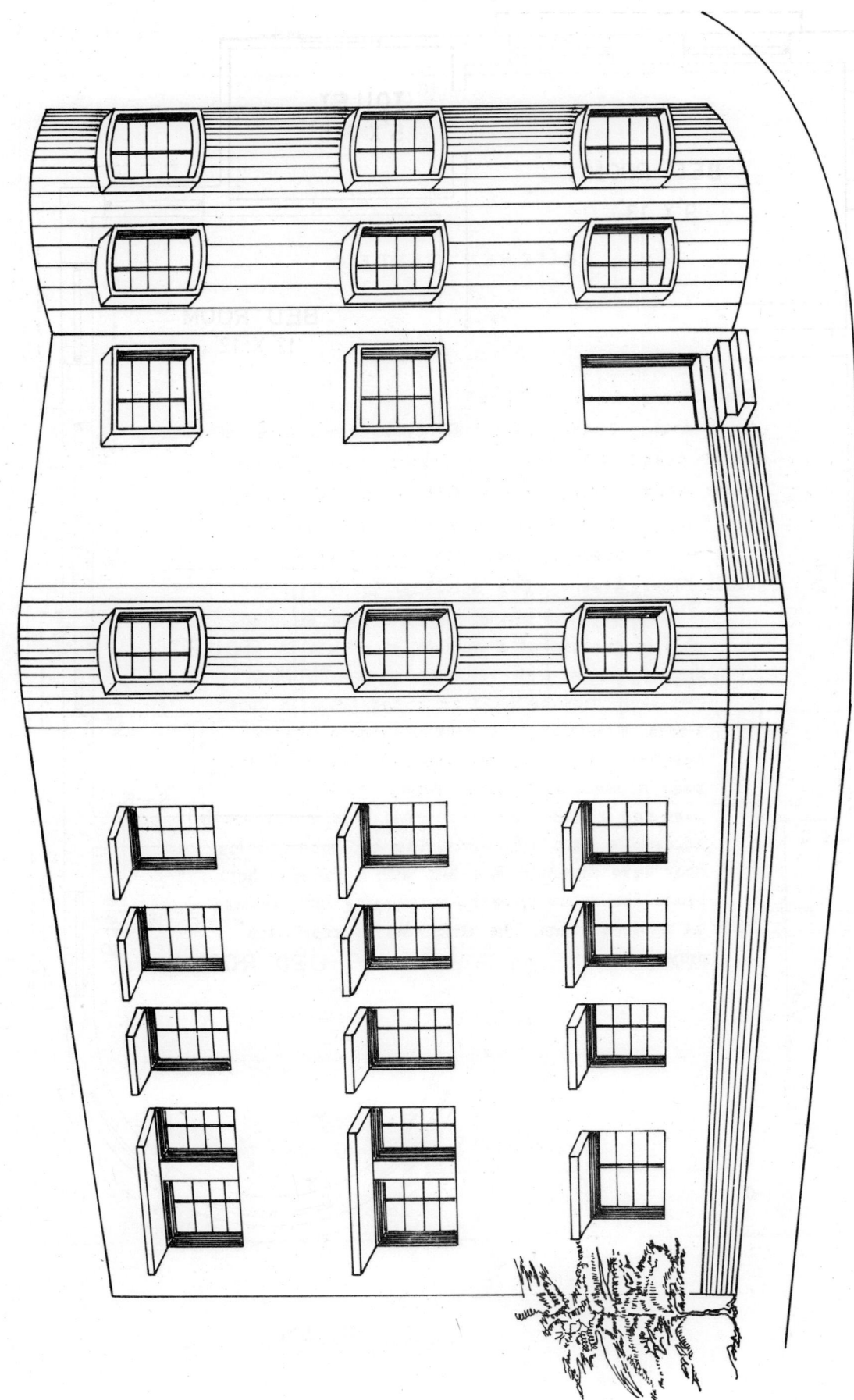

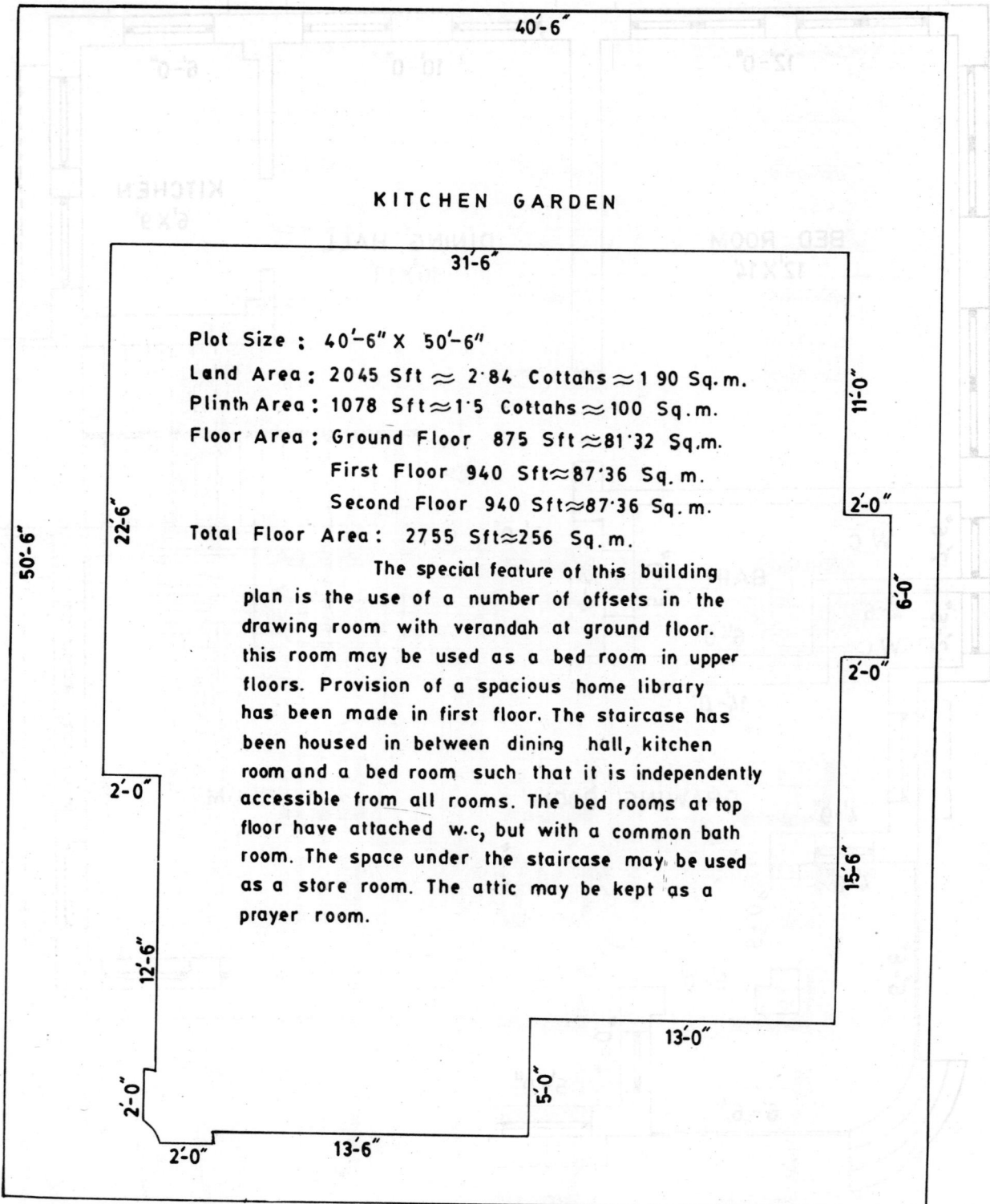

KITCHEN GARDEN

40'-6"

31'-6"

50'-6"

22'-6"

11'-0"

2'-0"

6'-0"

2'-0"

15'-6"

2'-0"

12'-6"

13'-0"

2'-0"

2'-0"

13'-6"

5'-0"

Plot Size : 40'-6" X 50'-6"

Land Area : 2045 Sft ≈ 2·84 Cottahs ≈ 1·90 Sq.m.

Plinth Area : 1078 Sft ≈ 1·5 Cottahs ≈ 100 Sq.m.

Floor Area : Ground Floor 875 Sft ≈ 81·32 Sq.m.

First Floor 940 Sft ≈ 87·36 Sq.m.

Second Floor 940 Sft ≈ 87·36 Sq.m.

Total Floor Area : 2755 Sft ≈ 256 Sq.m.

The special feature of this building plan is the use of a number of offsets in the drawing room with verandah at ground floor. this room may be used as a bed room in upper floors. Provision of a spacious home library has been made in first floor. The staircase has been housed in between dining hall, kitchen room and a bed room such that it is independently accessible from all rooms. The bed rooms at top floor have attached w.c, but with a common bath room. The space under the staircase may be used as a store room. The attic may be kept as a prayer room.

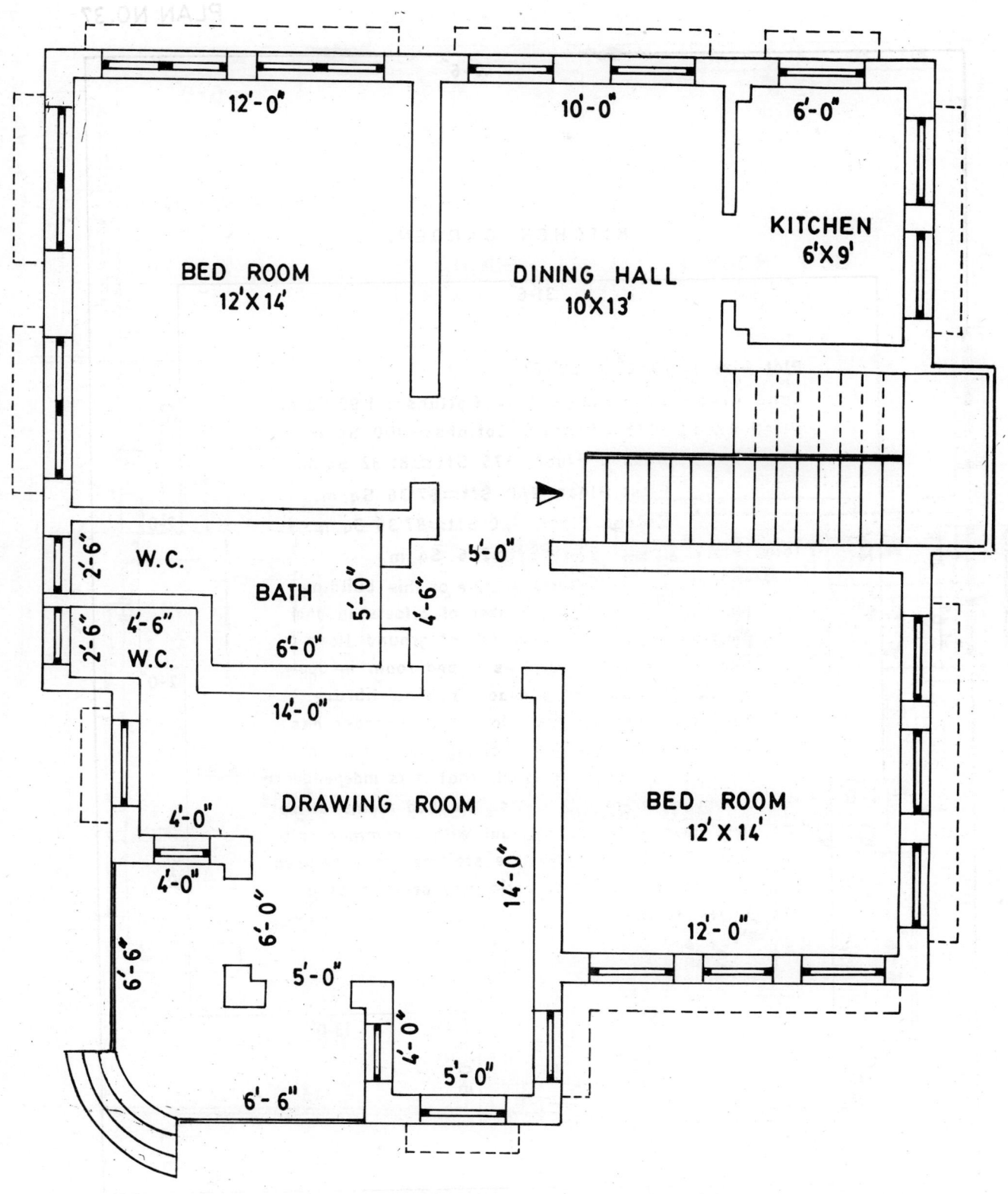

GROUND FLOOR PLAN

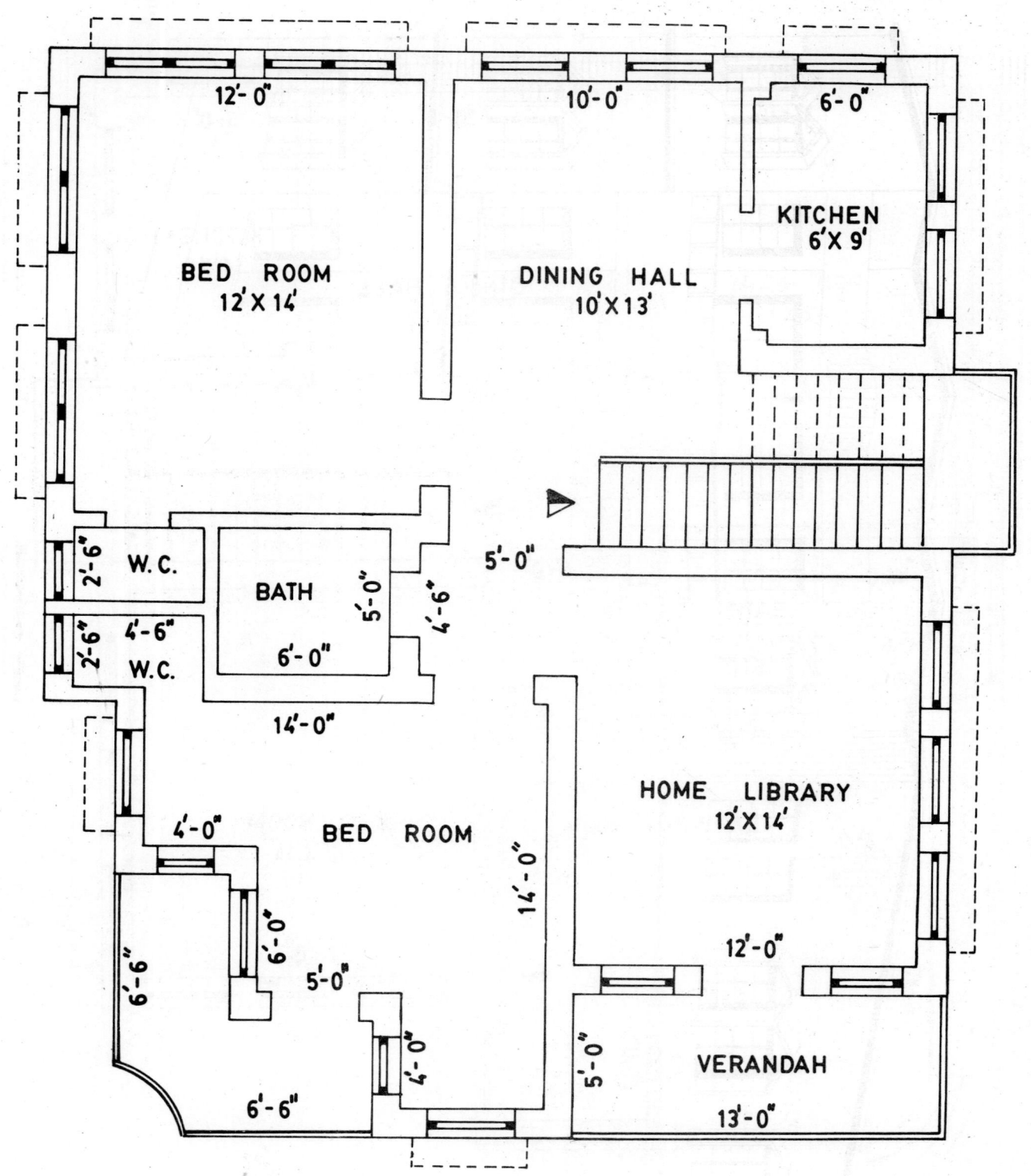

FIRST FLOOR PLAN

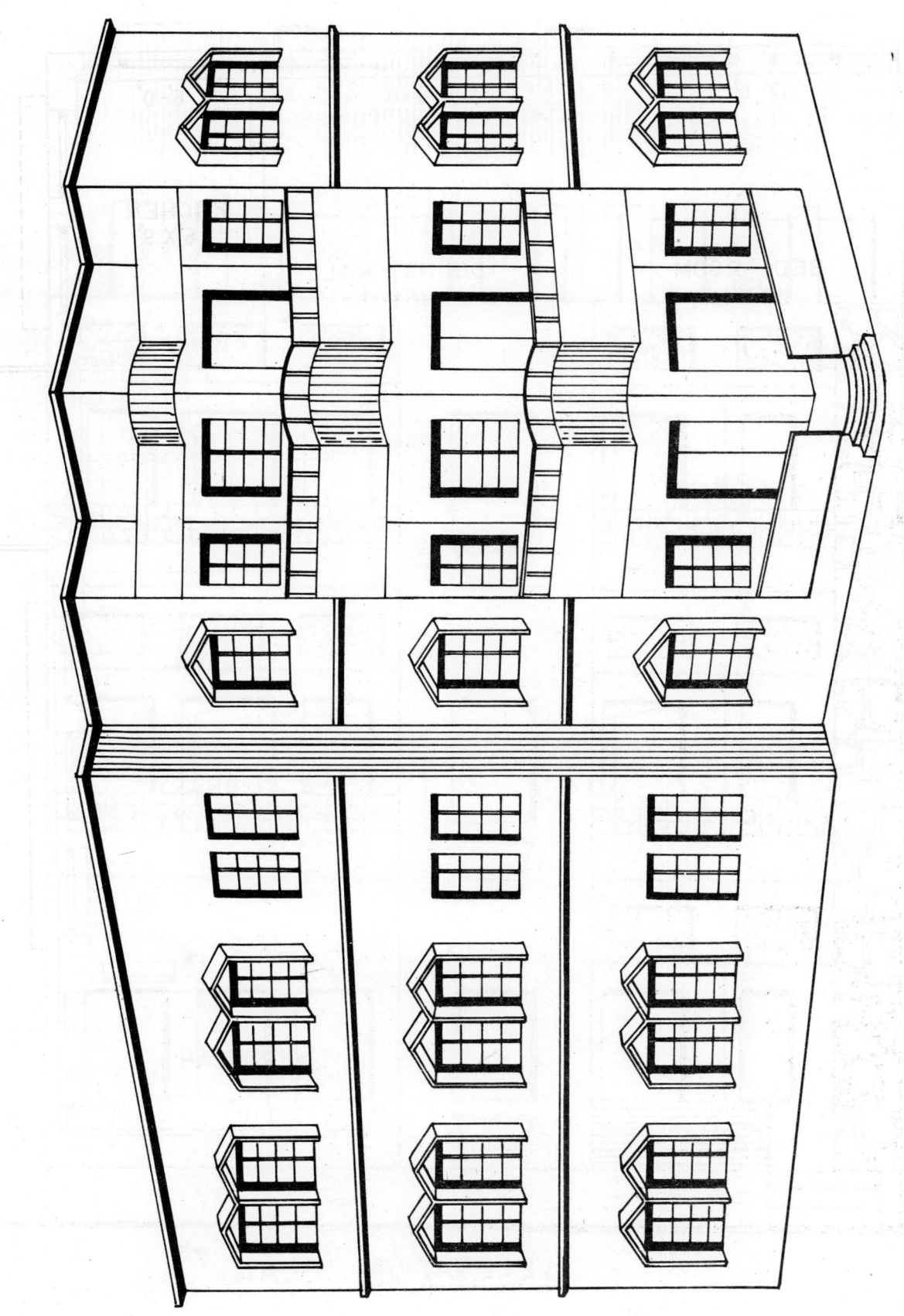

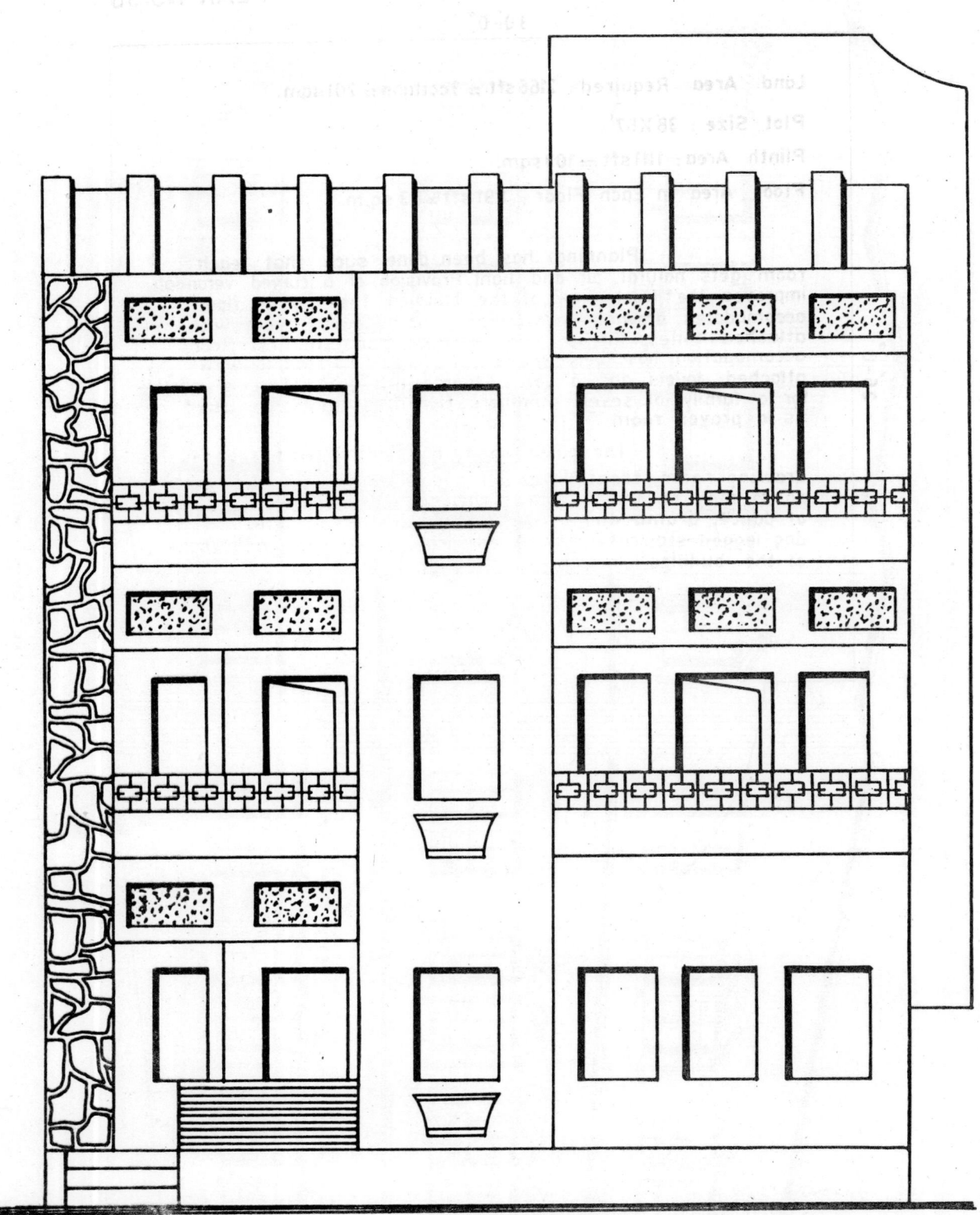

ELEVATION

PLAN NO. 38

30'-0"

Land Area Required : 2166 sft ≈ 3 cottahs ≈ 201 sqm.

Plot Size : 38' X 57'

Plinth Area : 1111 sft = 103 sqm.

Floor Area In Each Floor : 891 sft ≈ 83 sq.m.

Planning has been done such that each room gets natural air and light. Provision of a curved verandah improves the front view of the building. The ground floor accomodates drawing room, dining hall, guest room with attached toilet, kitchen and servant's room. In first floor, accomodation has been made for three bed rooms with attached toilets and a study room. Such a building is suitable for a family of seven members. The attic may be used as a prayer room.

The guest room has individuel access to drawing room, front verandah and dinning hall. The guest room may be used for recreational purposes like practice of dance, drama and music, when there is no guest. A dog legged staircase has been provided at one back corner of the building.

26'-0"

42'-0"

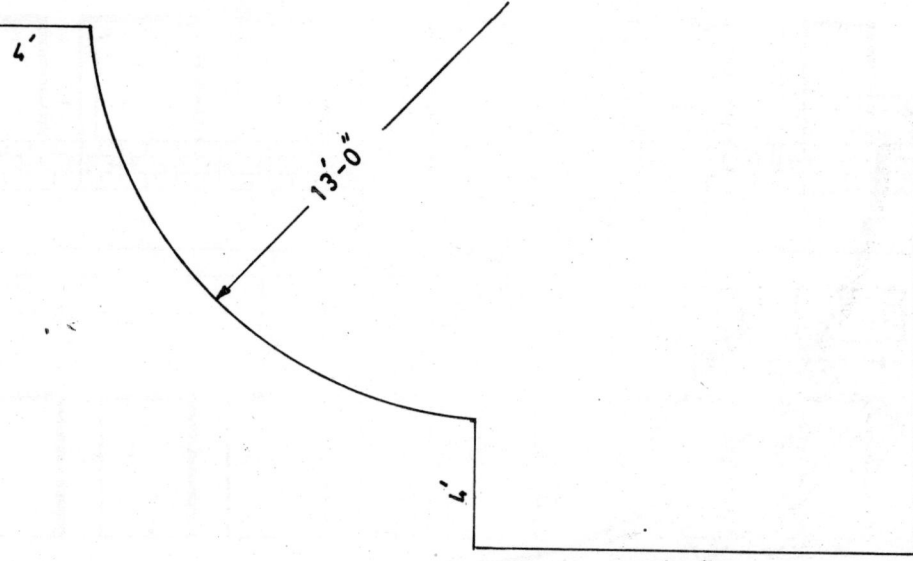

4'

13'-0"

4'

14'-0"

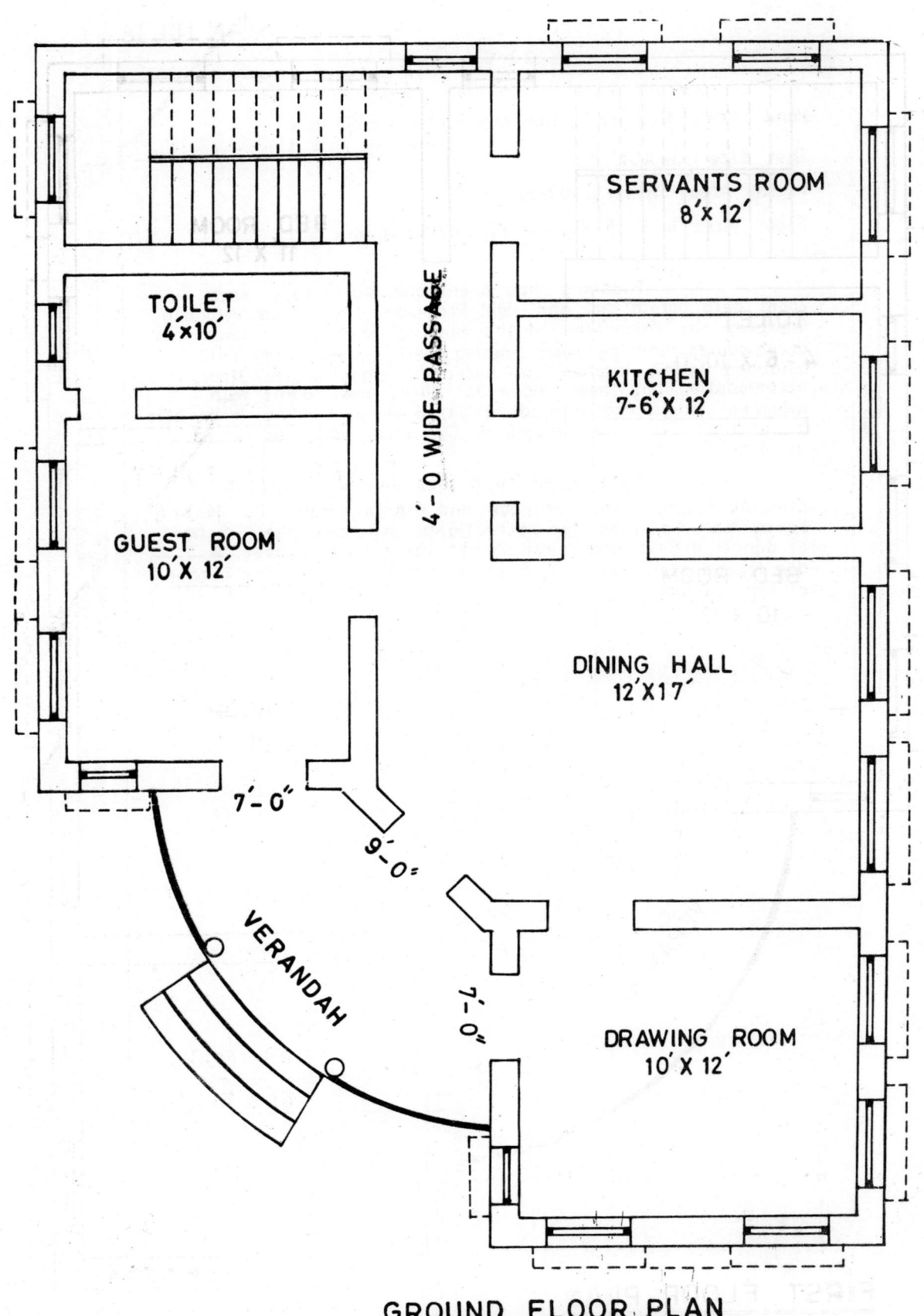

GROUND FLOOR PLAN

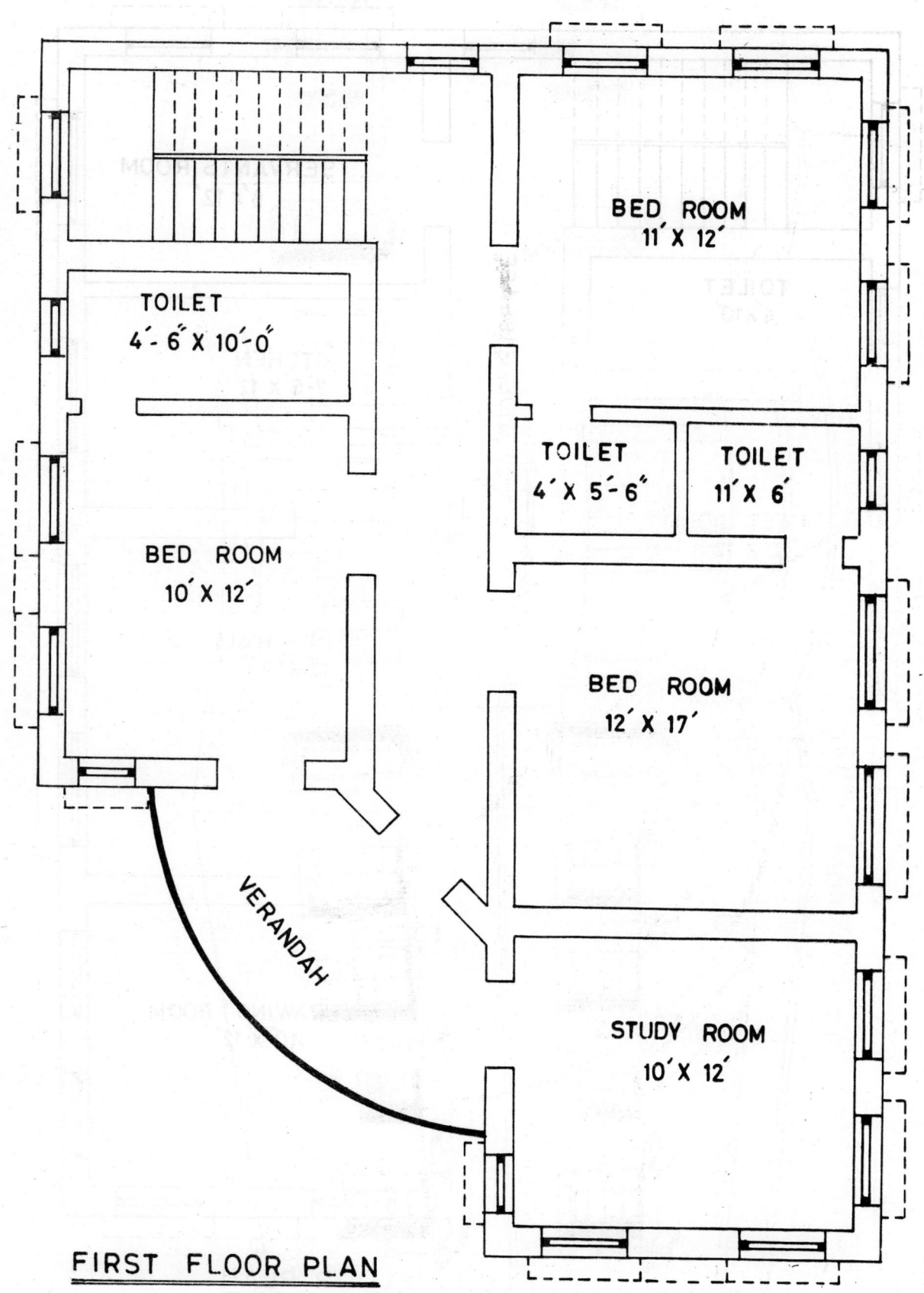

BED ROOM
11' X 12'

TOILET
4'- 6" X 10'-0"

TOILET
4' X 5'-6"

TOILET
11' X 6'

BED ROOM
10' X 12'

BED ROOM
12' X 17'

VERANDAH

STUDY ROOM
10' X 12'

FIRST FLOOR PLAN

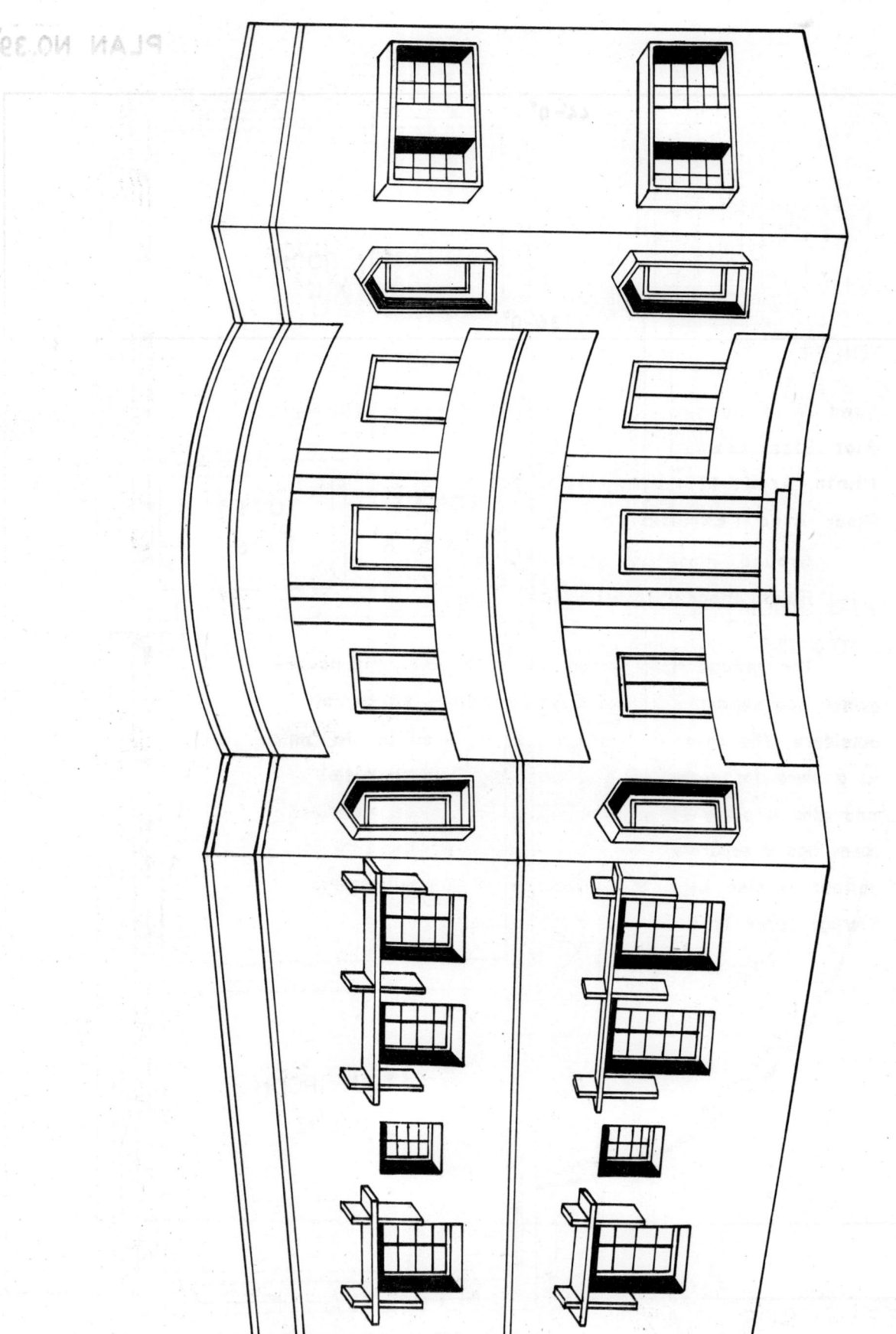

PLAN NO.39

44'-0"

36'-0"

Land Area Required: 2156 Sft ≈ 3 Cottahs ≈ 200 Sq.m.

Plot Size: 44'x 49'

Plinth Area: 1251 Sft ≈ 116·3 Sq.m.

Floor Area (Excluding Stair Case)

    Ground Floor 852 Sft ≈ 79·2 Sq.m.

    First Floor 875 Sft ≈ 81·4 Sq.m.

    The ground floor is for rental purpose. The house-owner has separate entrance with a sitting space for outsiders. The covered verandah may be used by the tenant as a place for reception and discussion with outsiders and also a place for relaxation. In first floor, each bed room has a separate verandah. A home library cum parlour is also kept in addition to other requirements. Storage space is provided in the kitchen.

12'-0"

4'-6"

15'-0"

4'-6"

14'-0"

4'

4'

49'-0"

36'-0"

3'

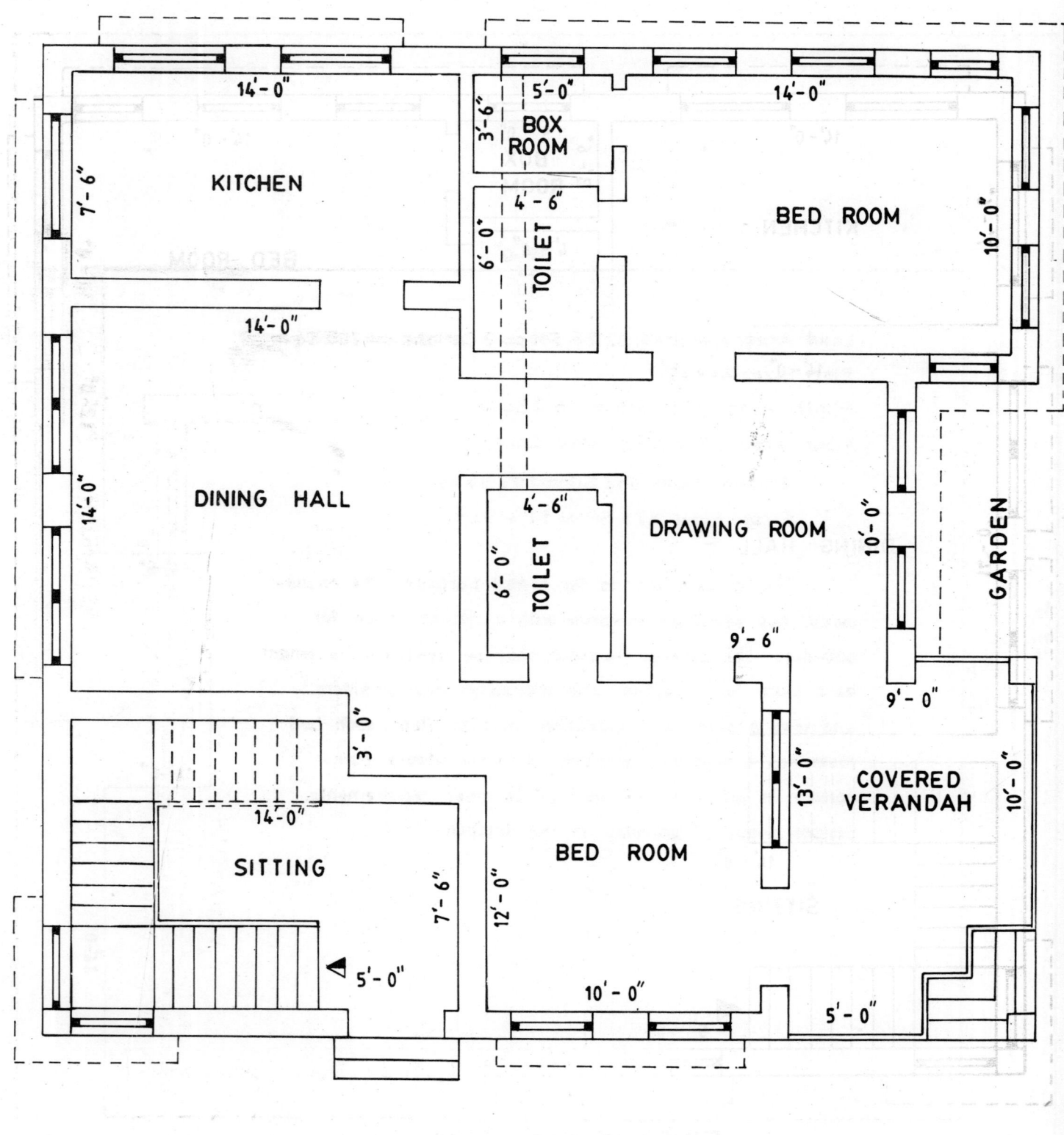

GROUND FLOOR PLAN

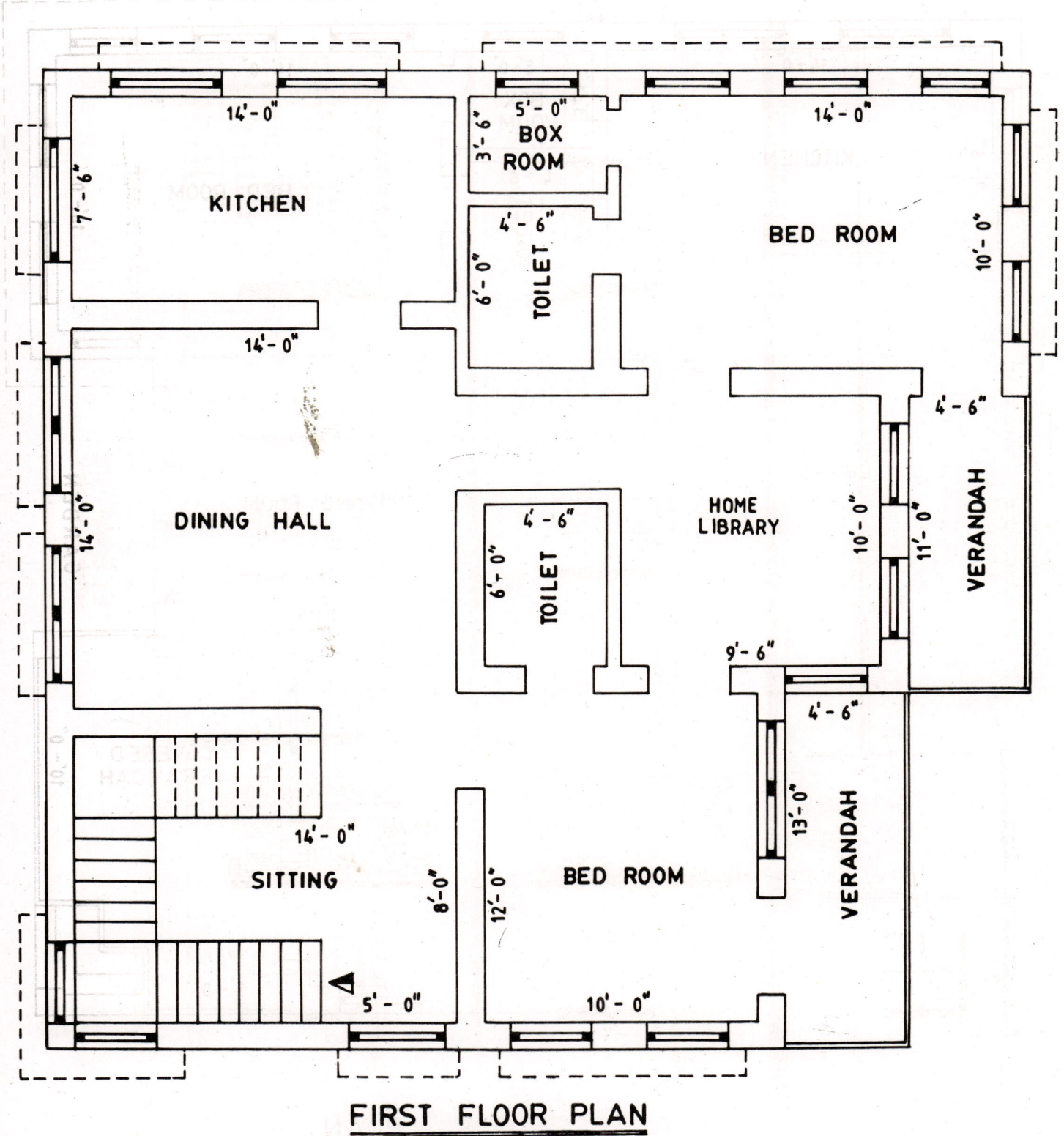

FIRST FLOOR PLAN

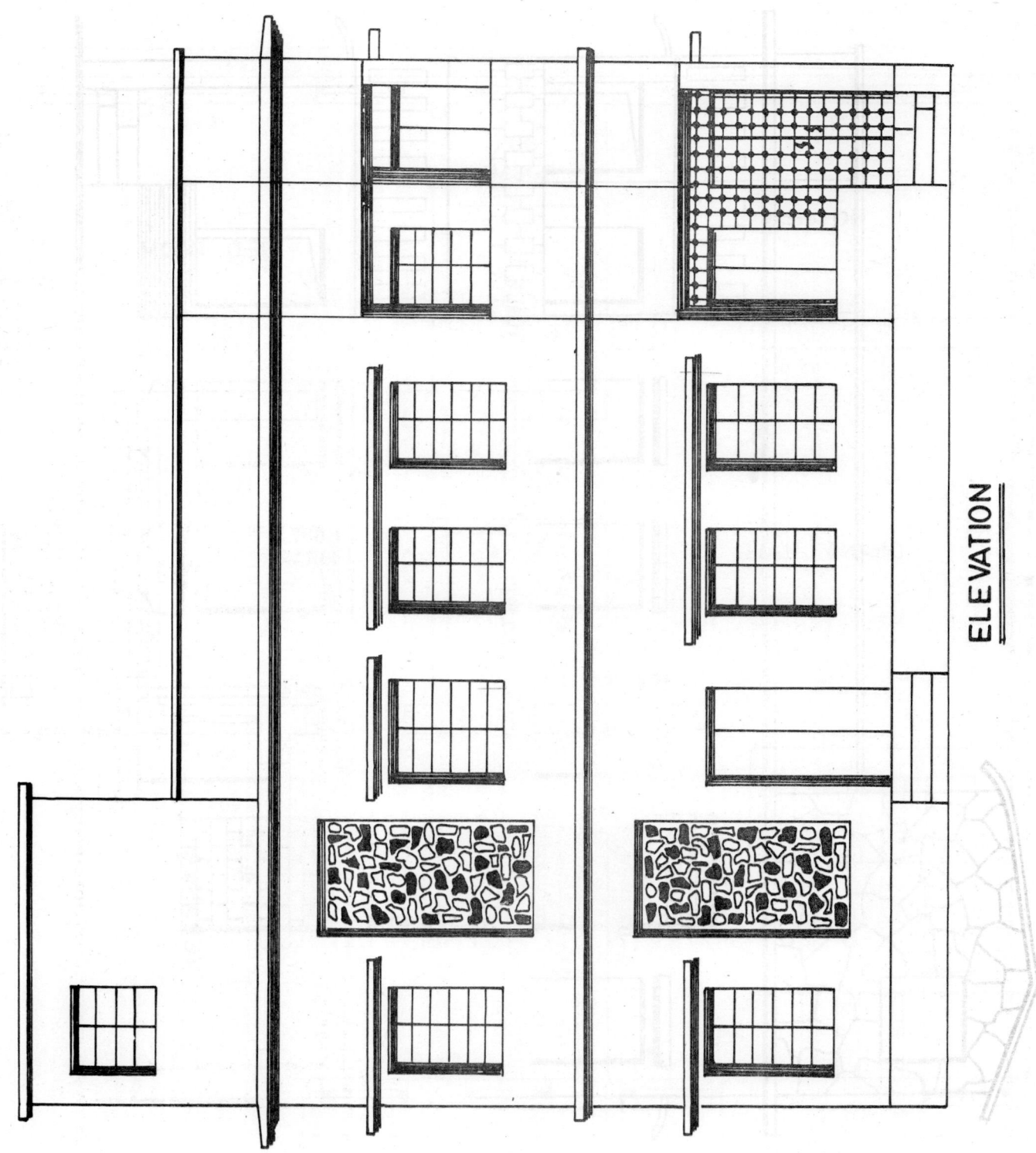

ELEVATION

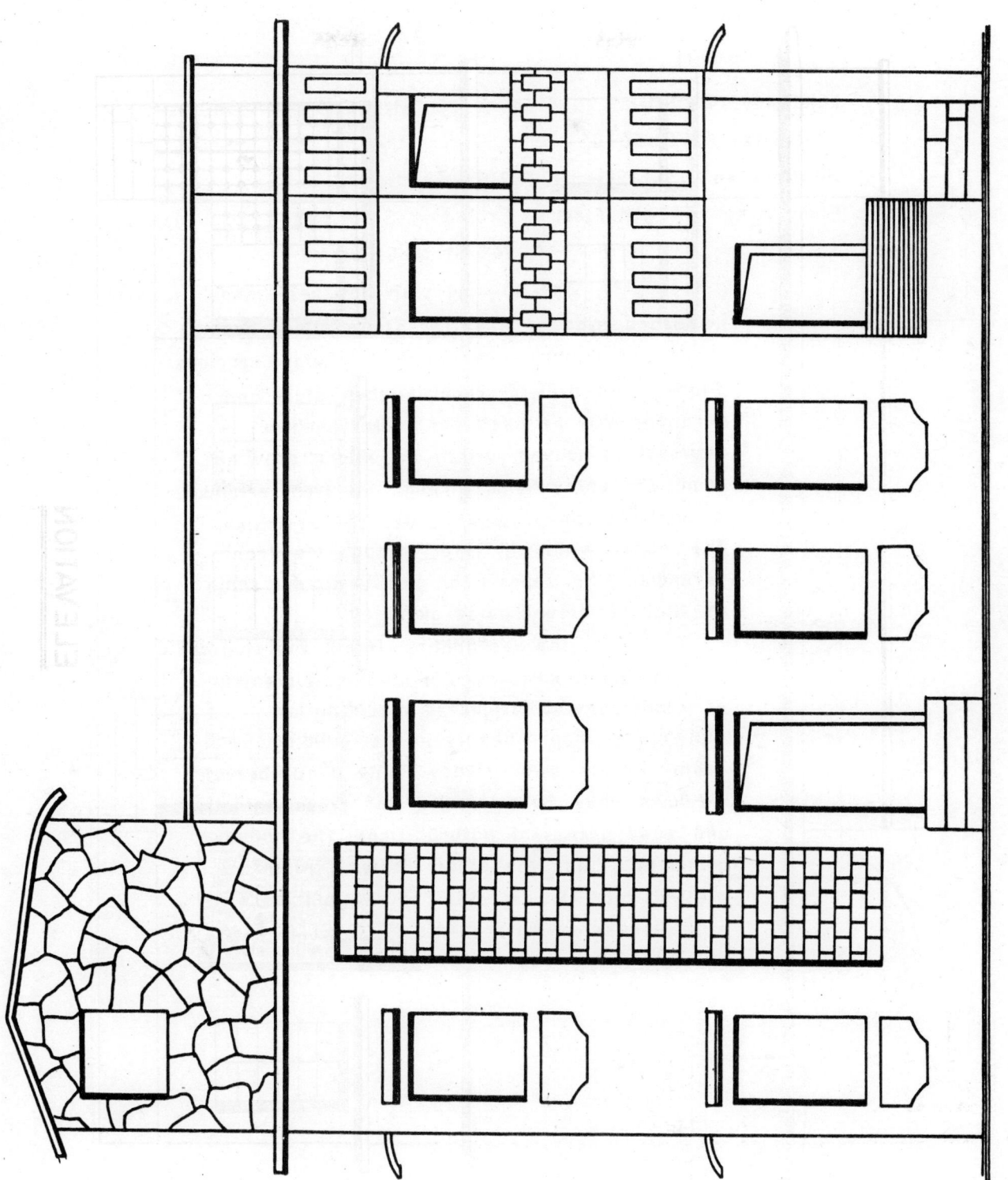

ELEVATION

34'-6"

12'-0"

Land Area Required: 2520 Sft ≈ 3˙5 Cottahs ≈ 234 Sq.m.

Plot Size: 42'-0" x 60'-0"

Plinth Area: 1386 Sft ≈ 127˙62 Sq.m.

Floor Area In Each Floor: (Excluding Stair Case)

1034˙5 Sft ≈ 96 Sq.m.

The building plan has been made in such a manner that the ground floor having three bed rooms with attached toilets, a spacious dining cum drawing room, kitchen, store and a verandah may be used for rental purpose. A staircase is housed within an oblong drawing room, the entry being through the front verandah. A toilet may be provided under the staircase. The house-owner will enter through the front verandah. The upper floor plan is almost same as that of ground floor plan.

The verandahs, offsets, the staircase with projected semi-circular landing, placement of windows with projected moulding of sunshades impart beauty to the building. Each room has its own privacy. Quite a number of windows have been provided for cross-ventilation and easy access of natural light. The house-owner may use the small drawing room at ground floor for reception of the outsiders and discussion with them.

5'-0"

20'-6"

8'-0"

45'-6"

5'-6"

7'-6"

2'-6"

19'-6"

4'-6"

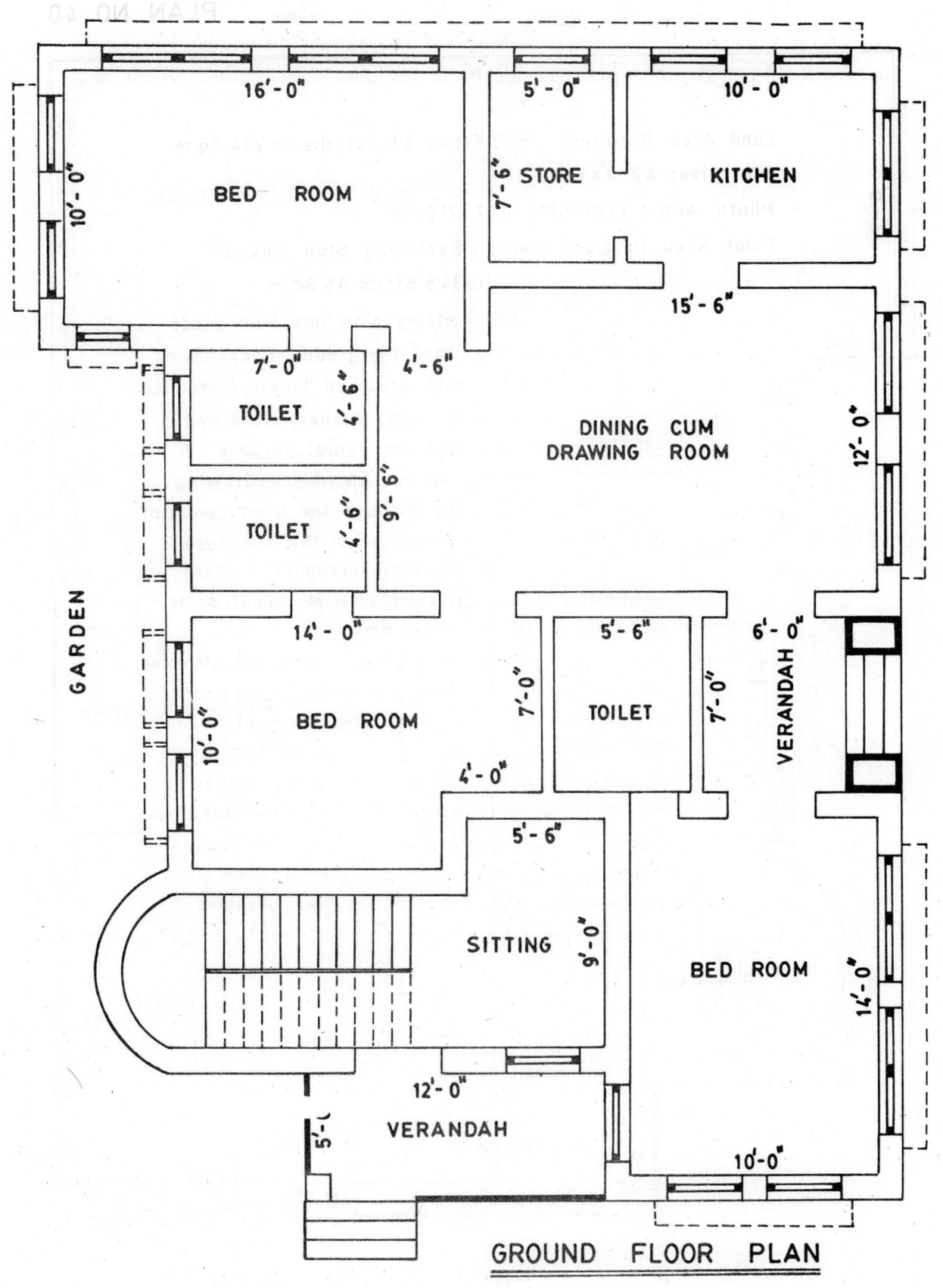

GROUND FLOOR PLAN

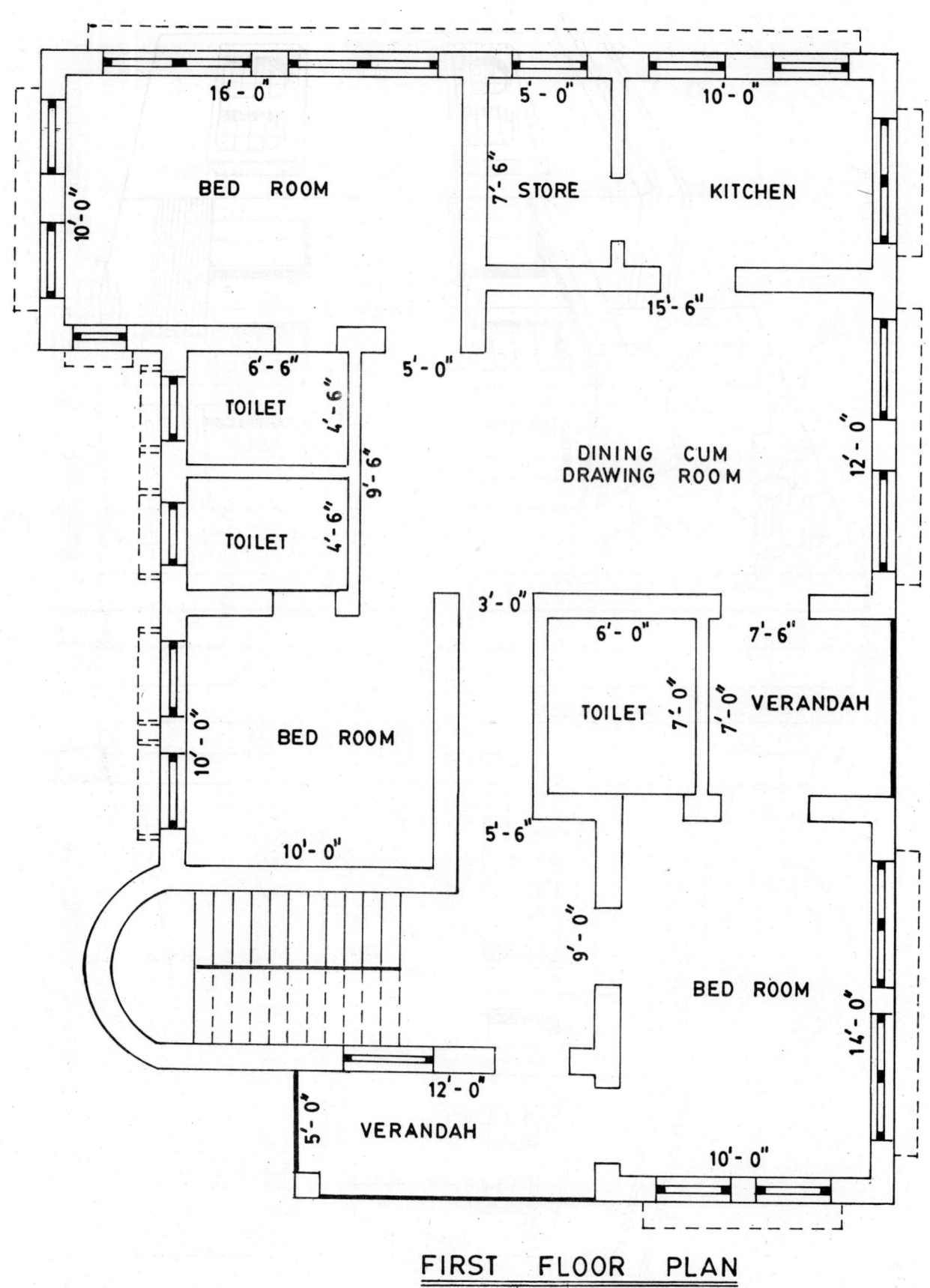

FIRST FLOOR PLAN

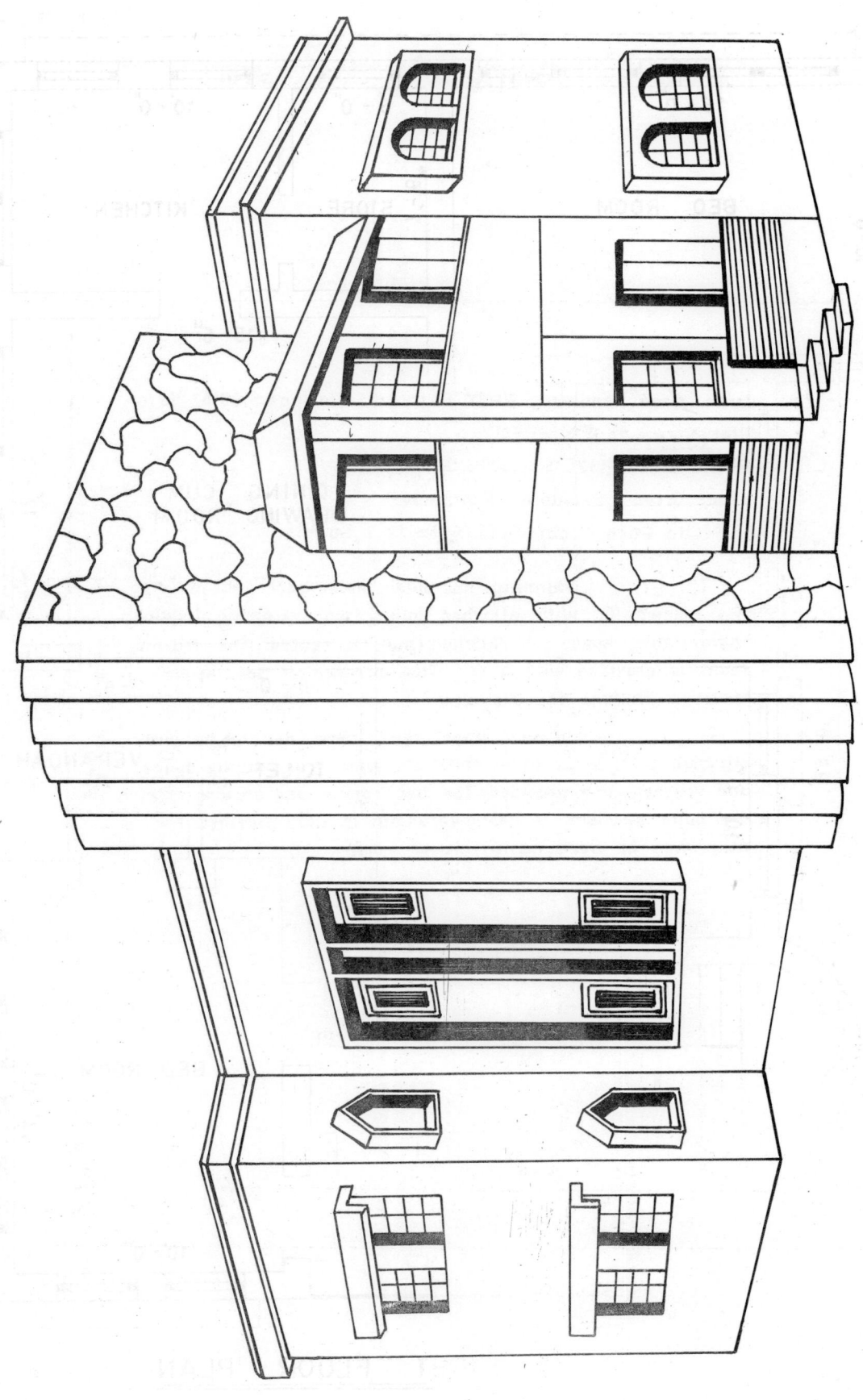

PLAN NO. 41

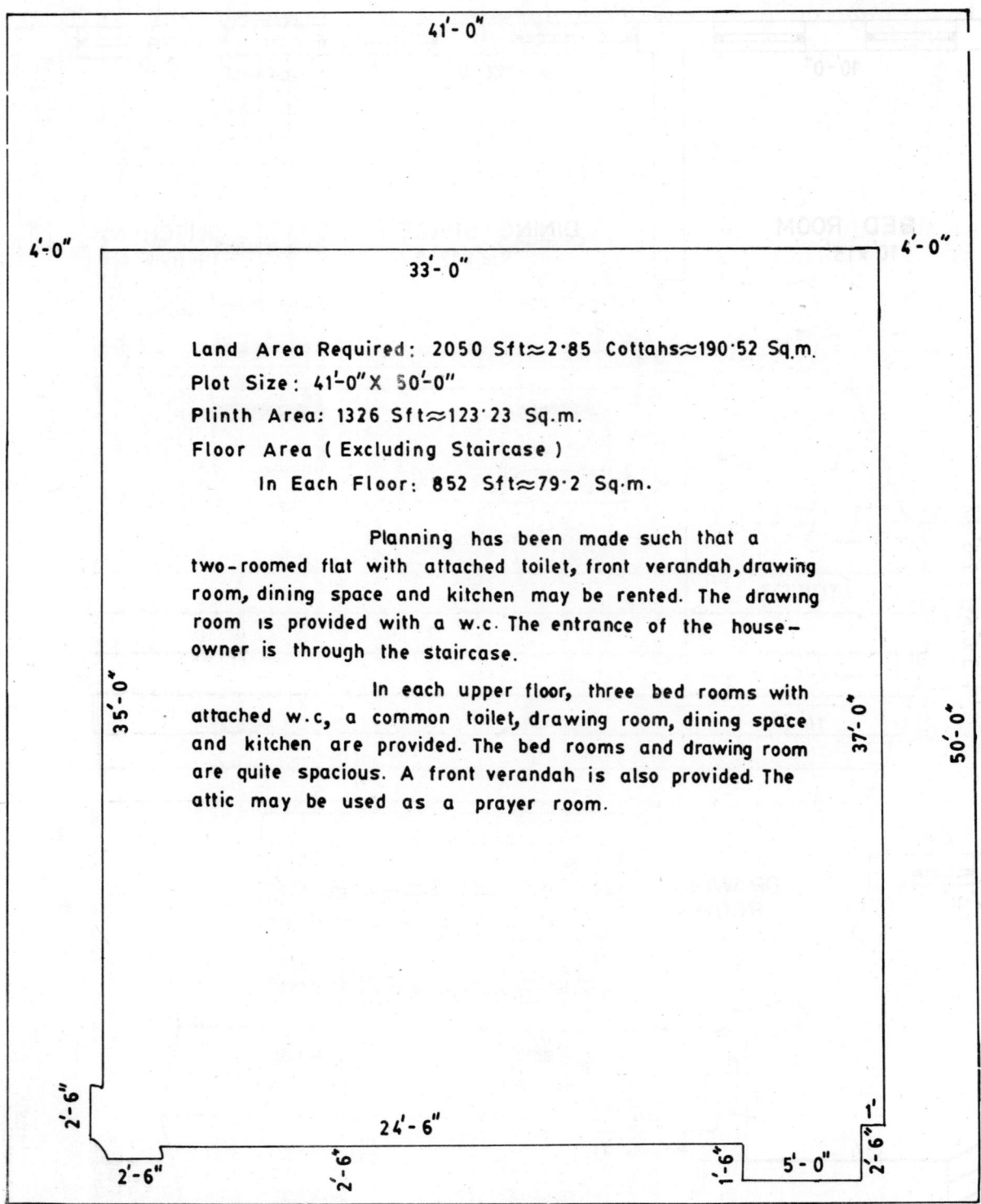

41'- 0"

33'- 0"

4'-0"

4'-0"

Land Area Required: 2050 Sft≈2·85 Cottahs≈190·52 Sq.m.

Plot Size: 41'-0" X 50'-0"

Plinth Area: 1326 Sft≈123·23 Sq.m.

Floor Area (Excluding Staircase)

In Each Floor: 852 Sft≈79·2 Sq·m.

Planning has been made such that a two-roomed flat with attached toilet, front verandah, drawing room, dining space and kitchen may be rented. The drawing room is provided with a w.c. The entrance of the house-owner is through the staircase.

In each upper floor, three bed rooms with attached w.c, a common toilet, drawing room, dining space and kitchen are provided. The bed rooms and drawing room are quite spacious. A front verandah is also provided. The attic may be used as a prayer room.

35'- 0"

37'- 0"

50'- 0"

2'-6"

24'- 6"

2'- 6"

1'

2'- 6"

2'- 6"

1'- 6"

5'- 0"

2'- 6"

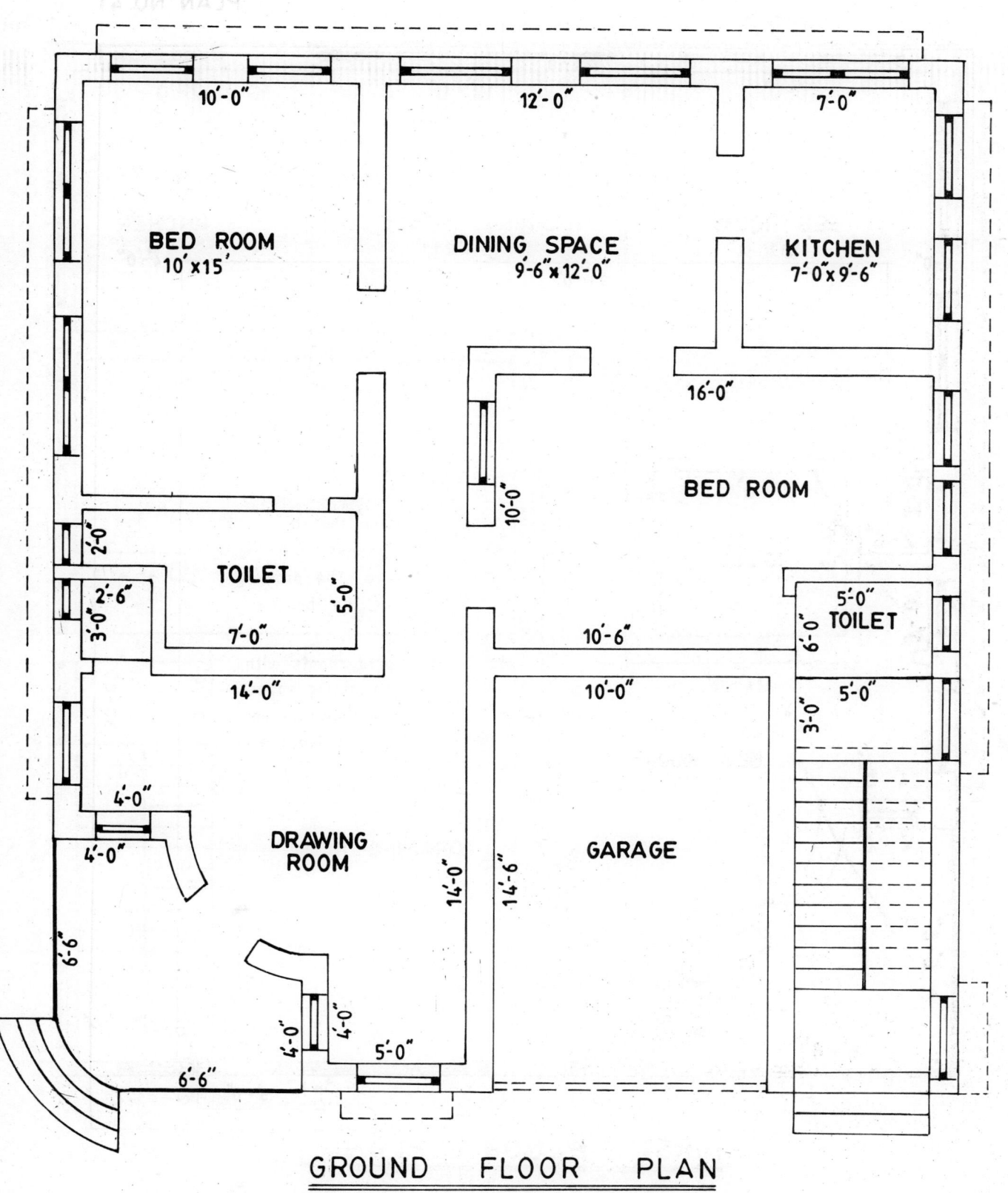

BED ROOM
10'x15'

DINING SPACE
9'-6"x12'-0"

KITCHEN
7'-0"x9'-6"

10'-0"

12'-0"

7'-0"

16'-0"

10'-0"

BED ROOM

TOILET

5'-0"

2'-0"

2'-6"

3'-0"

7'-0"

5'-0"
TOILET

6'-0"

5'-0"

3'-0"

14'-0"

10'-6"

4'-0"

10'-0"

4'-0"

DRAWING
ROOM

14'-0"

14'-6"

GARAGE

6'-6"

4'-0"

4'-0"

6'-6"

5'-0"

GROUND   FLOOR   PLAN

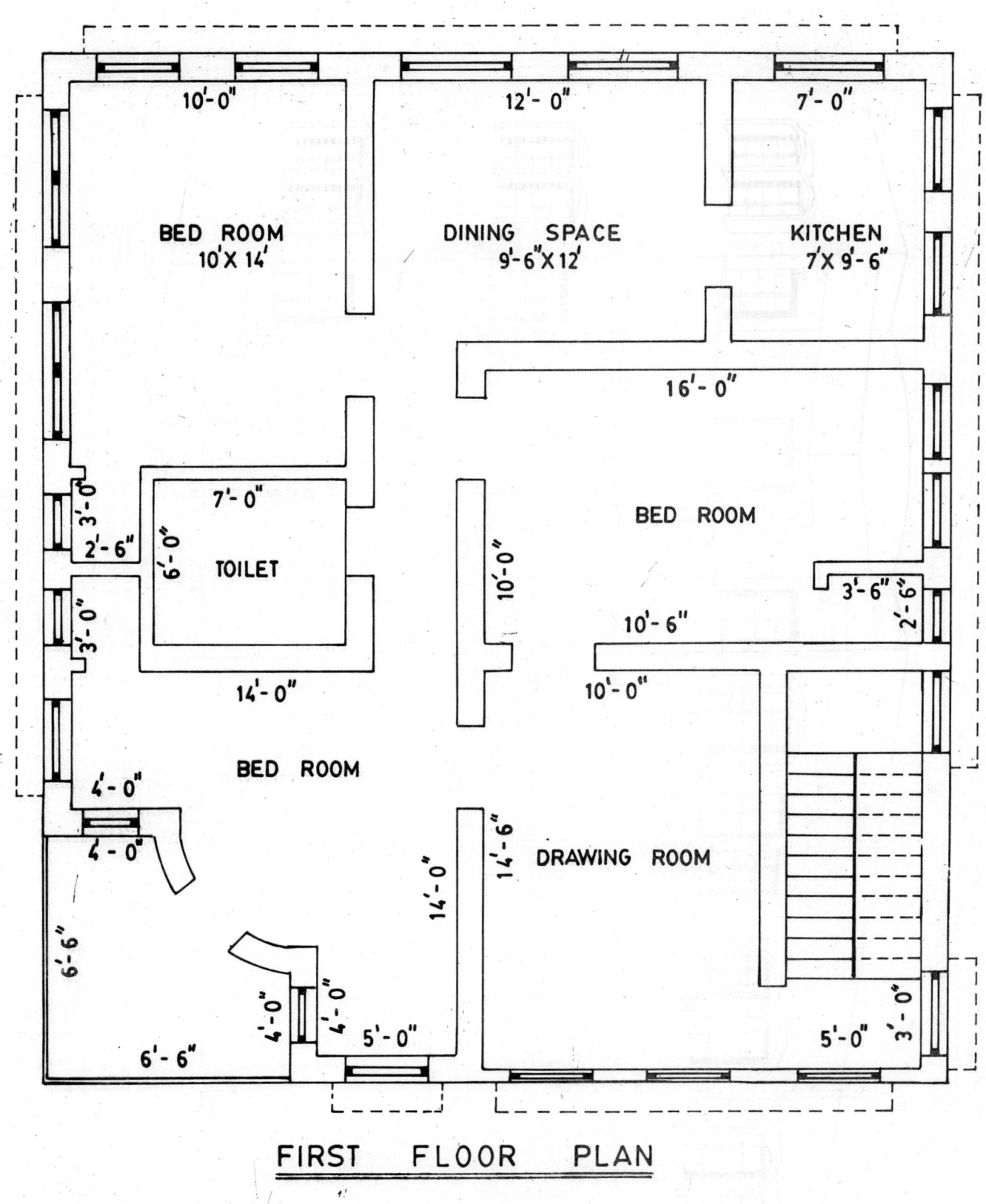

FIRST FLOOR PLAN

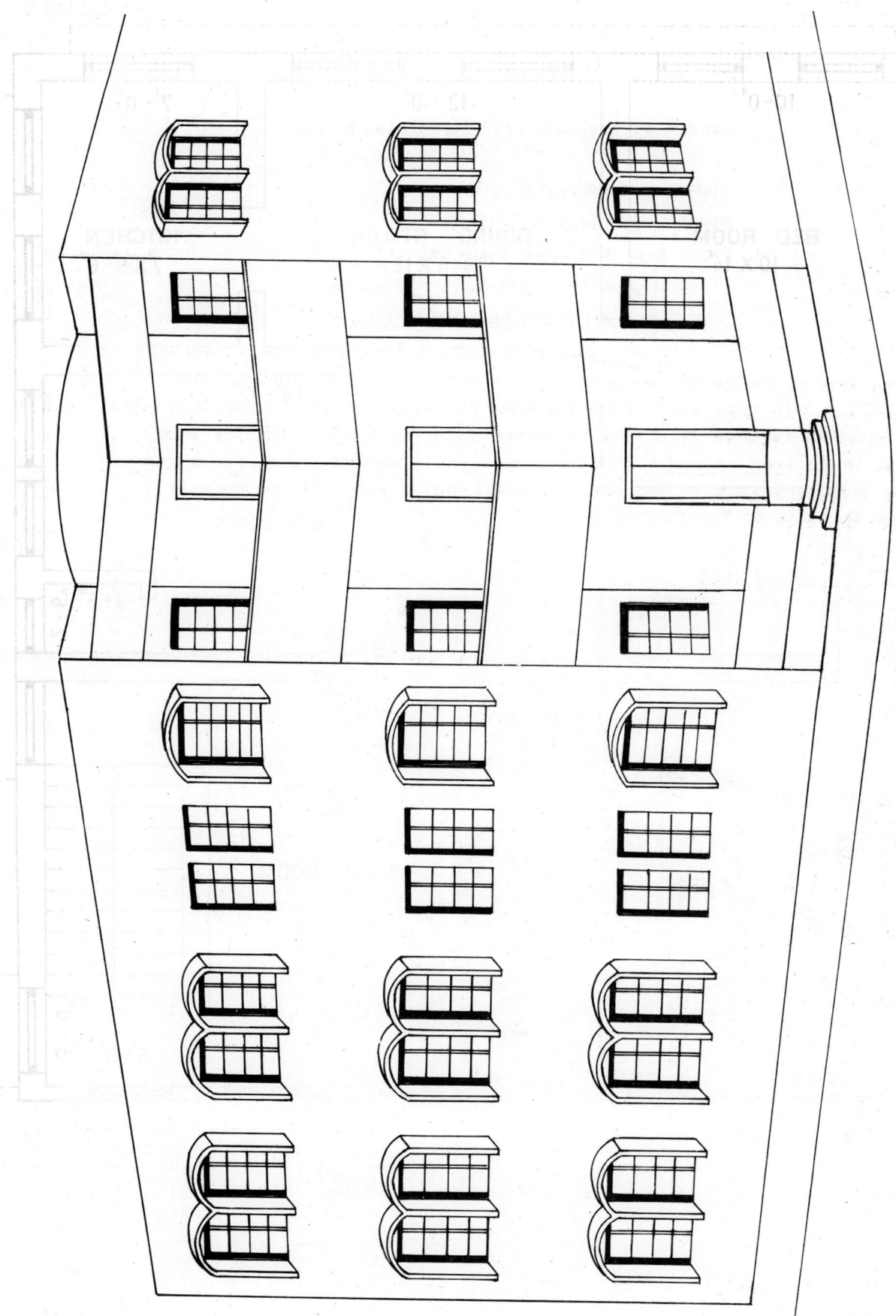

PLAN NO. 42

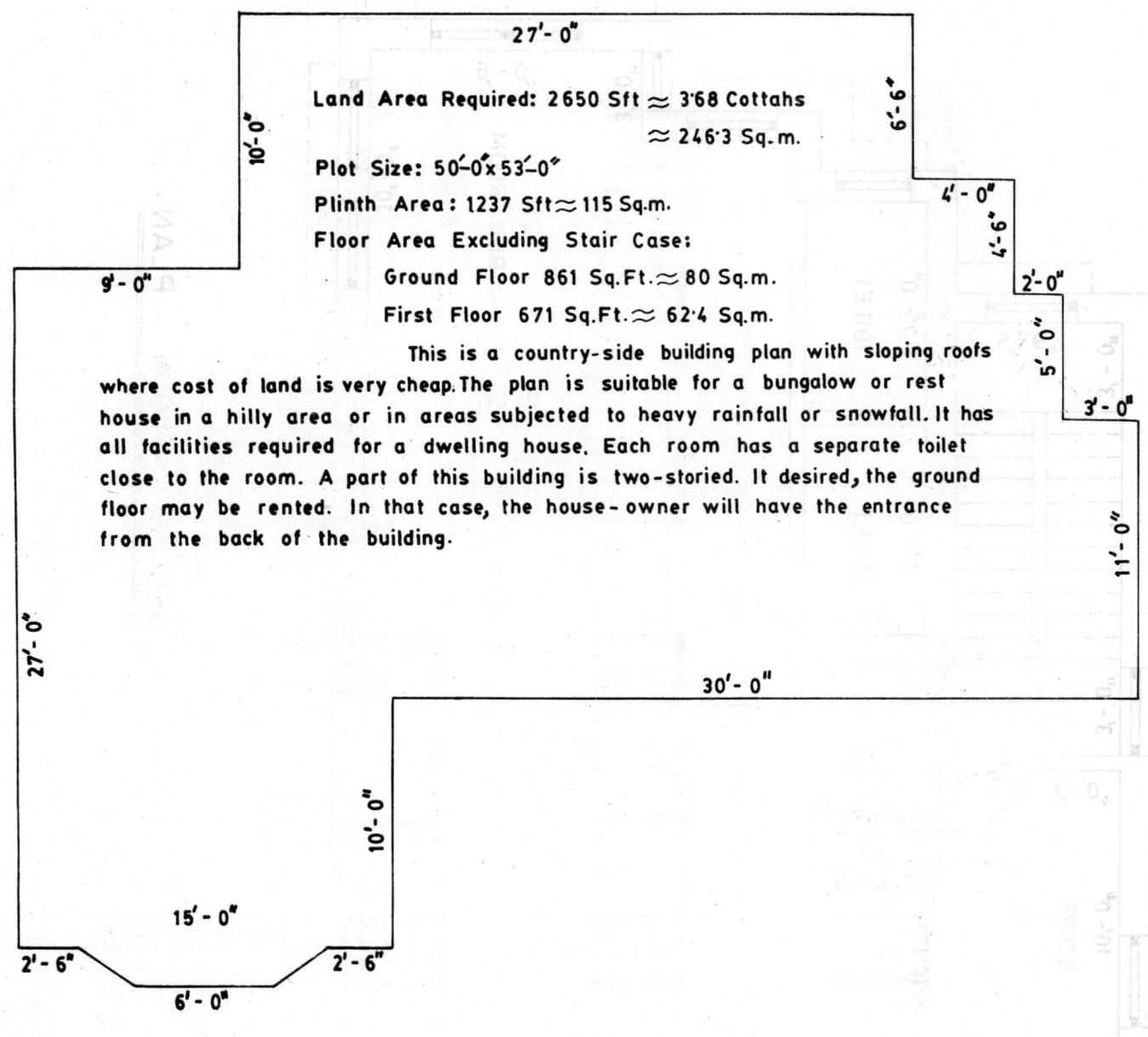

Land Area Required: 2650 Sft ≈ 3˙68 Cottahs

≈ 246˙3 Sq.m.

Plot Size: 50'-0" x 53'-0"

Plinth Area: 1237 Sft ≈ 115 Sq.m.

Floor Area Excluding Stair Case:

Ground Floor 861 Sq.Ft. ≈ 80 Sq.m.

First Floor 671 Sq.Ft. ≈ 62˙4 Sq.m.

This is a country-side building plan with sloping roofs where cost of land is very cheap. The plan is suitable for a bungalow or rest house in a hilly area or in areas subjected to heavy rainfall or snowfall. It has all facilities required for a dwelling house. Each room has a separate toilet close to the room. A part of this building is two-storied. If desired, the ground floor may be rented. In that case, the house-owner will have the entrance from the back of the building.

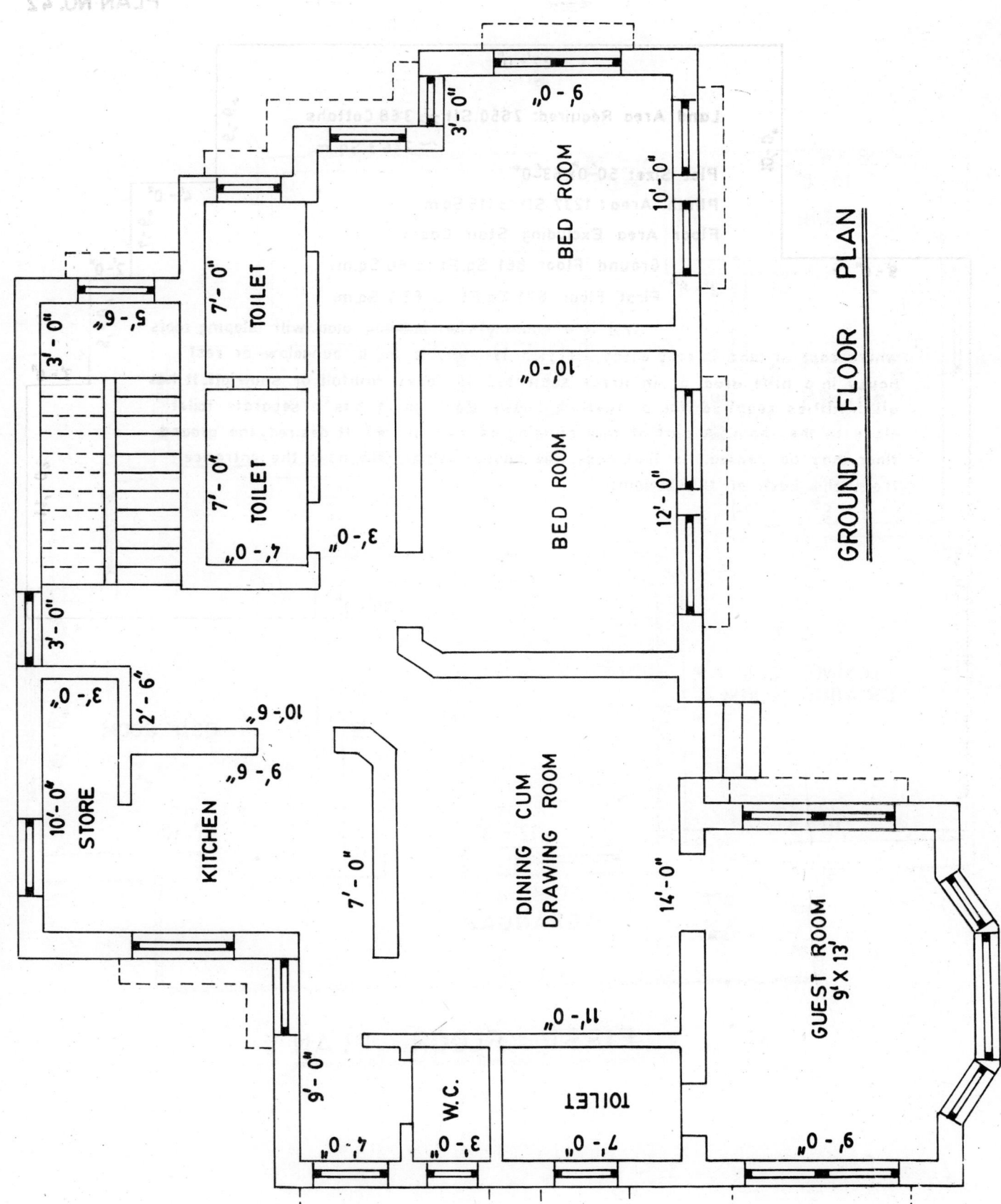

GROUND FLOOR PLAN

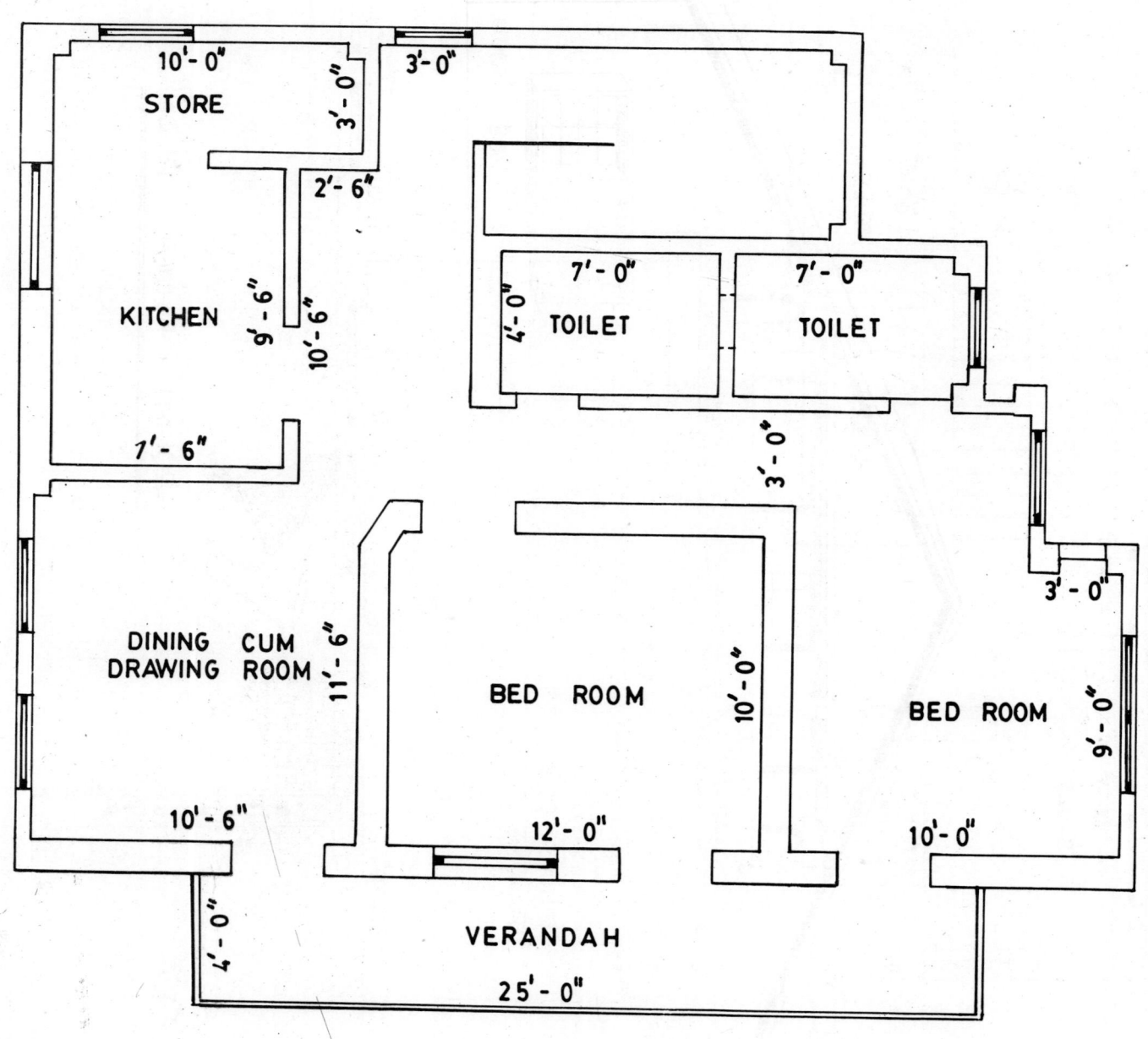

FIRST FLOOR PLAN

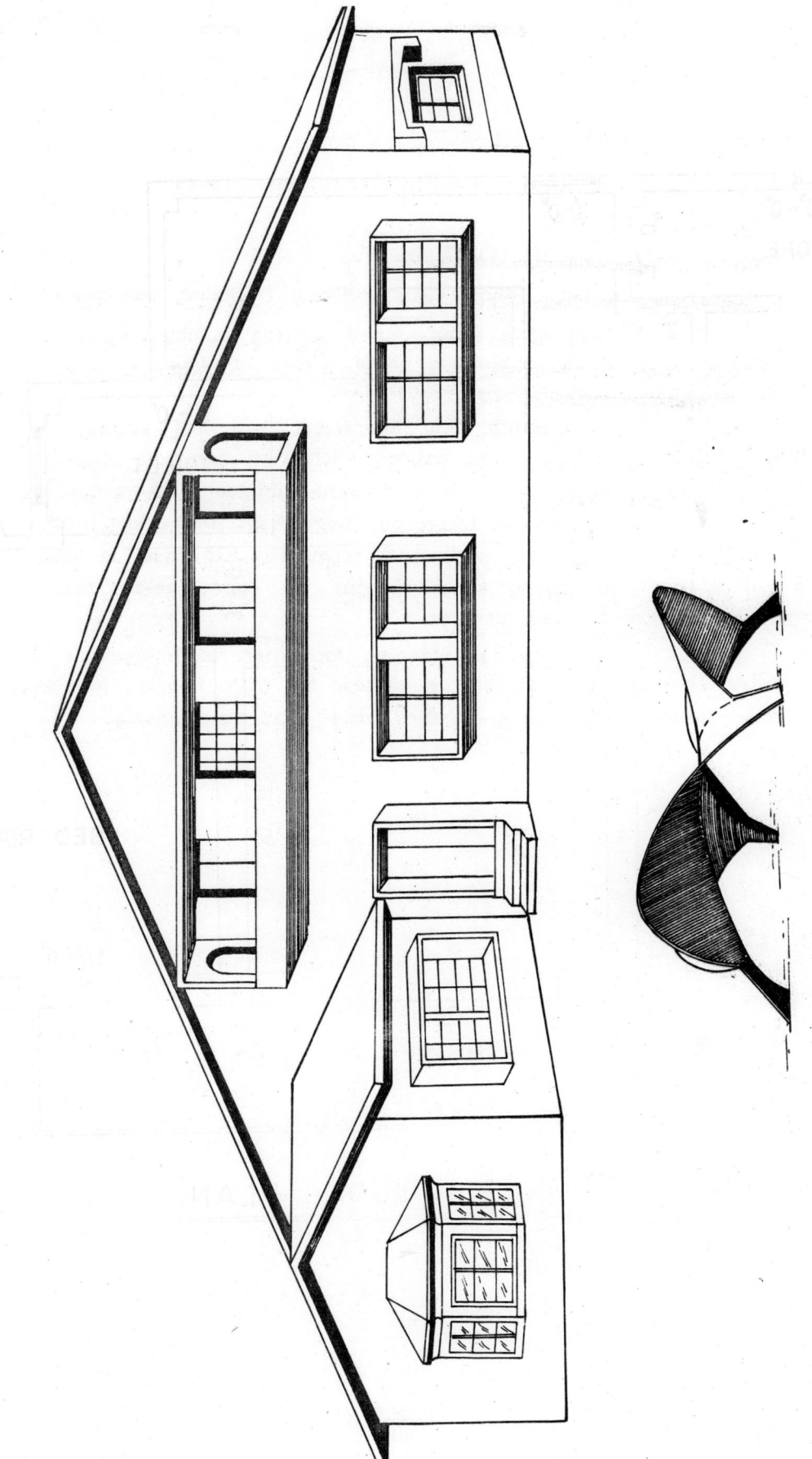

PLAN NO. 43

Land Area Required : 1680 sft = 2·33 cottahs = 156 sq.m.

Plot Size : 42' X 50'

Plinth Area : 1219 sft = 113·3 sq.m.

Floor Area In Each Floor : 1041 sft ≈ 97 sq.m. (Including Staircase)

   This is a three-storied building suitable for a divided large family (comprising three small families) residing in the same house.

   In ground floor, there are two bed rooms with attached toilets, dining space, kitchen, a common toilet, garage and drawing room with a front varandah. The mezzanine room may be used as a recreation room. Both first floor and second floor have one extra bed room. The staircase is housed within the dining space so that the space below the staircase can be well utilised.

   This type of planning facilitates in bringing down the capital cost to be shared by each family, provided they have mutual understanding and good relationship.

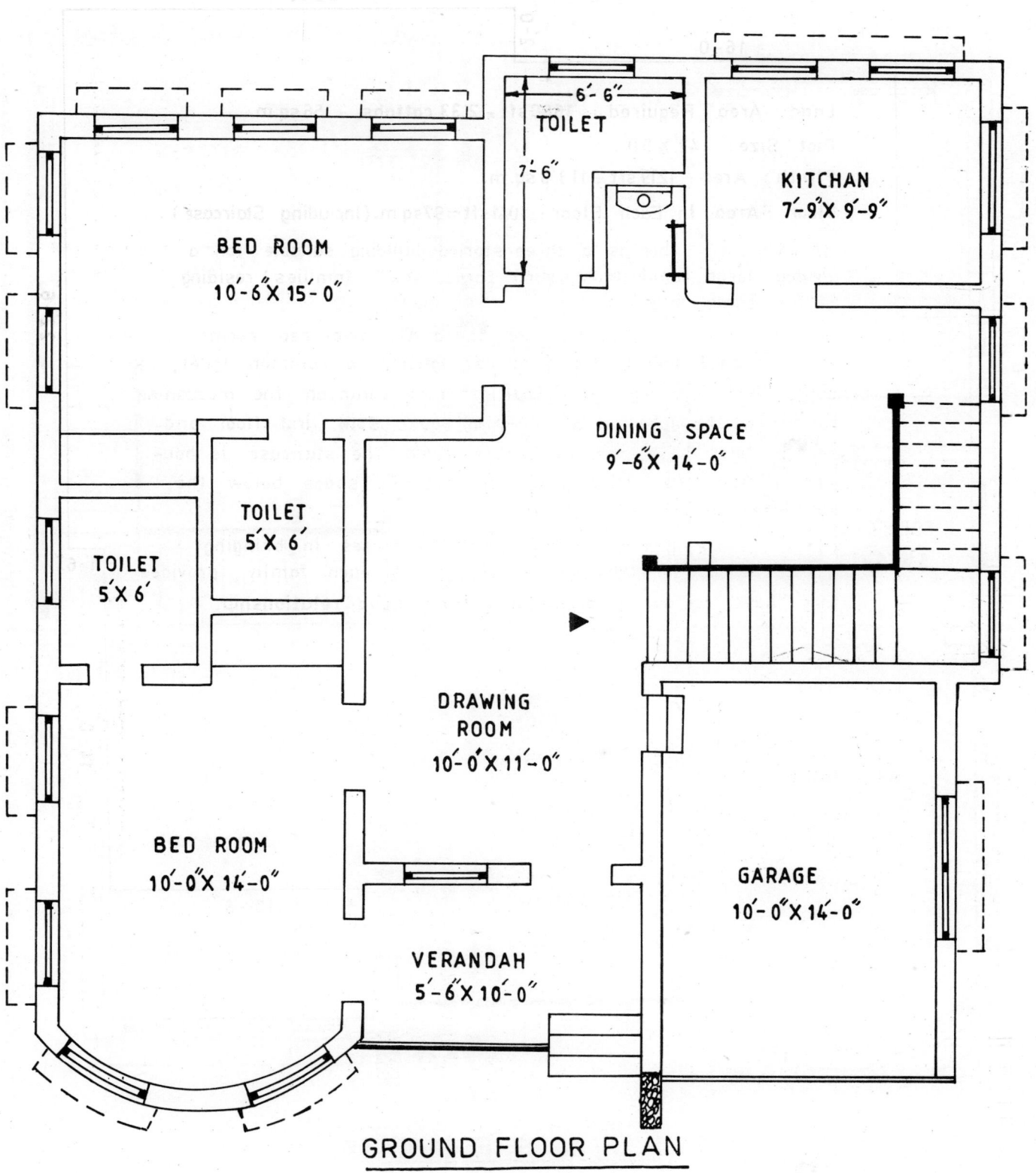

GROUND FLOOR PLAN

TOILET
6'-6"
7'-6"

KITCHAN
7'-9"X 9'-9"

BED ROOM
10'-6"X 15'-0"

DINING SPACE
9'-6"X 14'-0"

TOILET
5'X 6'

TOILET
5'X 6'

DRAWING
ROOM
10'-0"X 11'-0"

BED ROOM
10'-0"X 14'-0"

GARAGE
10'-0"X 14'-0"

VERANDAH
5'-6"X 10'-0"

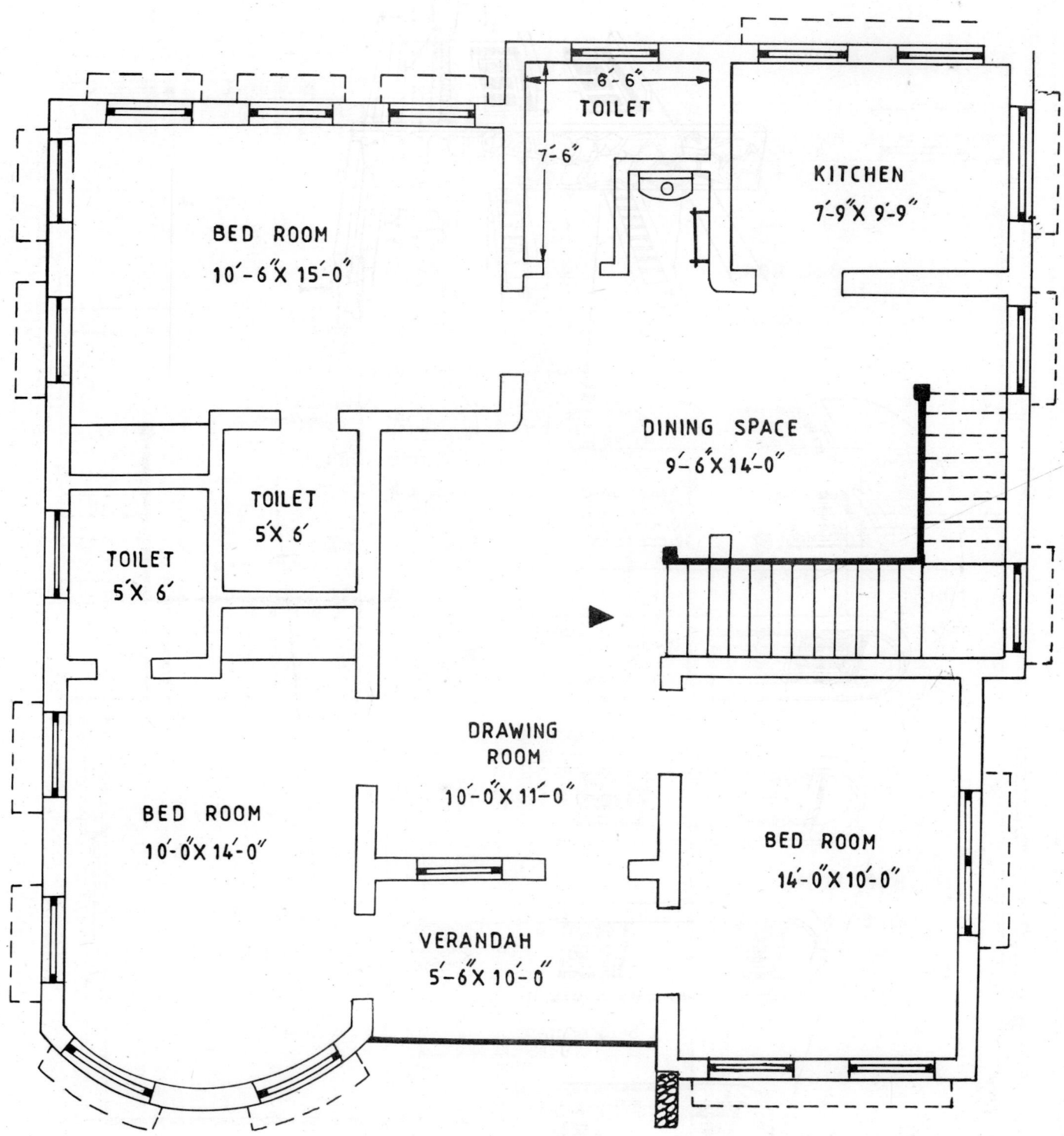

BED ROOM
10'-6" X 15'-0"

TOILET
6'-6"
7'-6"

KITCHEN
7'-9" X 9'-9"

TOILET
5' X 6'

TOILET
5' X 6'

DINING SPACE
9'-6" X 14'-0"

BED ROOM
10'-0" X 14'-0"

DRAWING ROOM
10'-0" X 11'-0"

BED ROOM
14'-0" X 10'-0"

VERANDAH
5'-6" X 10'-0"

F. F. & S. F. PLAN

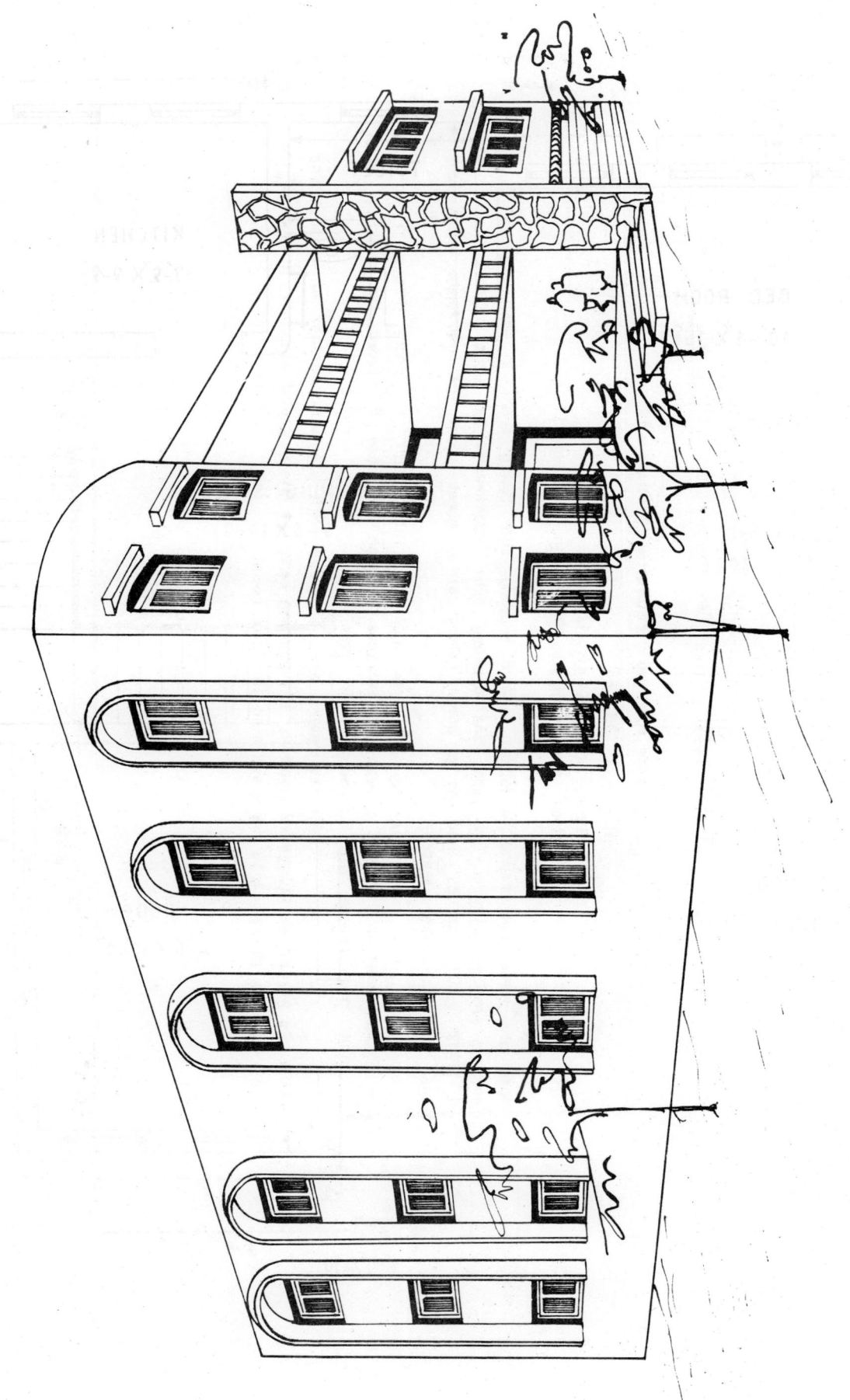

PLAN NO. 44

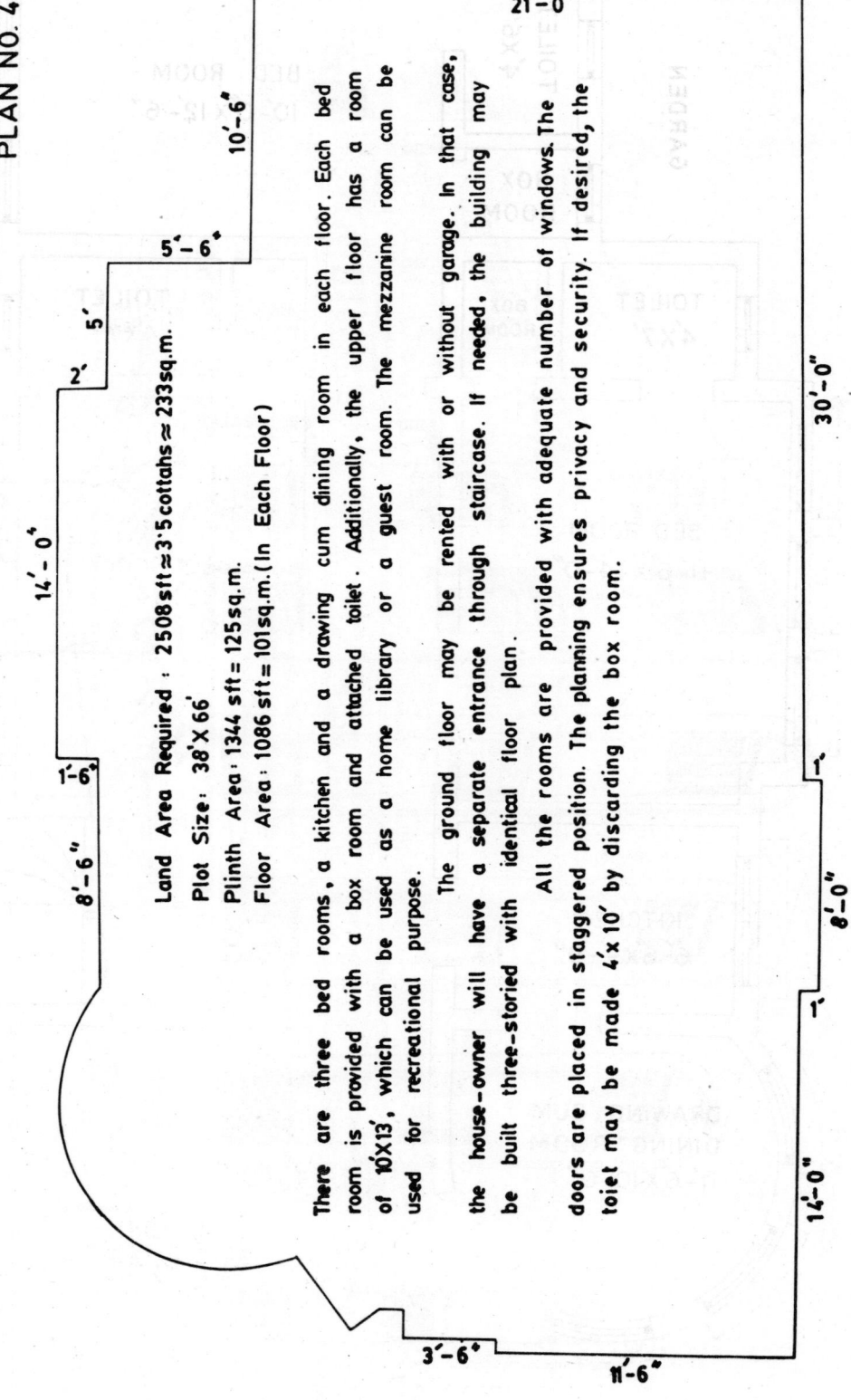

Land Area Required : 2508 sft ≈ 3·5 cottahs ≈ 233 sq.m.

Plot Size : 38' X 66'

Plinth Area : 1344 sft = 125 sq.m.

Floor Area : 1086 sft = 101 sq.m. (In Each Floor)

There are three bed rooms, a kitchen and a drawing cum dining room in each floor. Each bed room is provided with a box room and attached toilet. Additionally, the upper floor has a room of 10'X13', which can be used as a home library or a guest room. The mezzanine room can be used for recreational purpose.

The ground floor may be rented with or without garage. In that case, the house-owner will have a separate entrance through staircase. If needed, the building may be built three-storied with identical floor plan.

All the rooms are provided with adequate number of windows. The doors are placed in staggered position. The planning ensures privacy and security. If desired, the toilet may be made 4'x10' by discarding the box room.

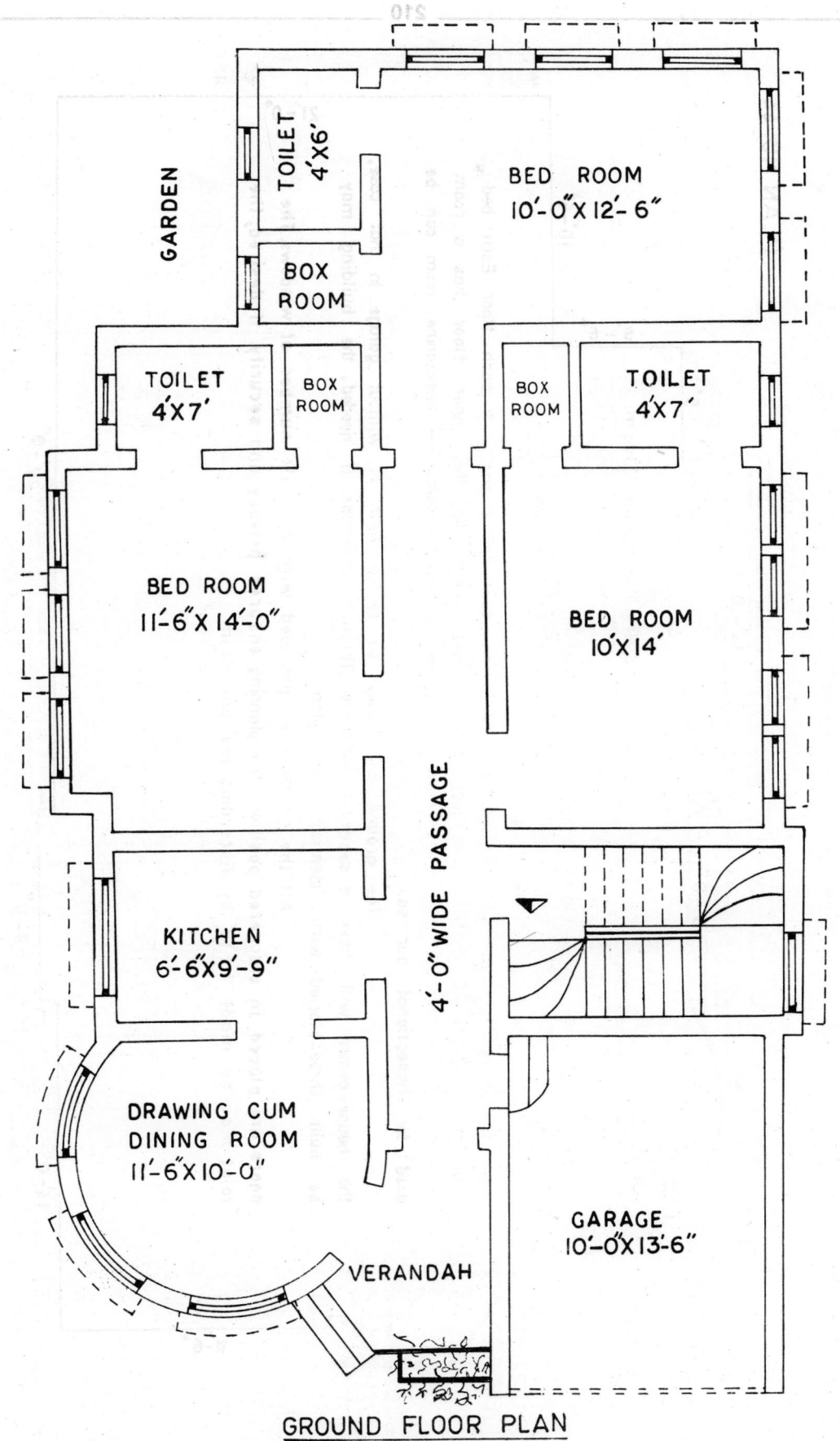

GROUND FLOOR PLAN

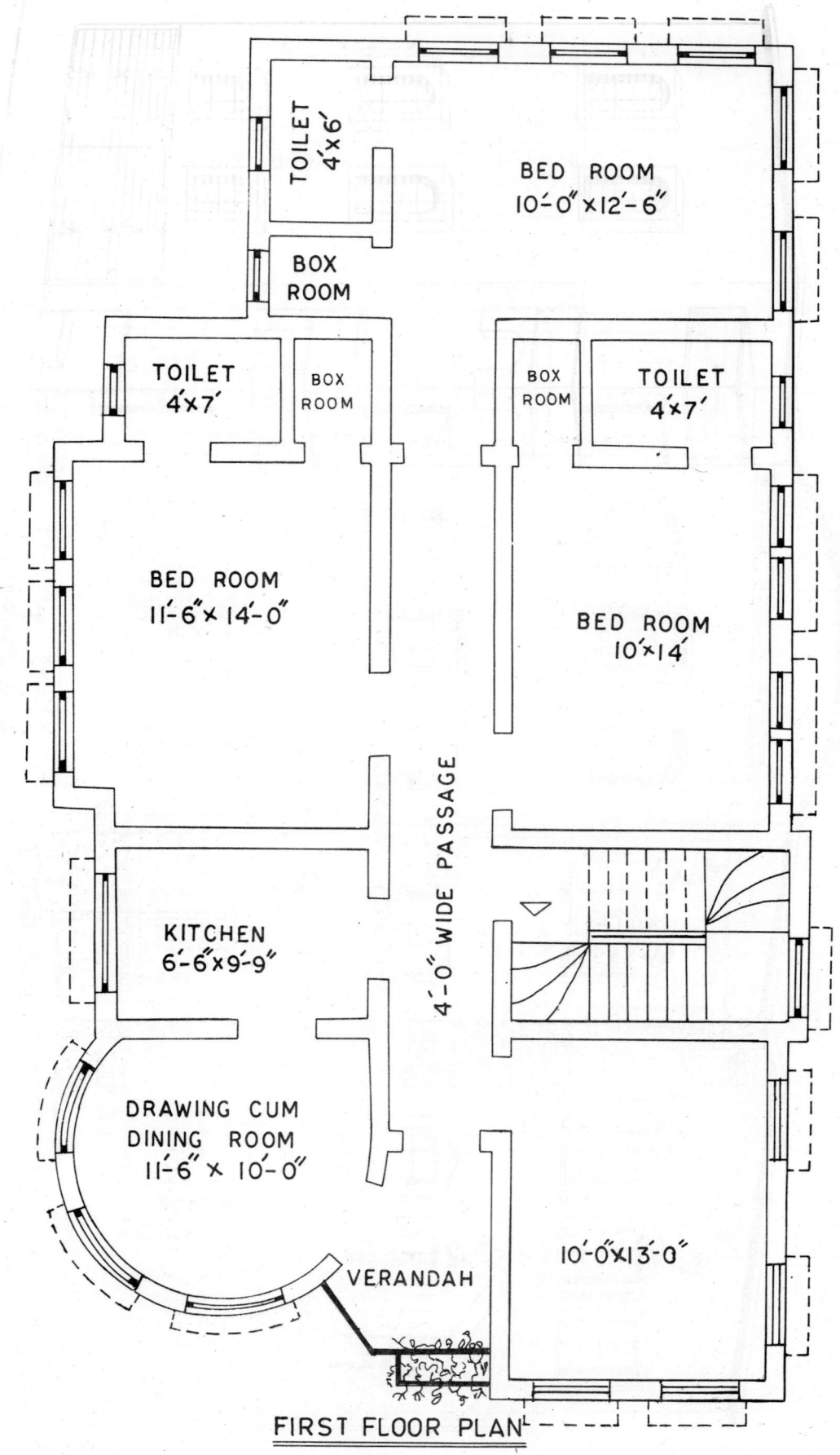

TOILET
4'x6'

BOX
ROOM

BED ROOM
10'-0"×12'-6"

TOILET
4'x7'

BOX
ROOM

BOX
ROOM

TOILET
4'x7'

BED ROOM
11'-6"×14'-0"

BED ROOM
10'×14'

4'-0" WIDE PASSAGE

KITCHEN
6'-6"×9'-9"

DRAWING CUM
DINING ROOM
11'-6" × 10'-0"

VERANDAH

10'-0"×13'-0"

FIRST FLOOR PLAN

PLAN NO. 45

Land Area Required : 2600 sft ≈ 3`6 cottahs ≈ 241`64 sq. m.

Plot Size : 40'×65'

Plinth Area : 1218 sft ≈ 113`2 sq. m.

Floor Area :

    Ground Floor : 945 sft ≈ 88 sq.m.
    First Floor : 400 sft ≈ 37 sq.m.

    This is a country-side building plan with sloping roof style. In ground floor, there are three bed rooms with attached toilets, a kitchen, staircase and a large hall that can serve as drawing cum dining space. In first floor, a multi-purpose room of 12'×20', a lobby and a toilet have been accomodated.

    Such type of building plan can be made, where cost of land is quite low and in areas of heavy rainfall. The type of building is suitable for making a guest house or rest house.

    The multipurpose room is meant for recreational purposes. A louvred partition or screen may be used to separate the dinning space from the drawing room. The rooms are well ventilated with access of natural light.

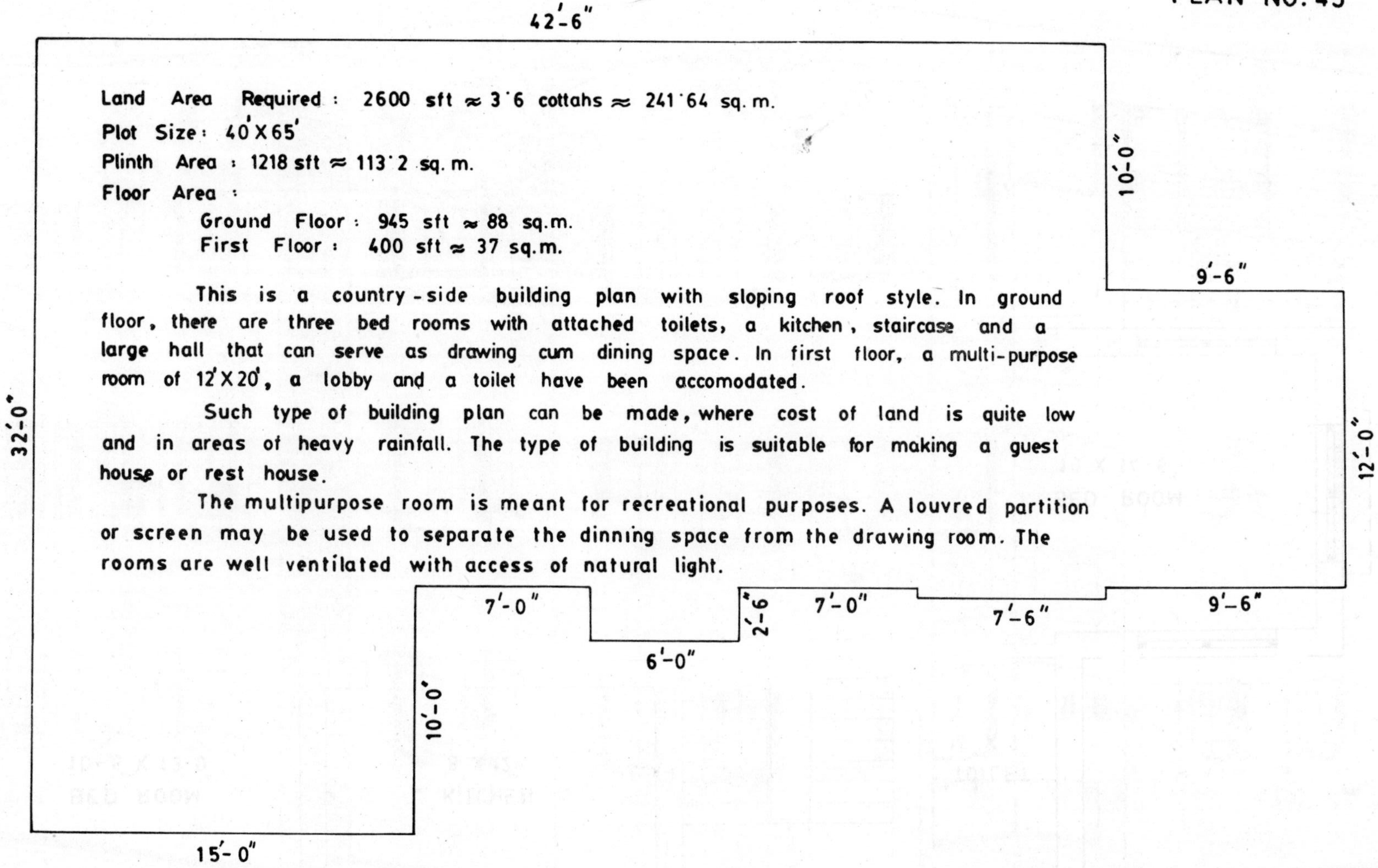

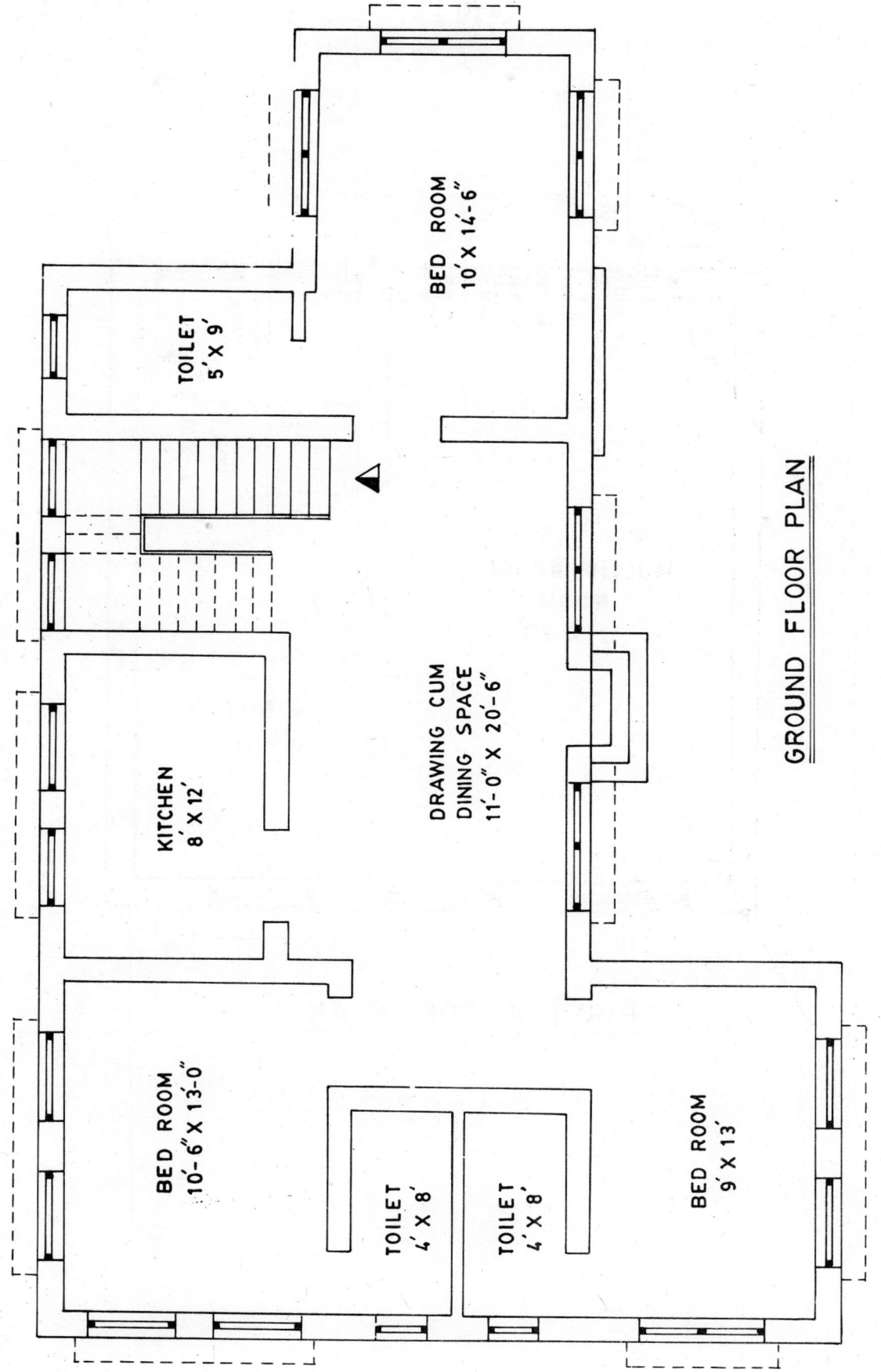

BED ROOM
10' X 14'-6"

TOILET
5' X 9'

KITCHEN
8' X 12'

DRAWING CUM
DINING SPACE
11'-0" X 20'-6"

BED ROOM
10'-6" X 13'-0"

TOILET
4' X 8'

TOILET
4' X 8'

BED ROOM
9' X 13'

GROUND FLOOR PLAN

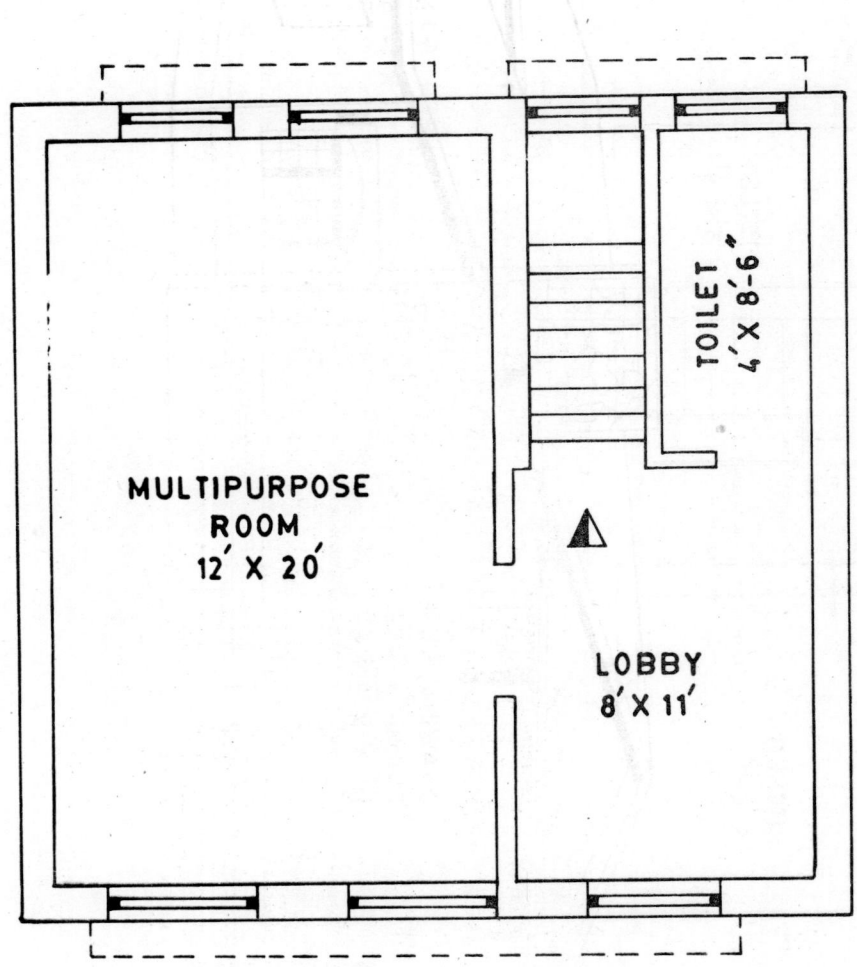

FIRST FLOOR PLAN

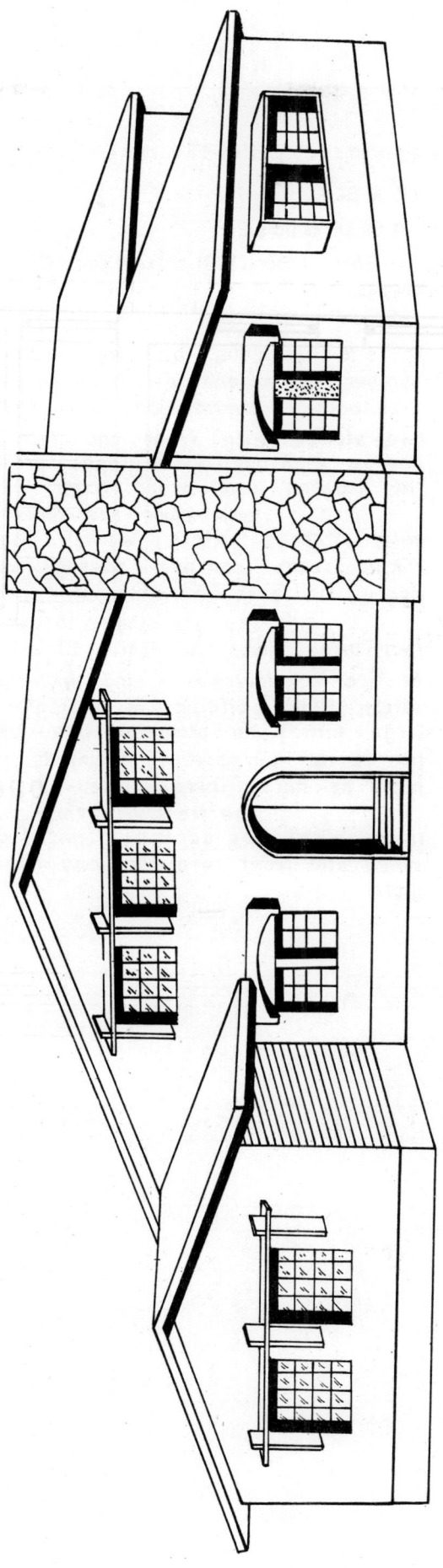

PLAN NO. 46

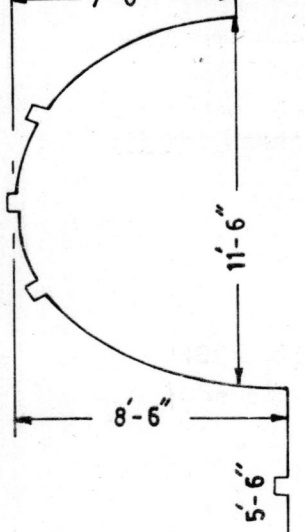

31'-6"

14'-6"

6'-0"

10'-6"

7'-0"

11'-6"

8'-6"

5'-6"

4'-0"

20'-0"

41'-6"

Land Area Required : 2538 sft ≈ 3·5 cottahs ≈ 236 sq.m.

Plot size : 47' X 54'

Plinth Area : 1184 sft ≈ 110 sq.m.

Floor Area In Each Floor = 1001 sft ≈ 93 sq.m.
(Including Staircase)

It is a luxurious building plan with a number of appendages to produce an architectural composition. The building is two-storied with front and rear verandahs and a staircase with finges projected from the drawing cum dining room.

Each floor has three bed rooms with attached toilets, kitchen, drawing cum dining room, common passage and a space for wash hand basin.

The planning although compact, facilitates good circulation of air and light and a comfortable living with privacy and safety. This building is suitable for a large family. No provision has been kept for rental purpose. If required, the building may be made three-storied.

The front bed room with attached toilet which has access to drawing cum dining room and front verandah, may be used as a guest room.

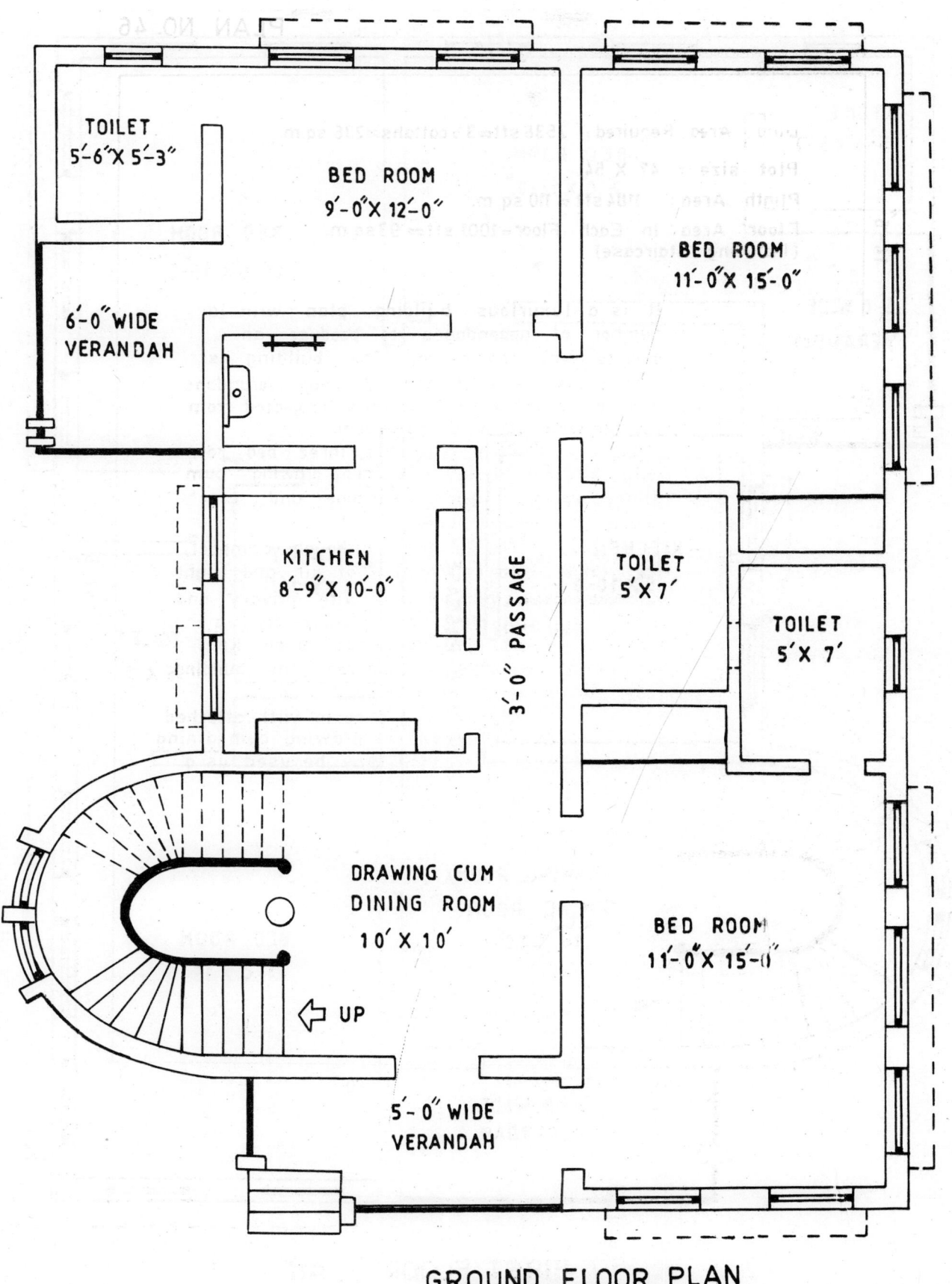

TOILET
5'-6"X 5'-3"

BED ROOM
9'-0"X 12'-0"

6'-0" WIDE
VERANDAH

BED ROOM
11'-0"X 15'-0"

KITCHEN
8'-9"X 10'-0"

3'-0" PASSAGE

TOILET
5'X 7'

TOILET
5'X 7'

DRAWING CUM
DINING ROOM
10'X 10'

BED ROOM
11'-0"X 15'-0"

UP

5'-0" WIDE
VERANDAH

GROUND FLOOR PLAN

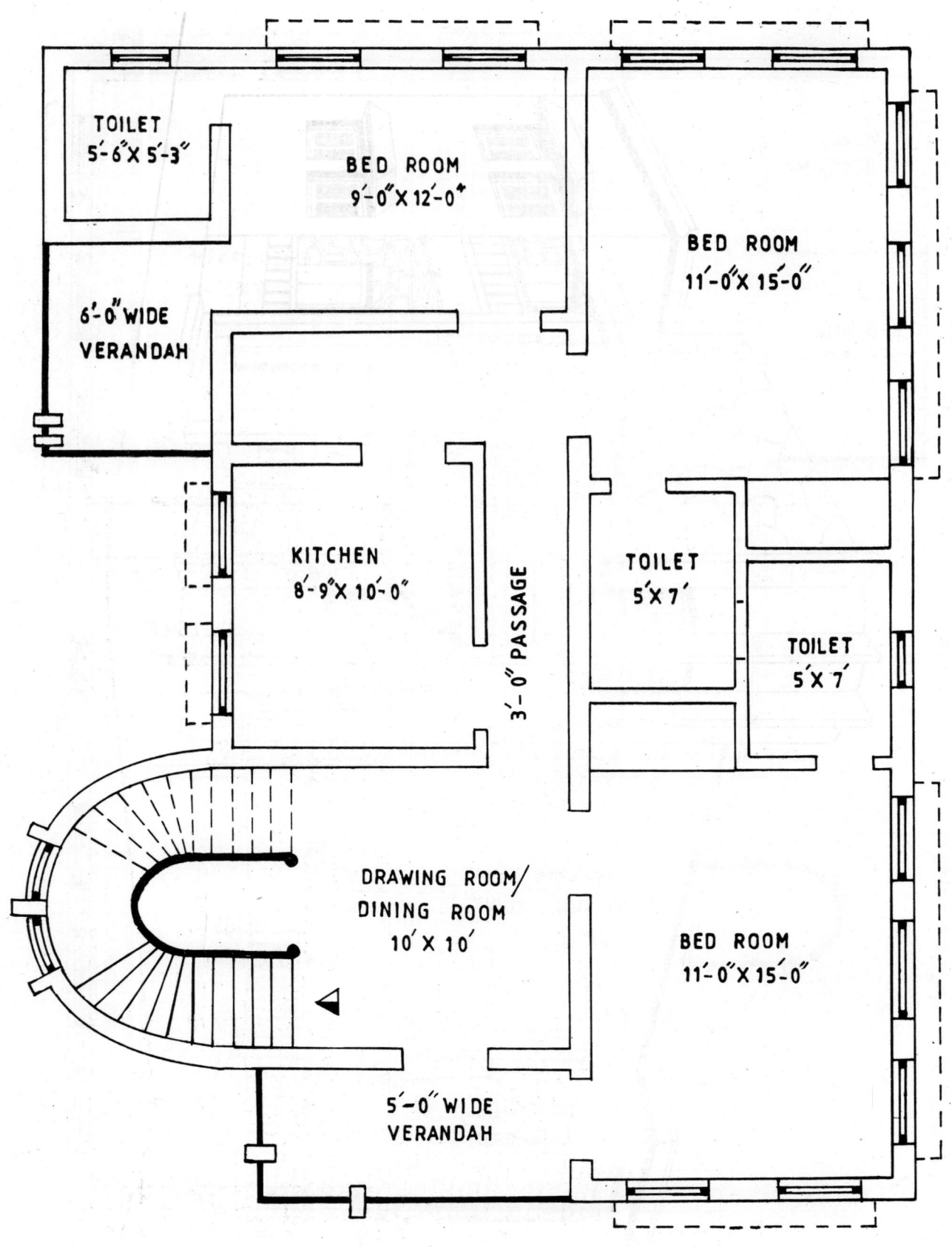

TOILET
5'-6"X 5'-3"

BED ROOM
9'-0"X 12'-0"

BED ROOM
11'-0"X 15'-0"

6'-0"WIDE
VERANDAH

KITCHEN
8'-9"X 10'-0"

3'-0" PASSAGE

TOILET
5'X 7'

TOILET
5'X 7'

DRAWING ROOM/
DINING ROOM
10'X 10'

BED ROOM
11'-0"X 15'-0"

5'-0"WIDE
VERANDAH

FIRST FLOOR PLAN

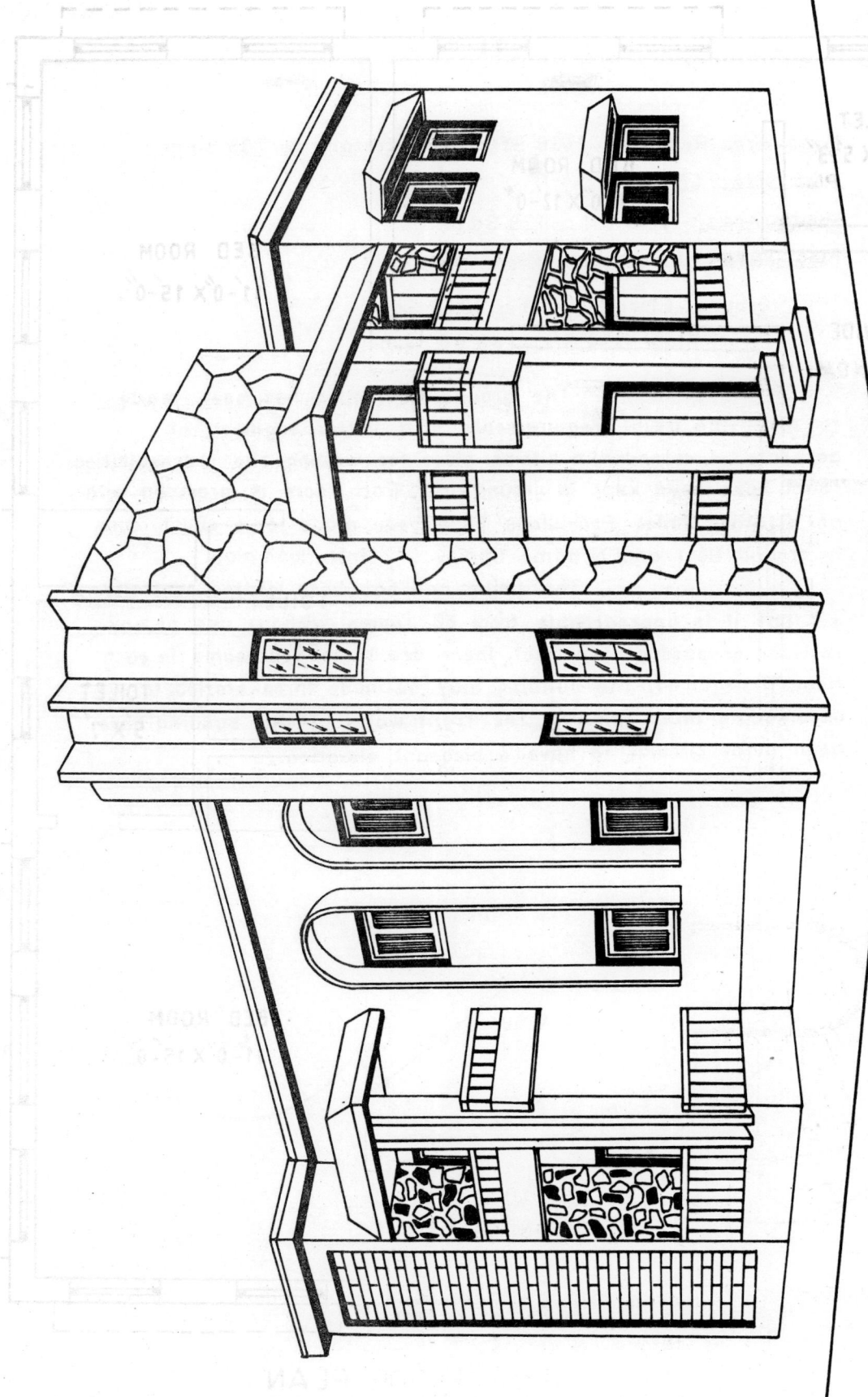

34'-0"

Land Area Required: 2436 Sft ≈ 3.38 Cottahs ≈ 235 Sq.m.

Plot Size: 42'x58'

Plinth Area : 1388 Sft ≈ 129 Sq.m.

Floor Area: (Excluding Stair Case):

Ground Floor 974 Sft ≈ 90.52 Sq.m.

First Floor 954 Sft ≈ 88.66 Sq.m.

   The ground floor plan has been made to meet the usual requirements of a doctor, engineer or advocate. A verandah, a sitting place for waiting and a consultation room have been kept in ground floor. Each room is provided with an attached toilet. Provisions have been made for a guest room in ground floor and a home library in first floor plan.

   The dining hall has been located centrally so that it is approachable from all rooms without use of any corridor or passage. In fact, there are three bed rooms in each floor. If required, the building may be made three-storied to accomodate three families. The front walls of the building have been made circular to have a pleasant elevation.

38'-0"

25'-6"

5'

6'

5'

6'

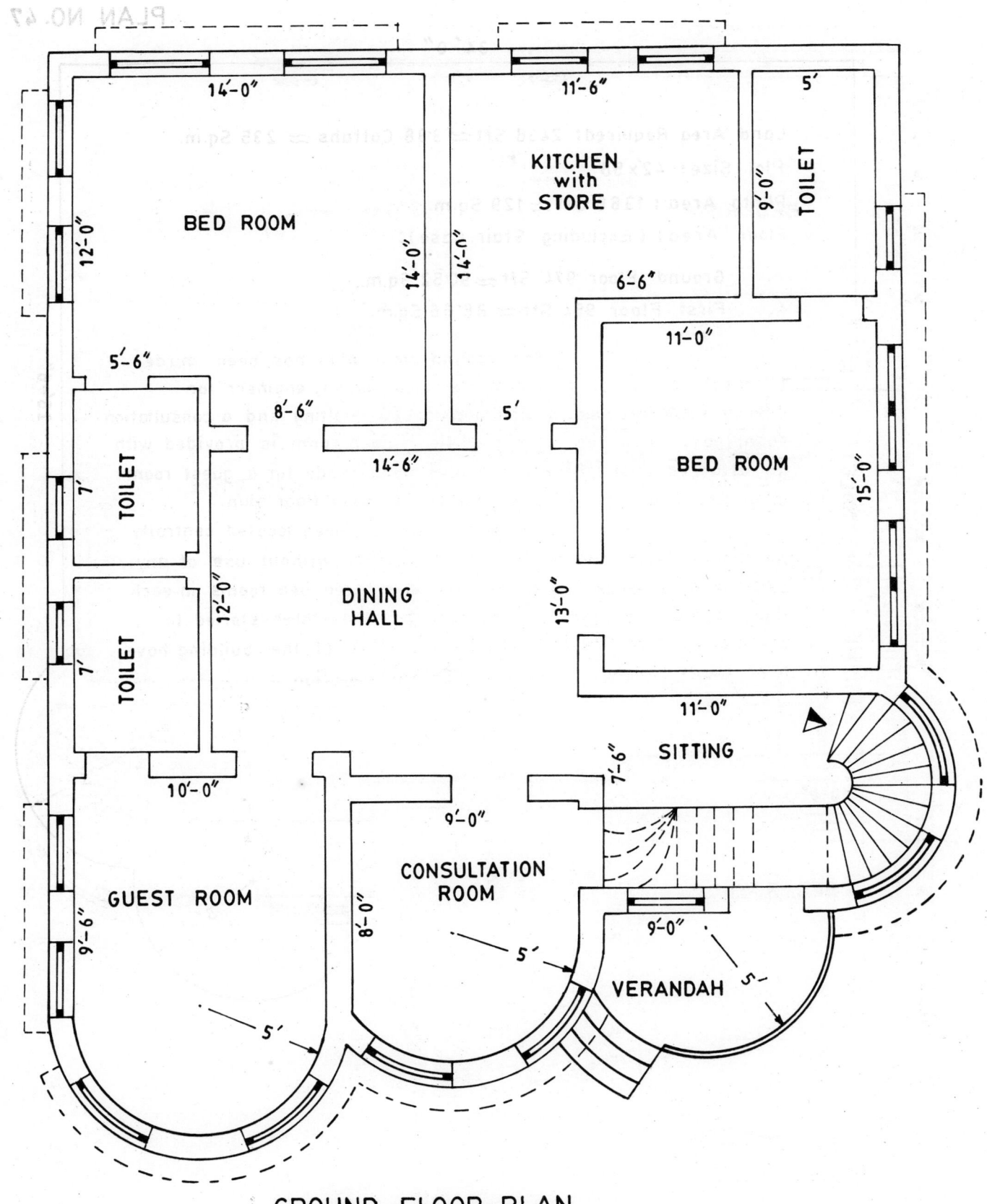

GROUND FLOOR PLAN

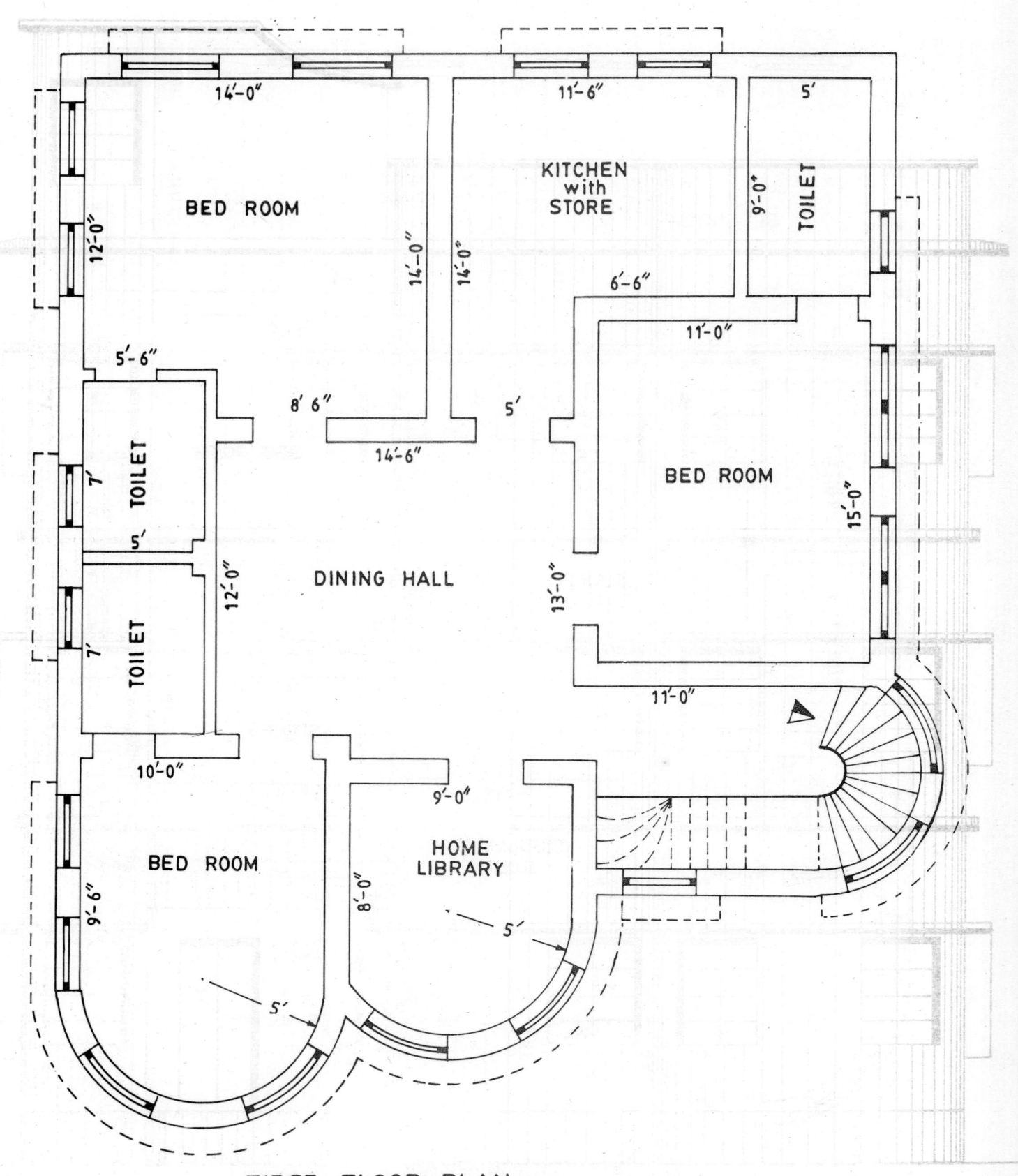

BED ROOM

14'-0"

12'-0"

KITCHEN
with
STORE

11'-6"

5'

9'-0"

TOILET

14'-0"

14'-0"

6'-6"

11'-0"

5'-6"

TOILET

7'

5'

8' 6"

5'

14'-6"

BED ROOM

15'-0"

TOILET

7'

12'-0"

DINING HALL

13'-0"

11'-0"

10'-0"

9'-0"

BED ROOM

9'-6"

8'-0"

HOME
LIBRARY

5'

5'

FIRST FLOOR PLAN

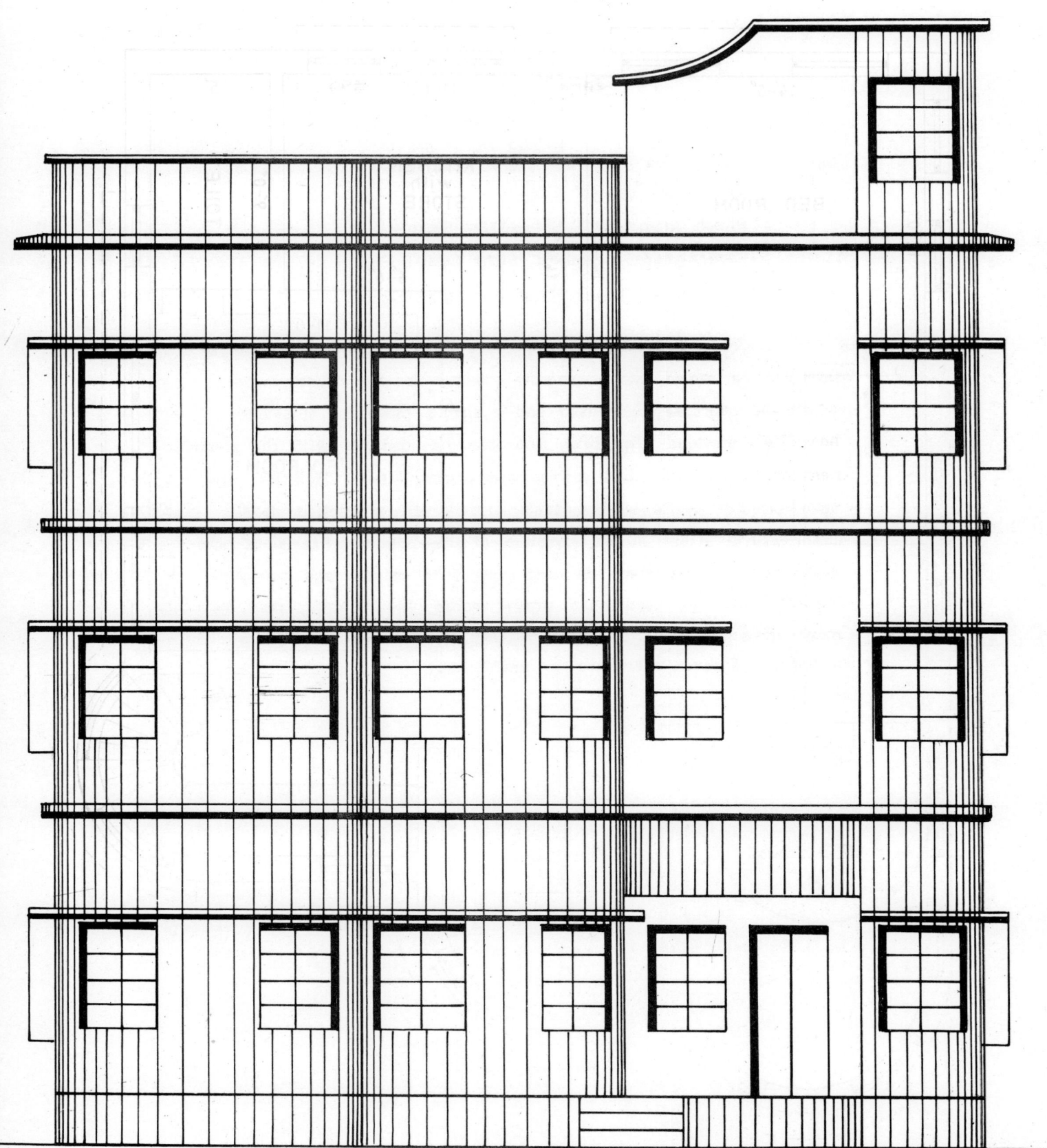

ELEVATION

PLAN NO. 48

32'-0"

8'

5'

25'-6"

21'-6"

3'

3'-6"

Land Area Required: 2385 Sft. ≈ 3·31 Cottahs ≈ 221·65 Sq. m.

Plot Size: 45' x 53'

Plinth Area: 1274 Sft ≈ 118·4 Sq. m.

Floor Area (Including Stair Case)

Each Floor 1004 Sft ≈ 93·31 Sq. m.

Planning has been done such that the entire ground floor can be rented with a separate entrance for the house-owner where the staircase is housed and a sitting place for outsiders have been provided. The first floor plan is identical with the ground floor plan. The three bed rooms have attached toilet facilities. The semi-circular walls with suitable offsets will produce a good architectural effect in the elevation of the building. Provisions have been made for adequate air and light with privacy and safety. The attic may be used as a prayer room. The building may be made three or four storied, if desired. It can well accomodate a family of seven members in each floor.

6'

5'

3'-6"

3'-6"

6'-6"

5'

3'-6"

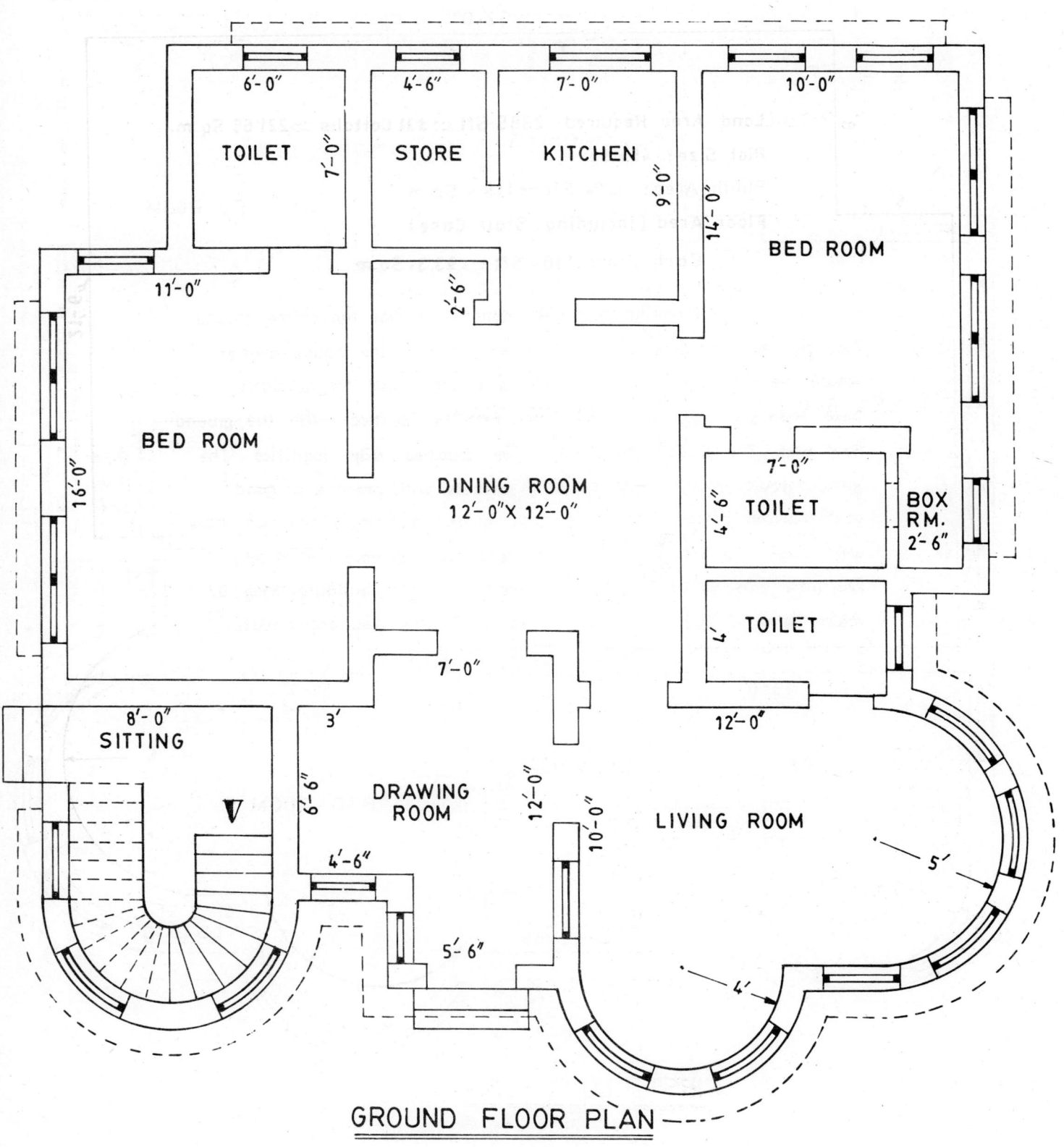

GROUND FLOOR PLAN

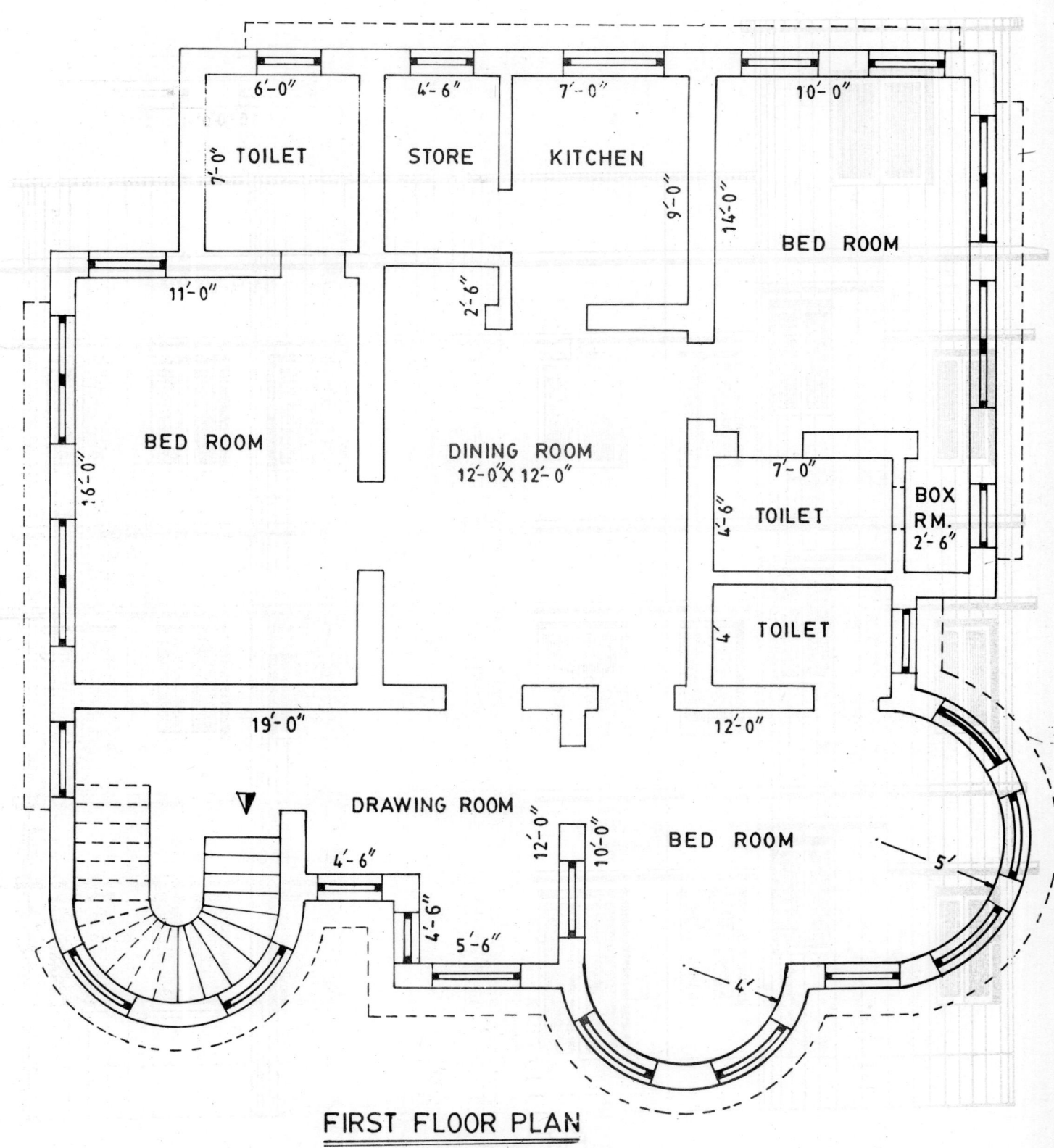

FIRST FLOOR PLAN

ELEVATION

LAND AREA REQUIRED: 2491 SFT ≈ 3·46 COTTAHS ≈ 231·5 SQ·M·

PLOT SIZE: 47'x 53'

PLINTH AREA: 1420 SFT ≈ 132 SQ·M·

FLOOR AREA (EXCLUDING STAIR CASE):

GROUND FLOOR 1046 SFT ≈ 97·21 SQ·M·

FIRST & SECOND FLOOR EACH 1100 SFT ≈ 102·23 SQ·M·

TOP FLOOR 981 SFT ≈ 91·2 SQ·M·

TERRACE GARDEN 154 SFT ≈ 14·3 SQ·M·

   The ground floor comprises two single-roomed flats with drawing room, dining, kitchen and attached toilet for rental purpose. Each upper floor has three bed rooms with attached toilet, a verandah, dining hall and kitchen to be used by the house owner. In first and second floors, provision has been kept for home library In top floor, terrace gardens have been provided on the front side of the building. The curved walls with offsets produce a good architectural effect in front elevation.

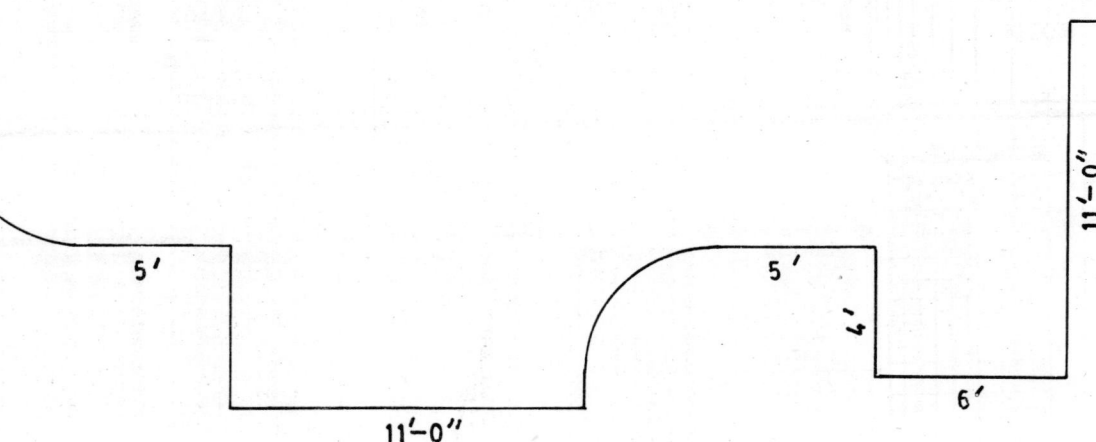

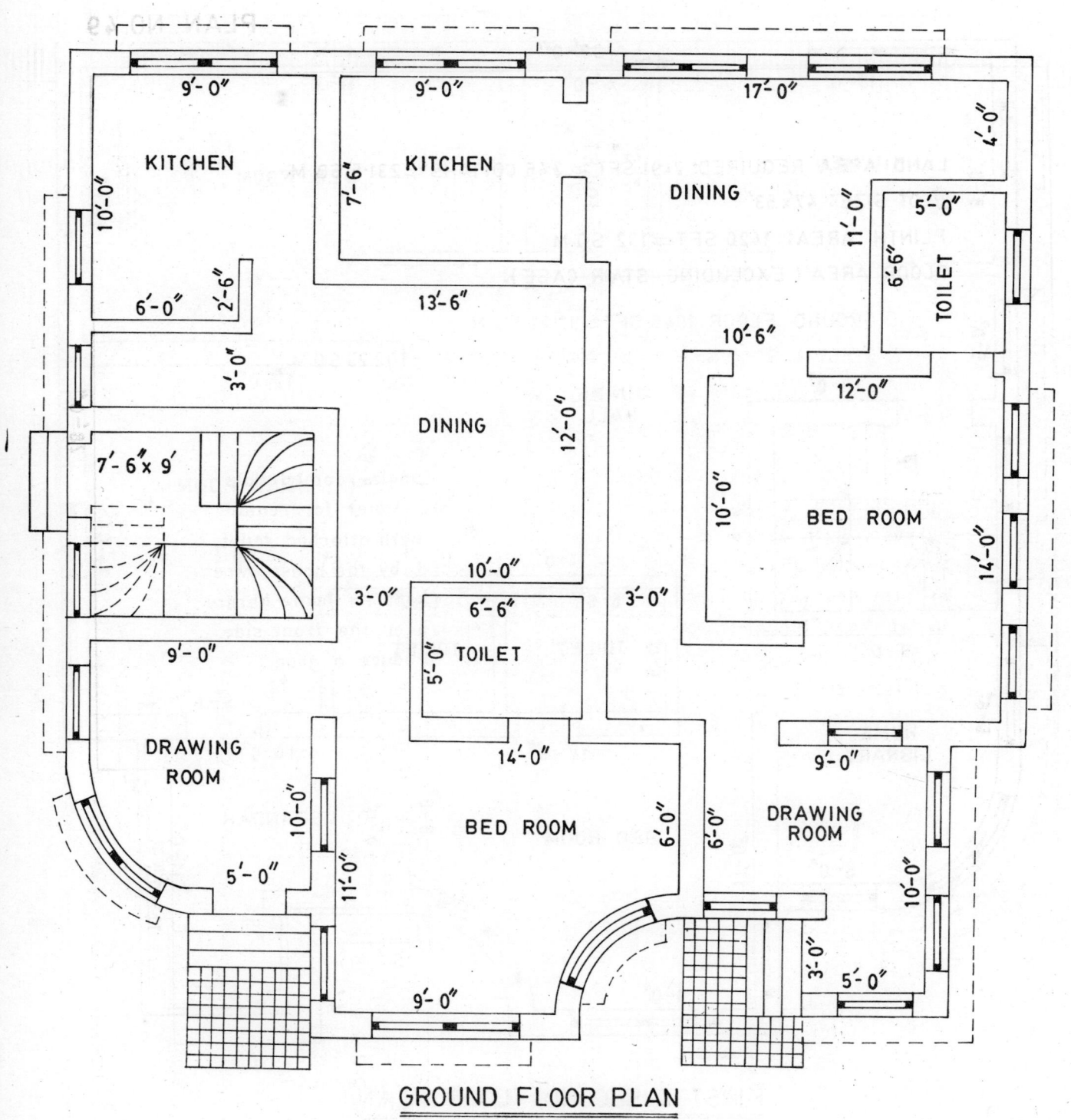

GROUND FLOOR PLAN

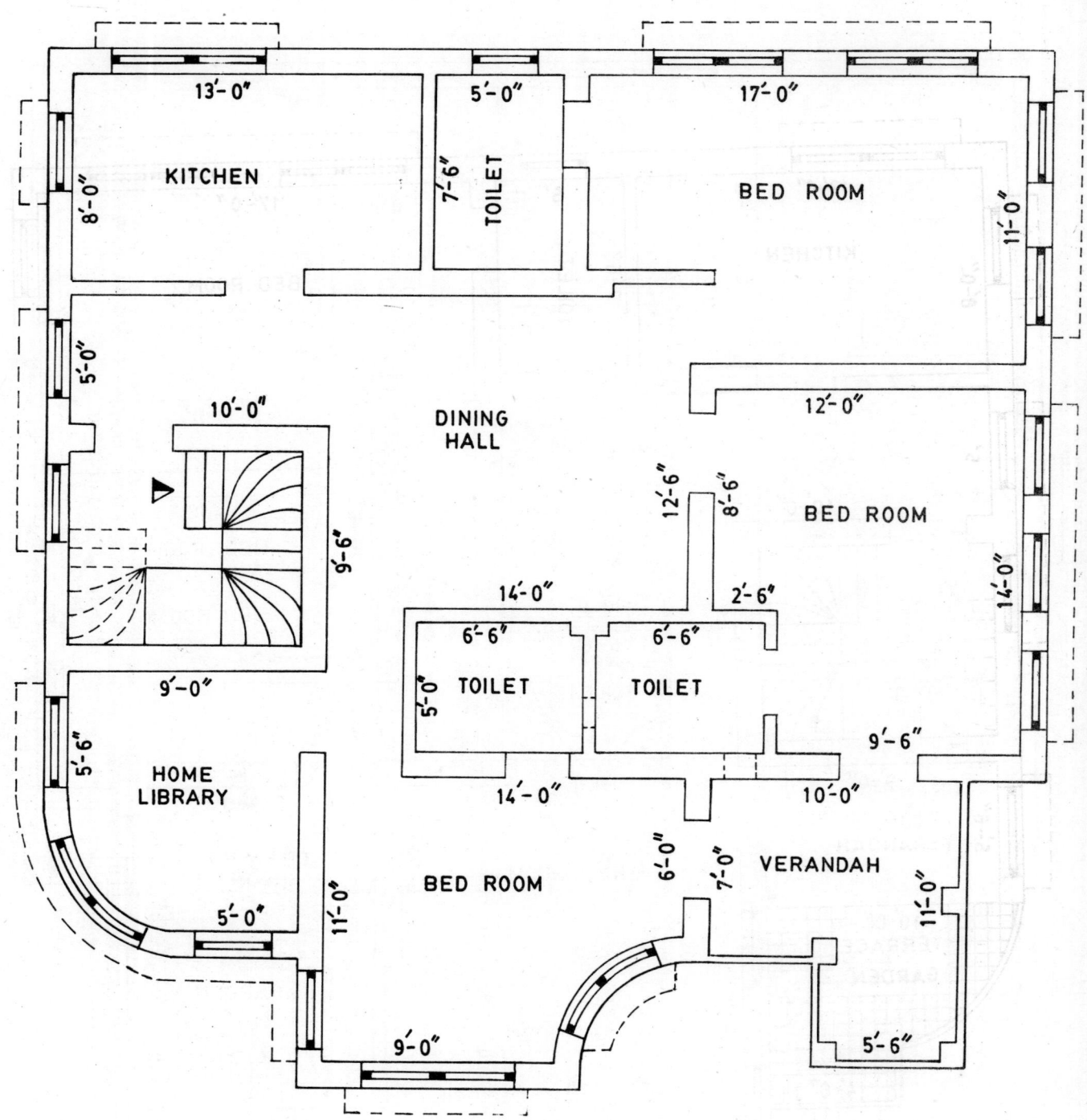

KITCHEN

13'-0"

5'-0"

7'-6"

TOILET

17'-0"

BED ROOM

8'-0"

11'-0"

5'-0"

10'-0"

DINING
HALL

9'-6"

12'-0"

12'-6"

8'-6"

BED ROOM

14'-0"

2'-6"

14'-0"

9'-0"

HOME
LIBRARY

6'-6"

5'-0"

TOILET

6'-6"

TOILET

14'-0"

5'-6"

5'-0"

11'-0"

BED ROOM

14'-0"

10'-0"

9'-6"

6'-0"

7'-0"

VERANDAH

11'-0"

9'-0"

5'-6"

FIRST & SECOND FLOOR PLAN

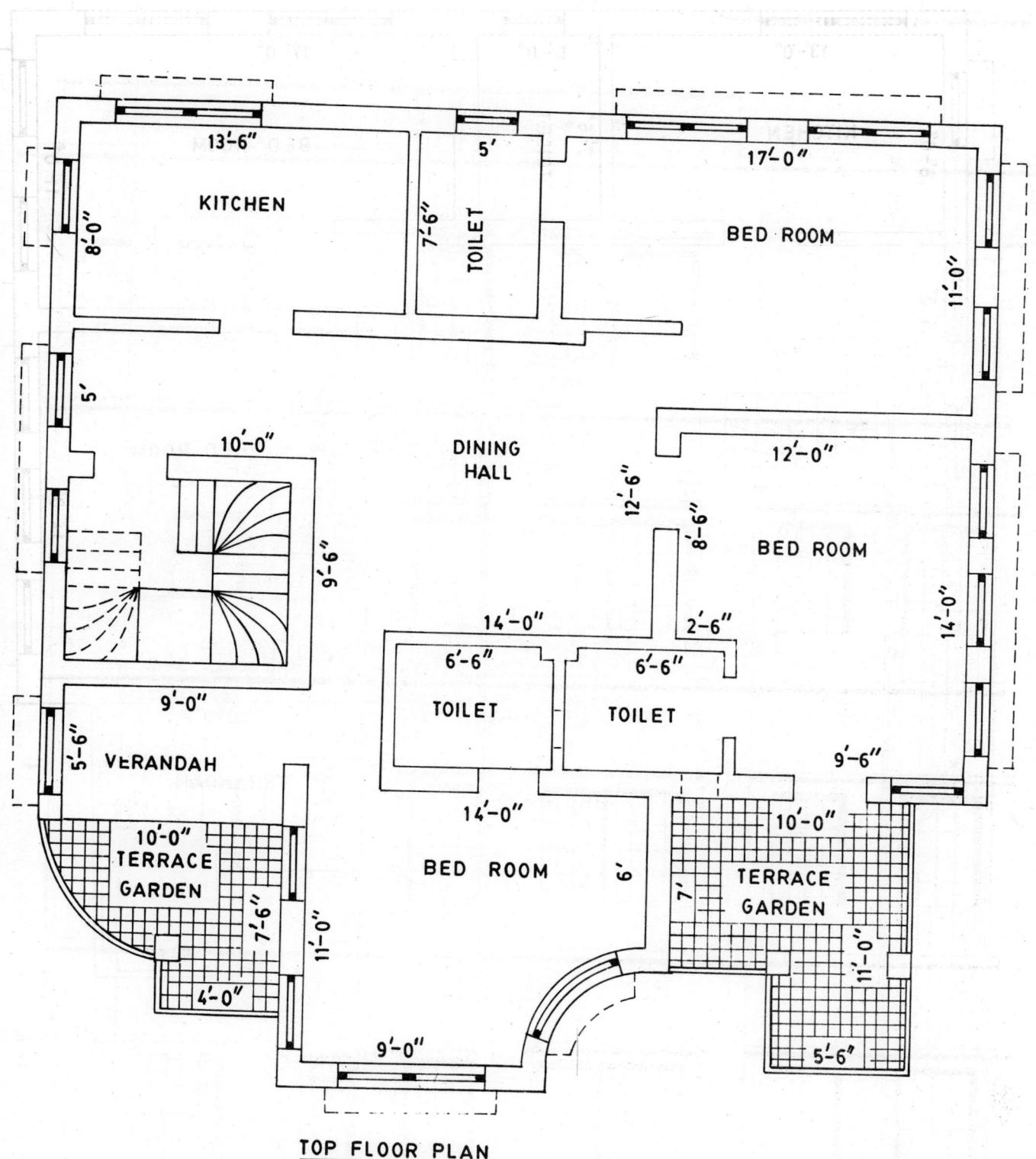

TOP FLOOR PLAN

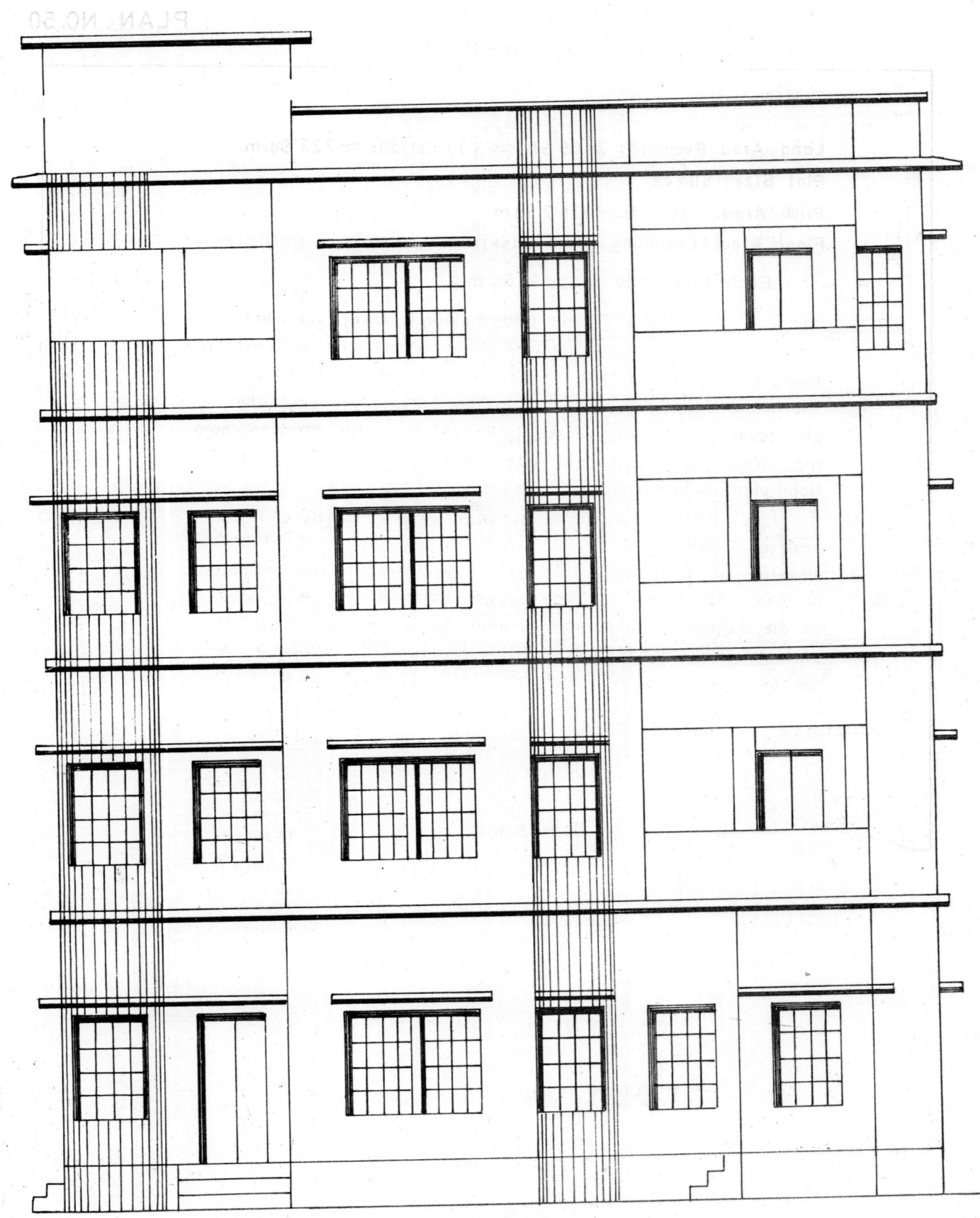

ELEVATION

PLAN NO. 50

39'-6"

18'-6"

Land Area Required: 2400 Sft ≈ 3·33 Cottahs ≈ 223 Sq.m.

Plot Size: 50'x48'

Plith Area: 1359 Sft ≈ 126·3 Sq.m.

Floor Area (Excluding Stair Case):

Each Floor 990 Sft ≈ 92 Sq.m.

The ground floor having a front verandah, drawing room, dining hall, kitchen, two bed rooms with attached toilet, a servant's room and one more toilet can be rented. The house-owner will have a separate entrance through the staircase cum drawing room. In ground floor, the drawing room may accomodate occassional guests of the tenant. The first floor plan is identical with the ground floor plan.

The curvatures used in front walls and the staircase wall are of special shape, which produce a beautiful elevation of the building. All the rooms have facilities of natural air and light. The kitchen has been located in the north-west of the building. The servant's room can be isolated from all parts of the building by providing a door. The staircase has the form of a spiral.

5'

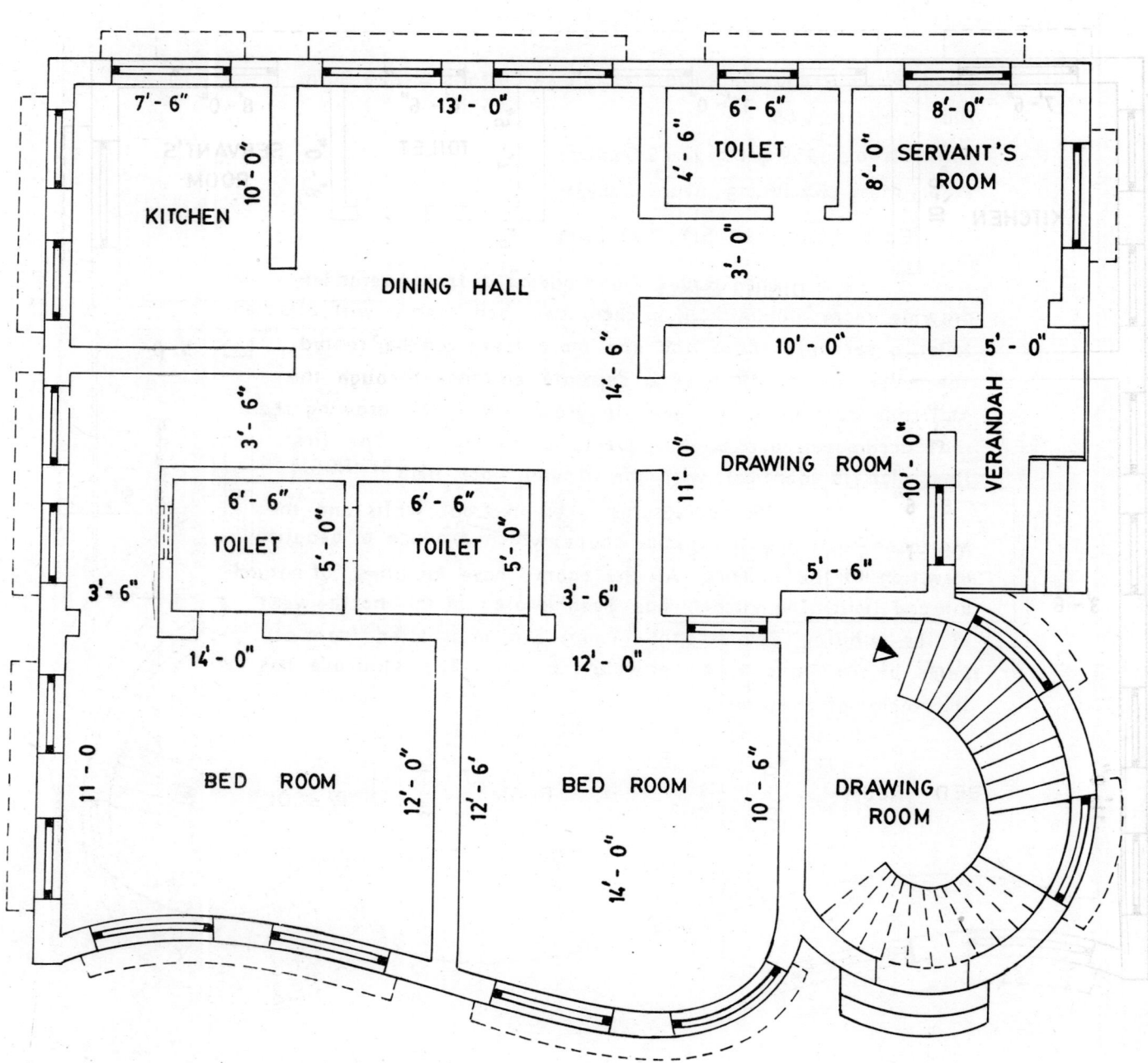

GROUND FLOOR PLAN

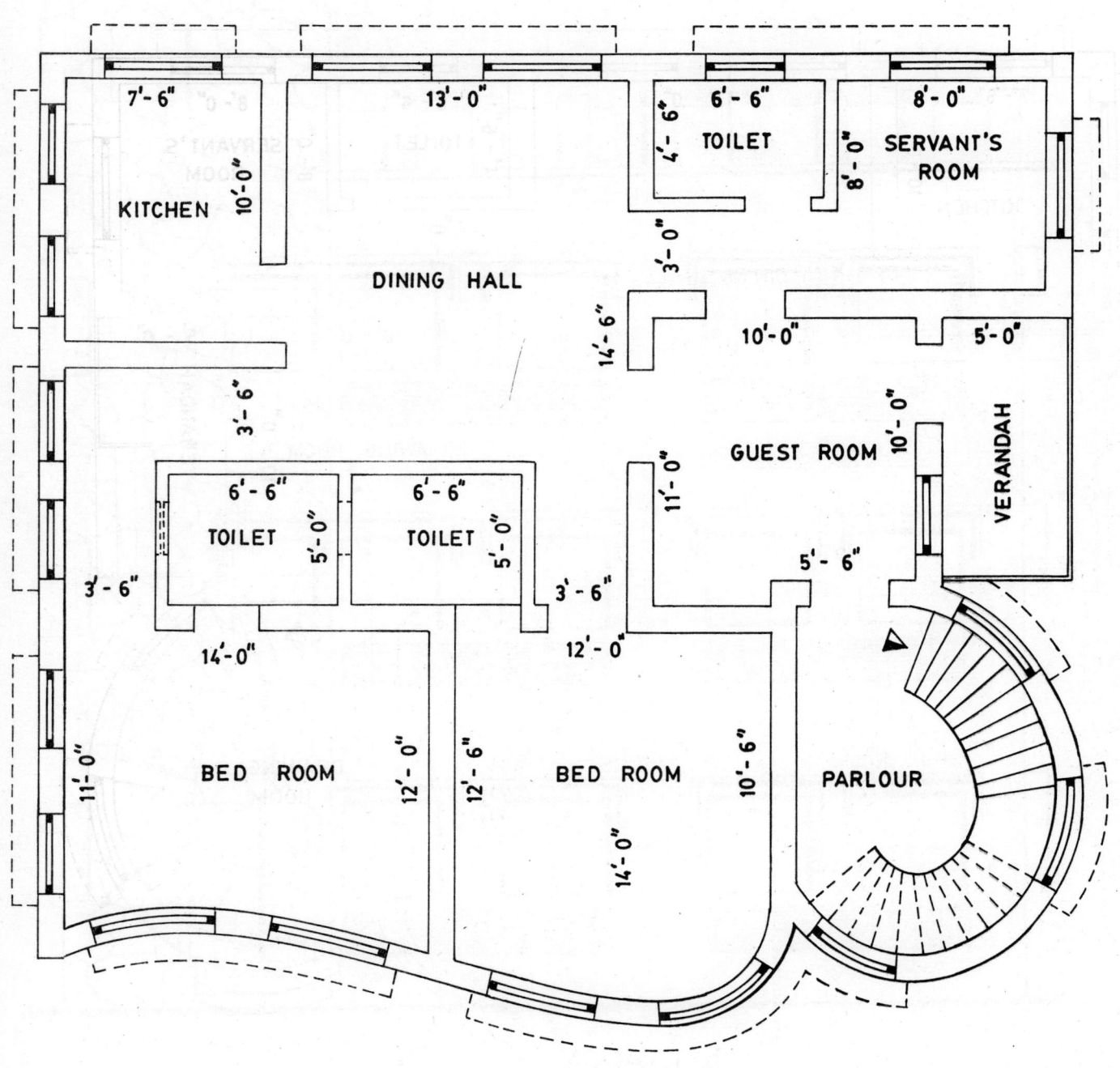

KITCHEN

7'-6"

10'-0"

13'-0"

DINING HALL

TOILET

6'-6"

4'-6"

SERVANT'S
ROOM

8'-0"

8'-0"

3'-0"

14'-6"

10'-0"

5'-0"

GUEST ROOM

10'-0"

11'-0"

VERANDAH

3'-6"

TOILET

6'-6"

5'-0"

TOILET

6'-6"

5'-0"

5'-6"

3'-6"

3'-6"

14'-0"

12'-0"

BED ROOM

12'-0"

12'-6"

BED ROOM

14'-0"

10'-6"

PARLOUR

11'-0"

FIRST FLOOR PLAN

ELEVATION

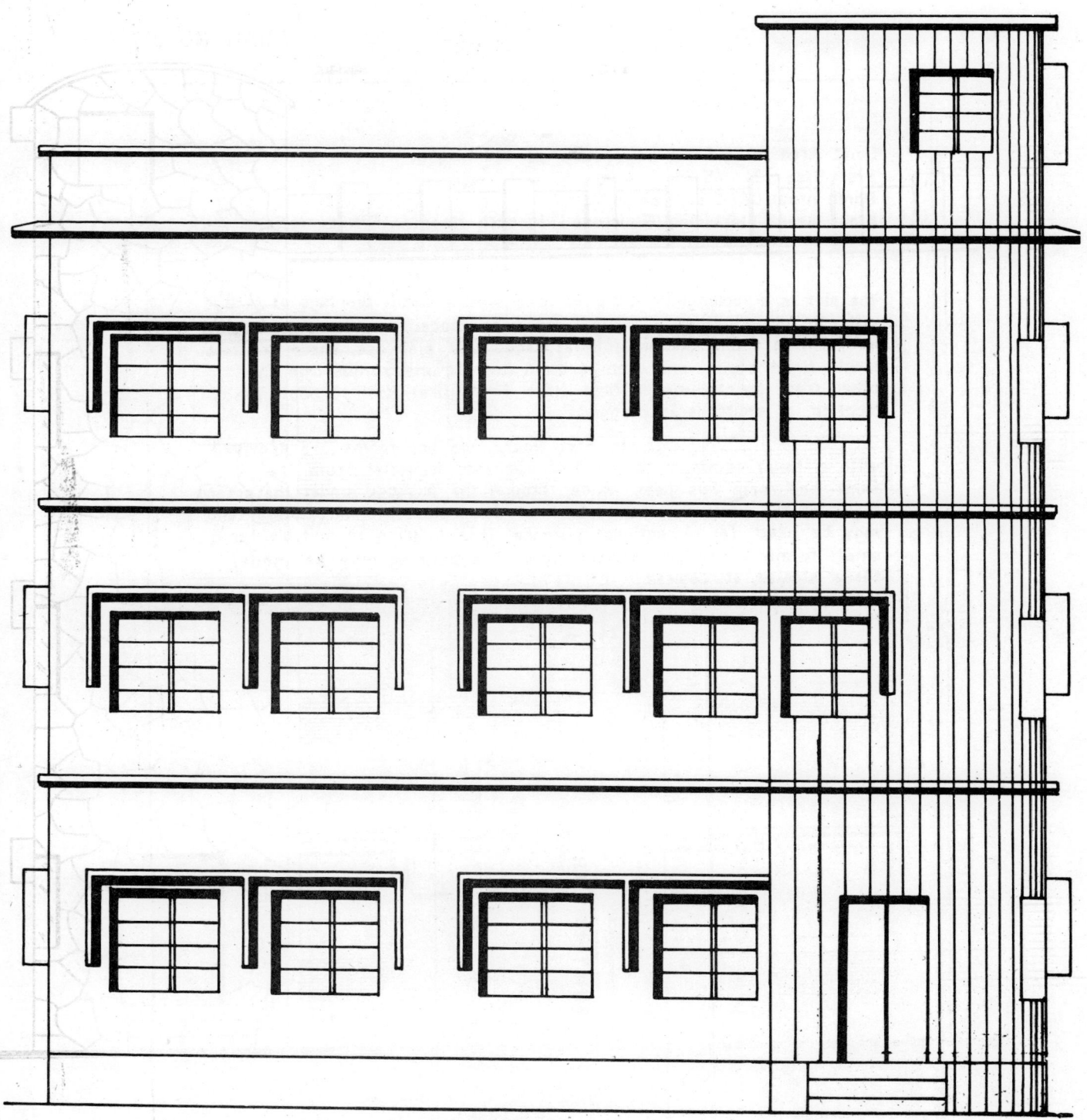

ELEVATION

Land Area: 1080 Sft = 1.5 Cottahs ≈ 100.4 Sq.m.
Plot Size : 24'x 45'
Plinth Area: 587 Sft ≈ 54.55 Sq.m.
Floor Area: (Including Staircase) In each floor: 441Sft. ≈ 41 Sq.m.

The plot is a rectangular strip of land with a road frontage of 24'-0" only. In ground floor, provision has been made to accomodate drawing cum dining room, kitchen, toilet, garage and staircase. There are two single-flight stairs - one from ground floor to mezzanine and the other from mezzanine to first floor. From first floor to roof, the location of staircase is changed.

In first floor, two bed rooms are provided with a toilet common to all. The approach to toilet from the front bed room has been made through the passage under the second flight of the stair as shown in first floor plan. The mezzanine floor may be used for recreational purpose. This building is suitable for a small family of middle income group. The building may be made three storied, if desired.

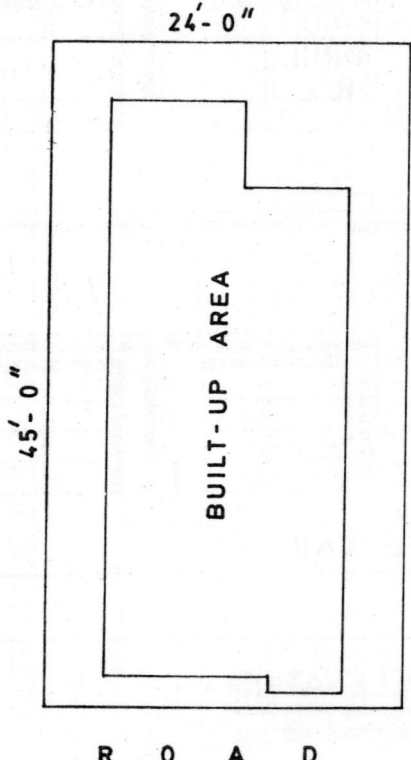

24'- 0"

45'- 0"

BUILT-UP AREA

R O A D

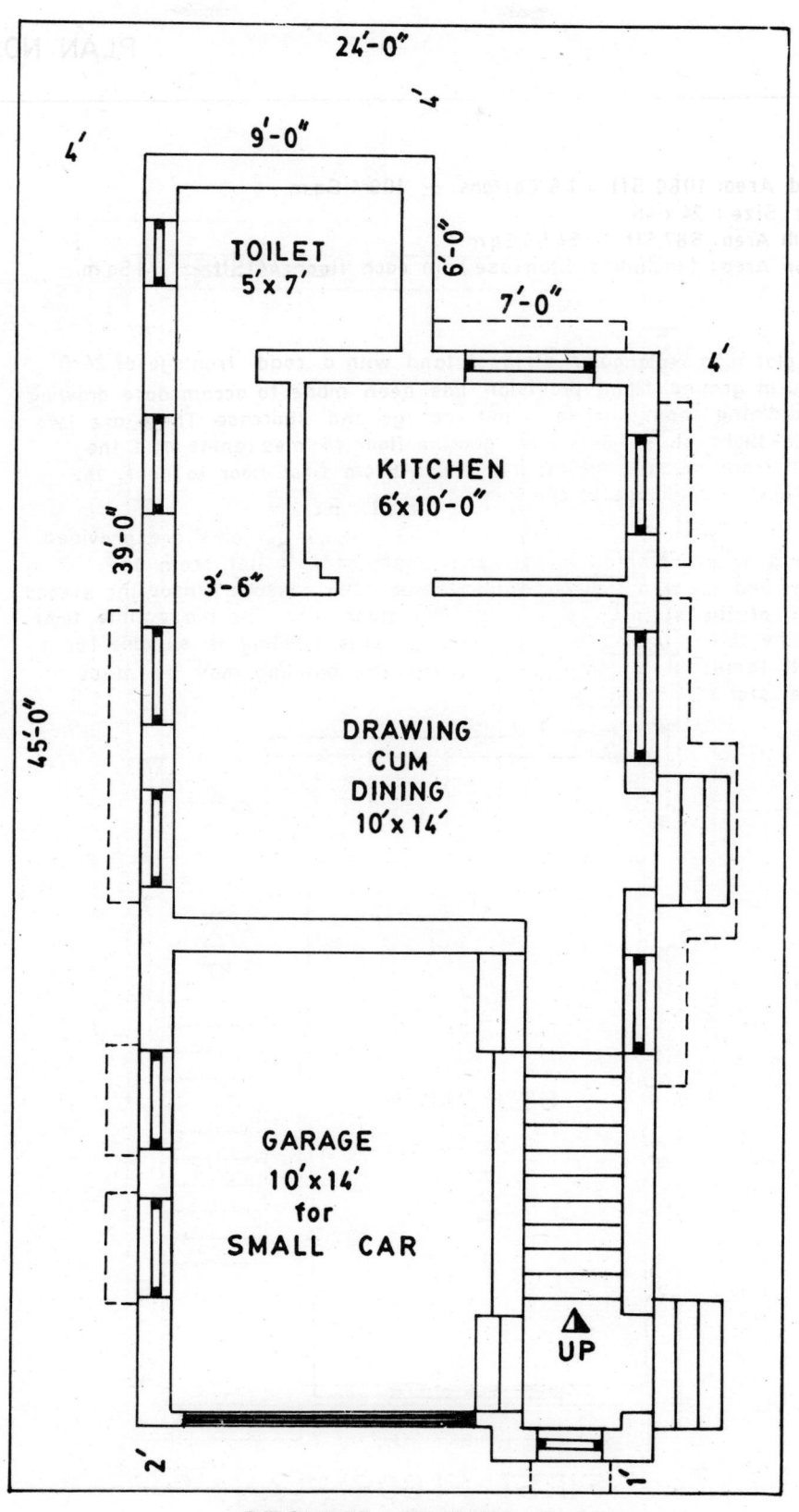

GROUND FLOOR PLAN

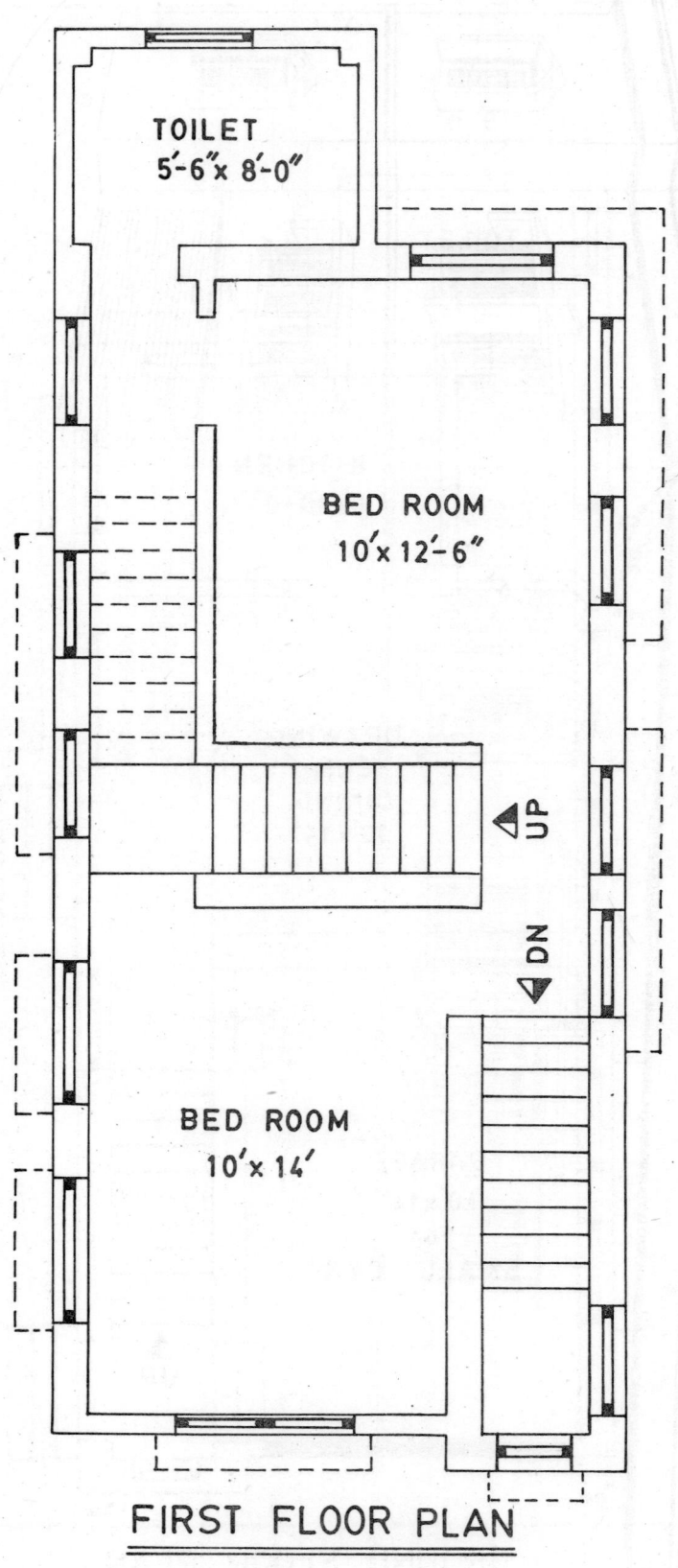

FIRST FLOOR PLAN

TOILET
5'-6" x 8'-0"

BED ROOM
10' x 12'-6"

UP

DN

BED ROOM
10' x 14'

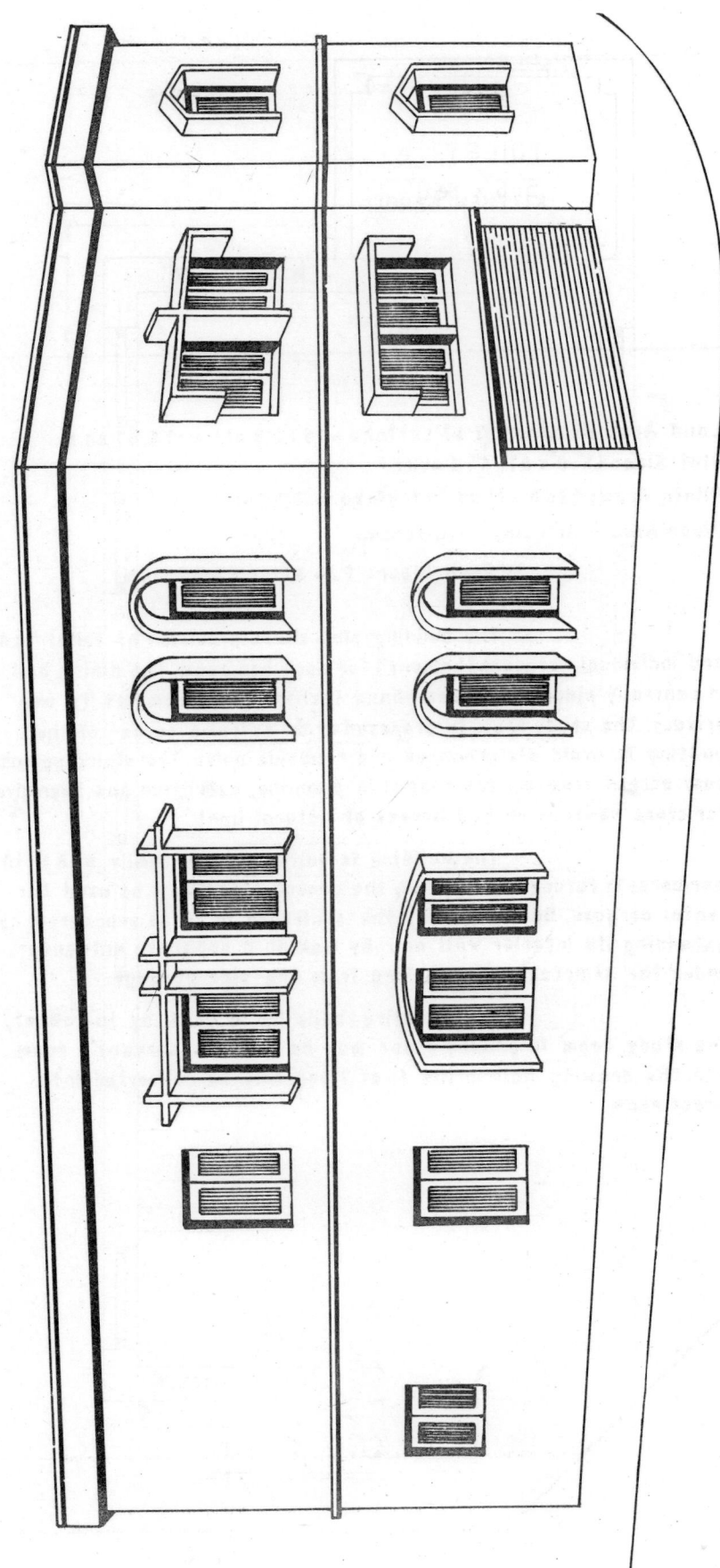

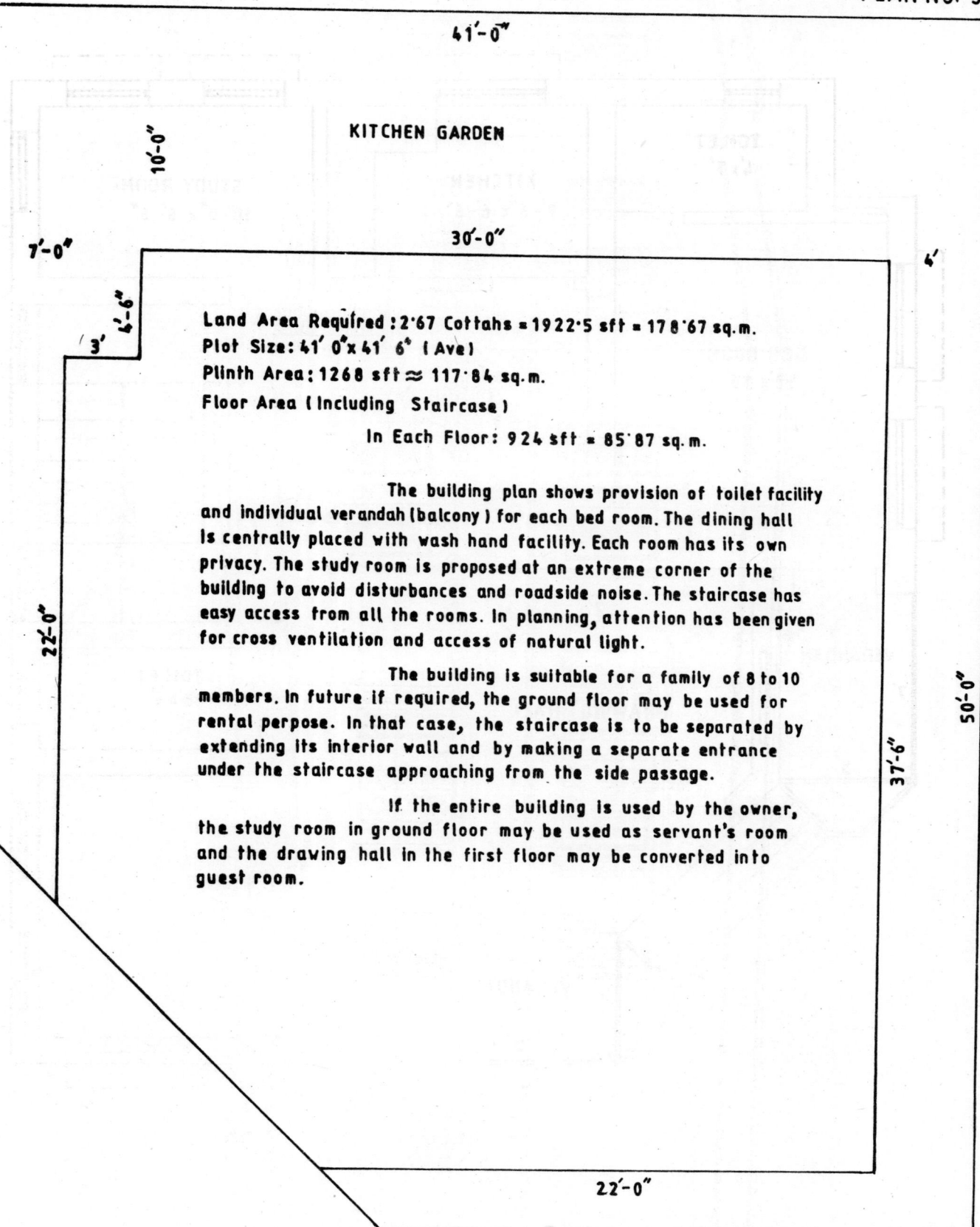

KITCHEN GARDEN

41'-0"

10'-0"

30'-0"

7'-0"

4'-6"

3'

22'-0"

37'-6"

50'-0"

4'

22'-0"

26'-0"

Land Area Required : 2·67 Cottahs = 1922·5 sft = 178·67 sq.m.
Plot Size : 41' 0"x 41' 6" (Ave)
Plinth Area : 1268 sft ≈ 117·84 sq.m.
Floor Area (Including Staircase)
        In Each Floor : 924 sft = 85·87 sq.m.

The building plan shows provision of toilet facility and individual verandah (balcony) for each bed room. The dining hall is centrally placed with wash hand facility. Each room has its own privacy. The study room is proposed at an extreme corner of the building to avoid disturbances and roadside noise. The staircase has easy access from all the rooms. In planning, attention has been given for cross ventilation and access of natural light.

The building is suitable for a family of 8 to 10 members. In future if required, the ground floor may be used for rental perpose. In that case, the staircase is to be separated by extending its interior wall and by making a separate entrance under the staircase approaching from the side passage.

If the entire building is used by the owner, the study room in ground floor may be used as servant's room and the drawing hall in the first floor may be converted into guest room.

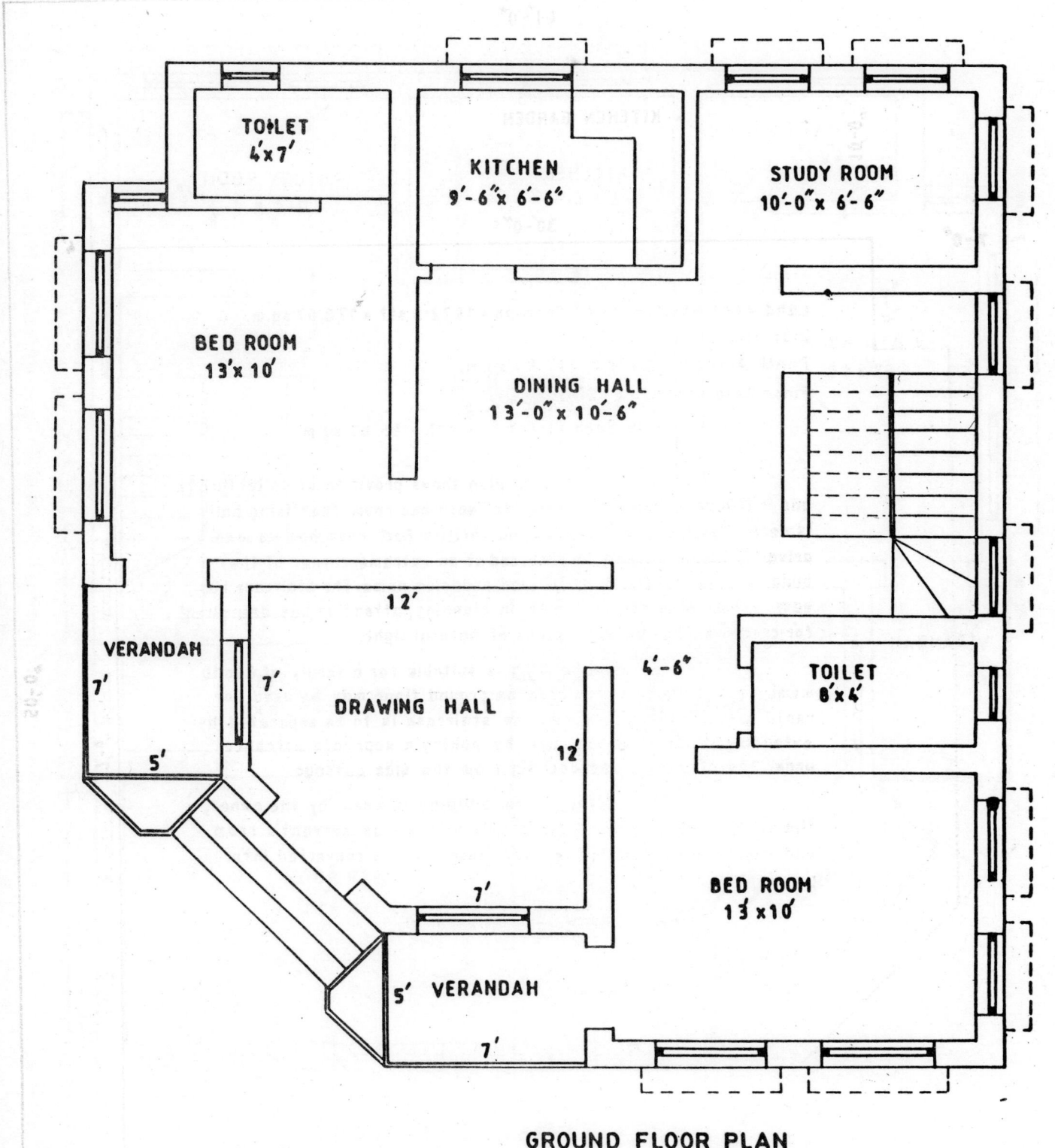

GROUND FLOOR PLAN

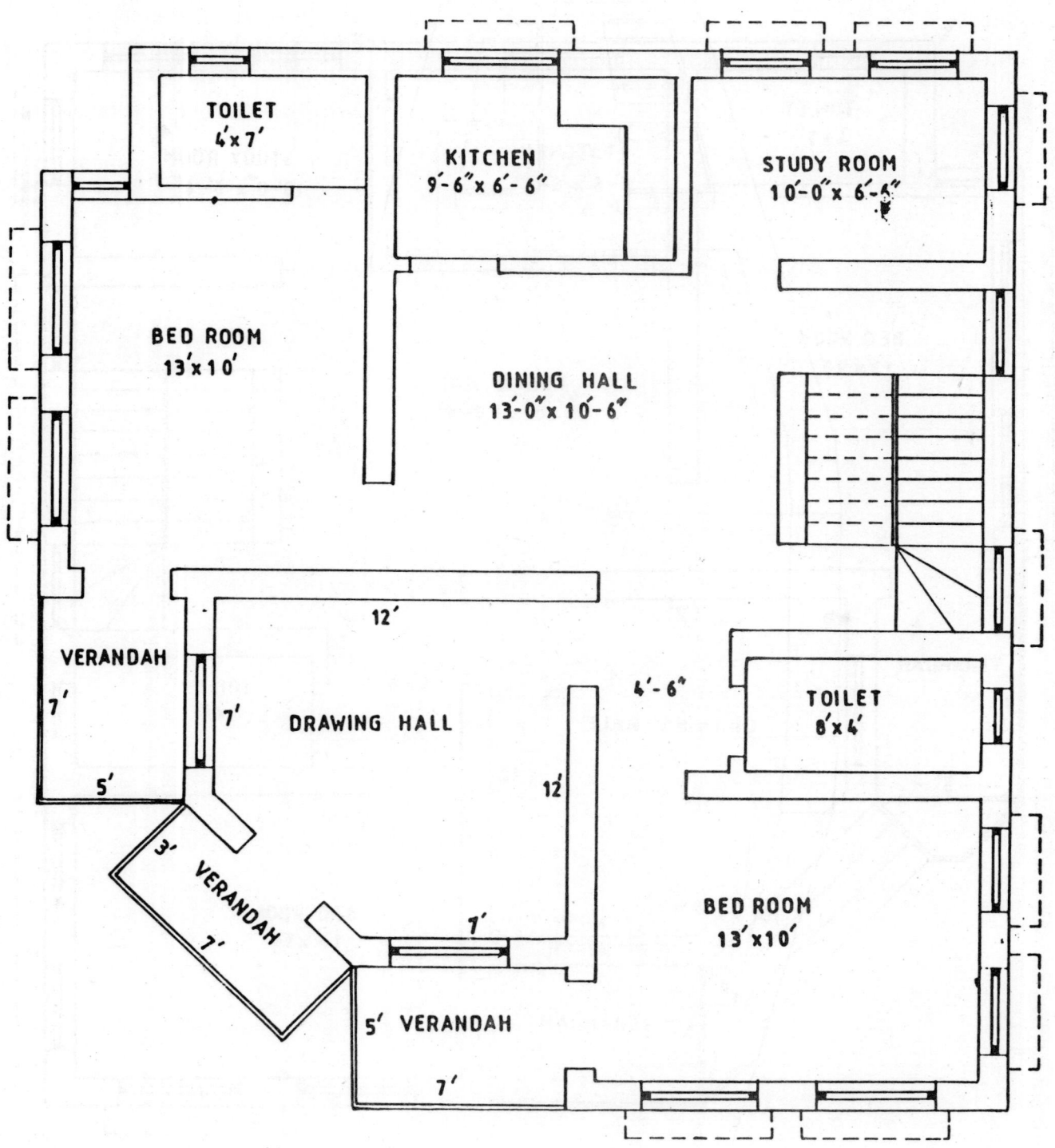

FIRST FLOOR PLAN

247

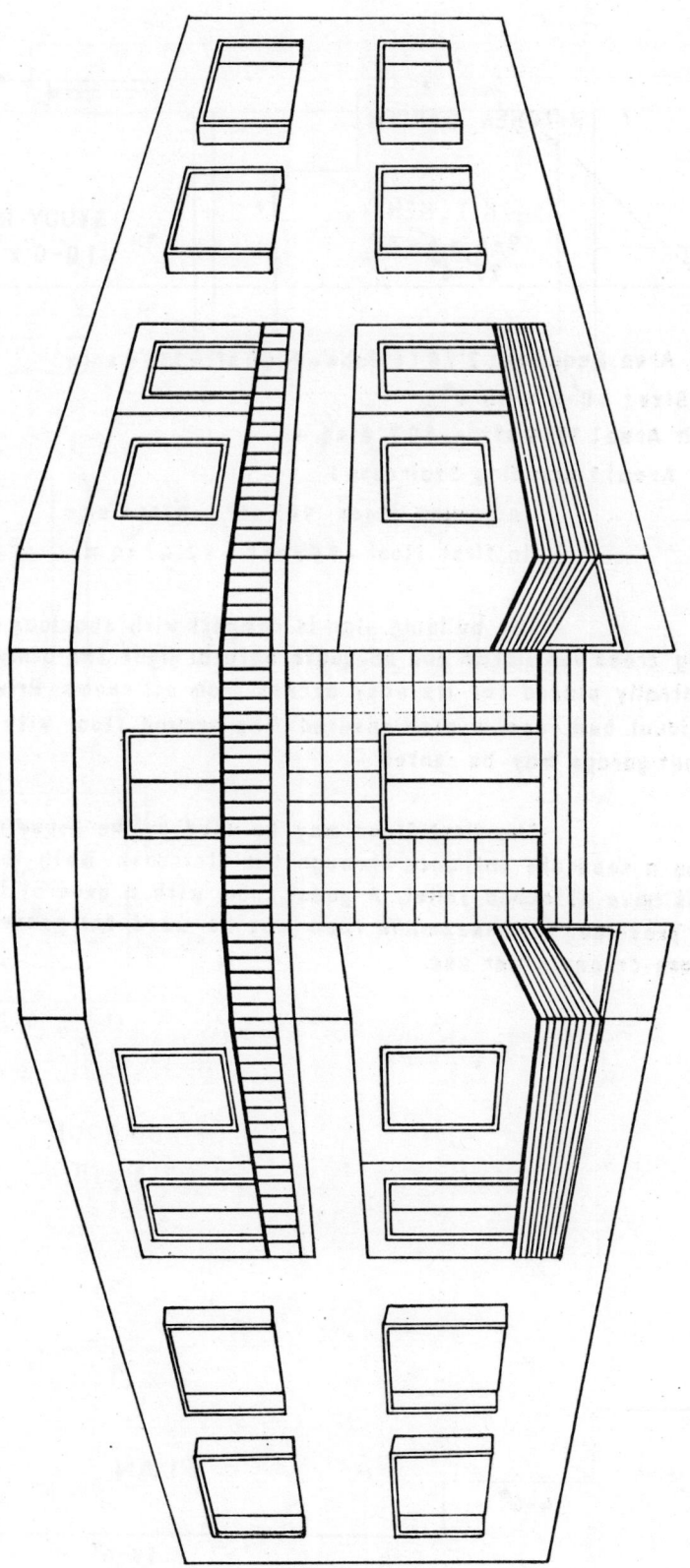

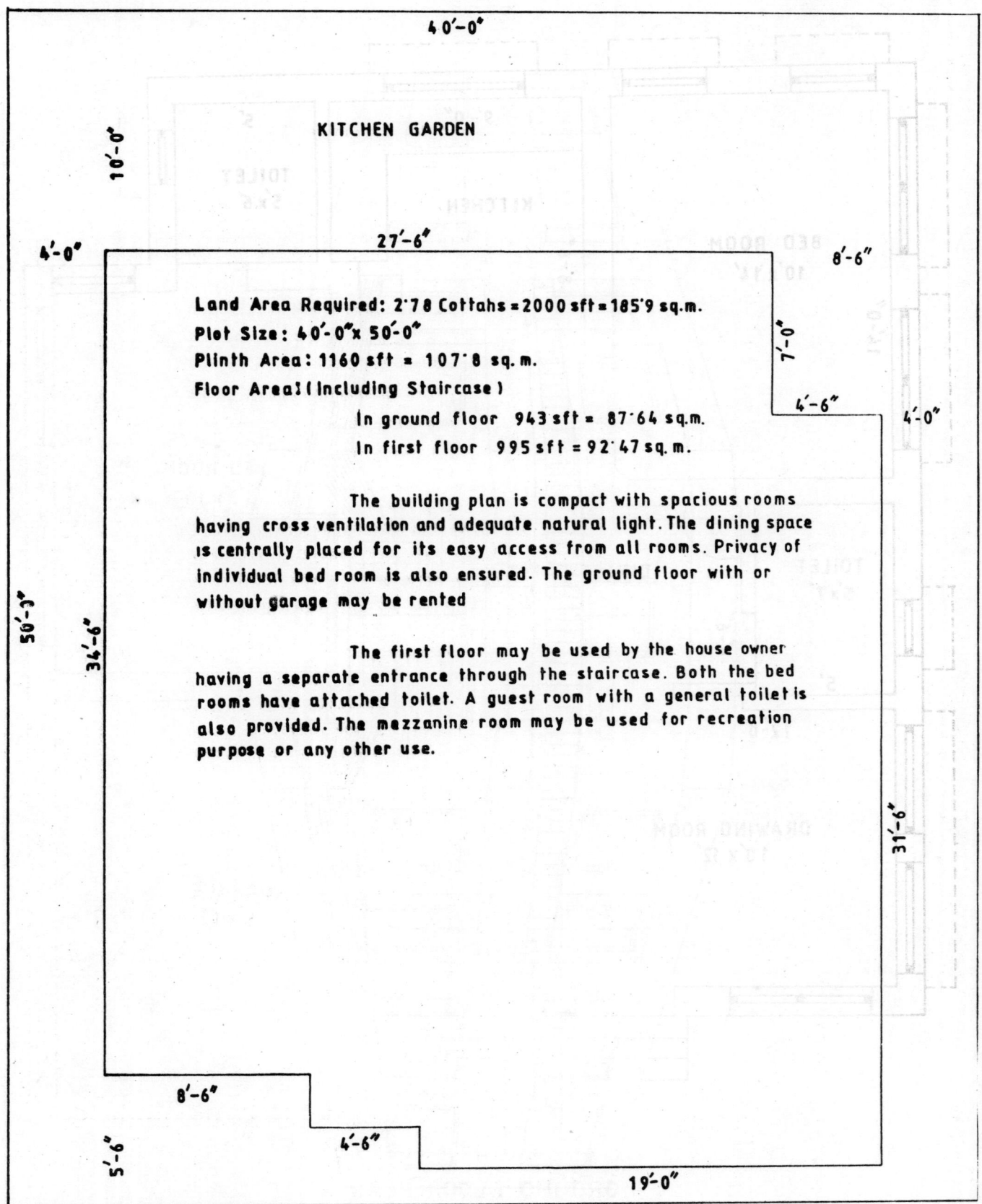

KITCHEN GARDEN

Land Area Required: 2'78 Cottahs = 2000 sft = 185'9 sq.m.
Plot Size: 40'-0" x 50'-0"
Plinth Area: 1160 sft = 107'8 sq.m.
Floor Area (Including Staircase)

    In ground floor 943 sft = 87'64 sq.m.
    In first floor 995 sft = 92'47 sq.m.

    The building plan is compact with spacious rooms having cross ventilation and adequate natural light. The dining space is centrally placed for its easy access from all rooms. Privacy of individual bed room is also ensured. The ground floor with or without garage may be rented

    The first floor may be used by the house owner having a separate entrance through the staircase. Both the bed rooms have attached toilet. A guest room with a general toilet is also provided. The mezzanine room may be used for recreation purpose or any other use.

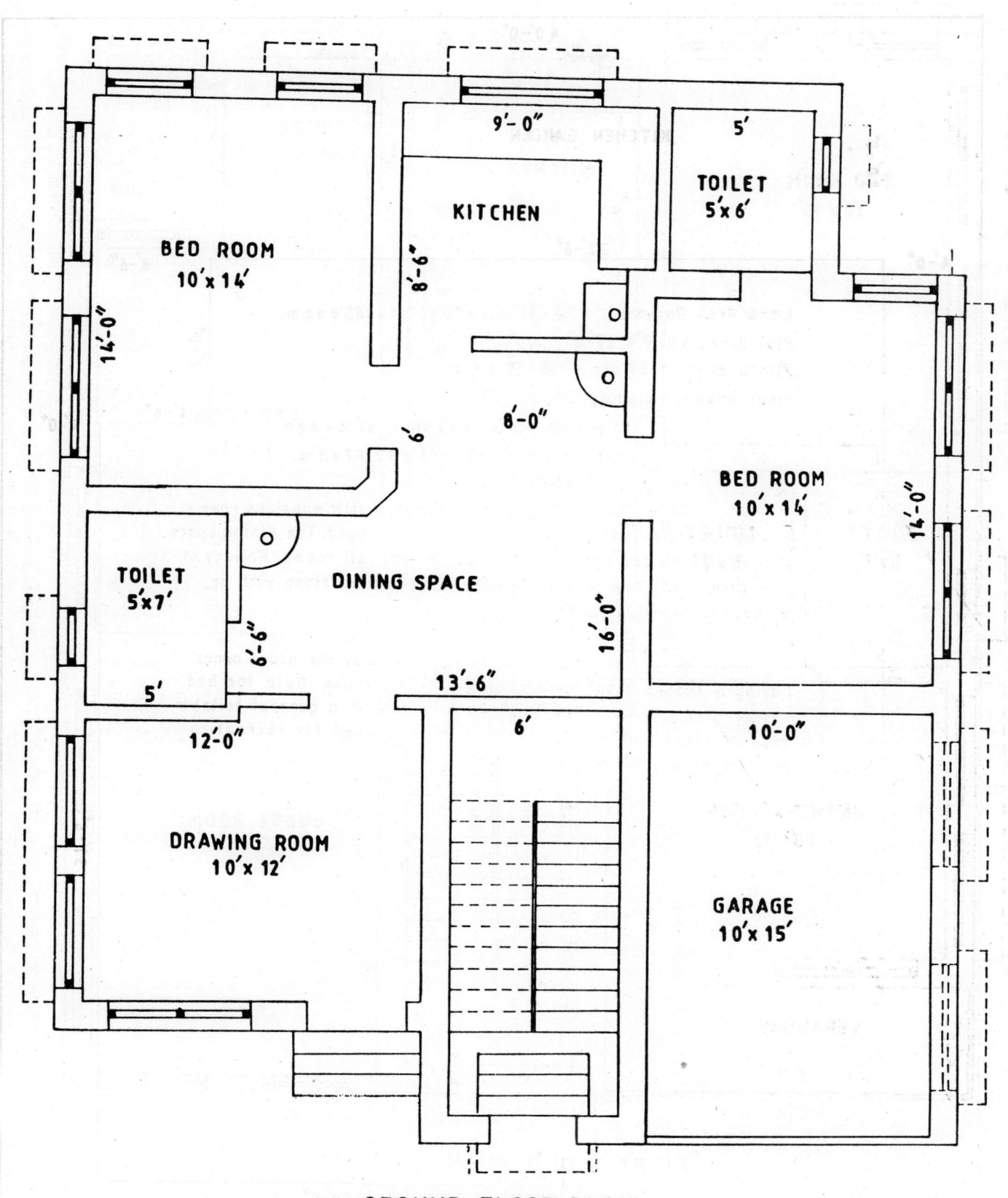

GROUND FLOOR PLAN

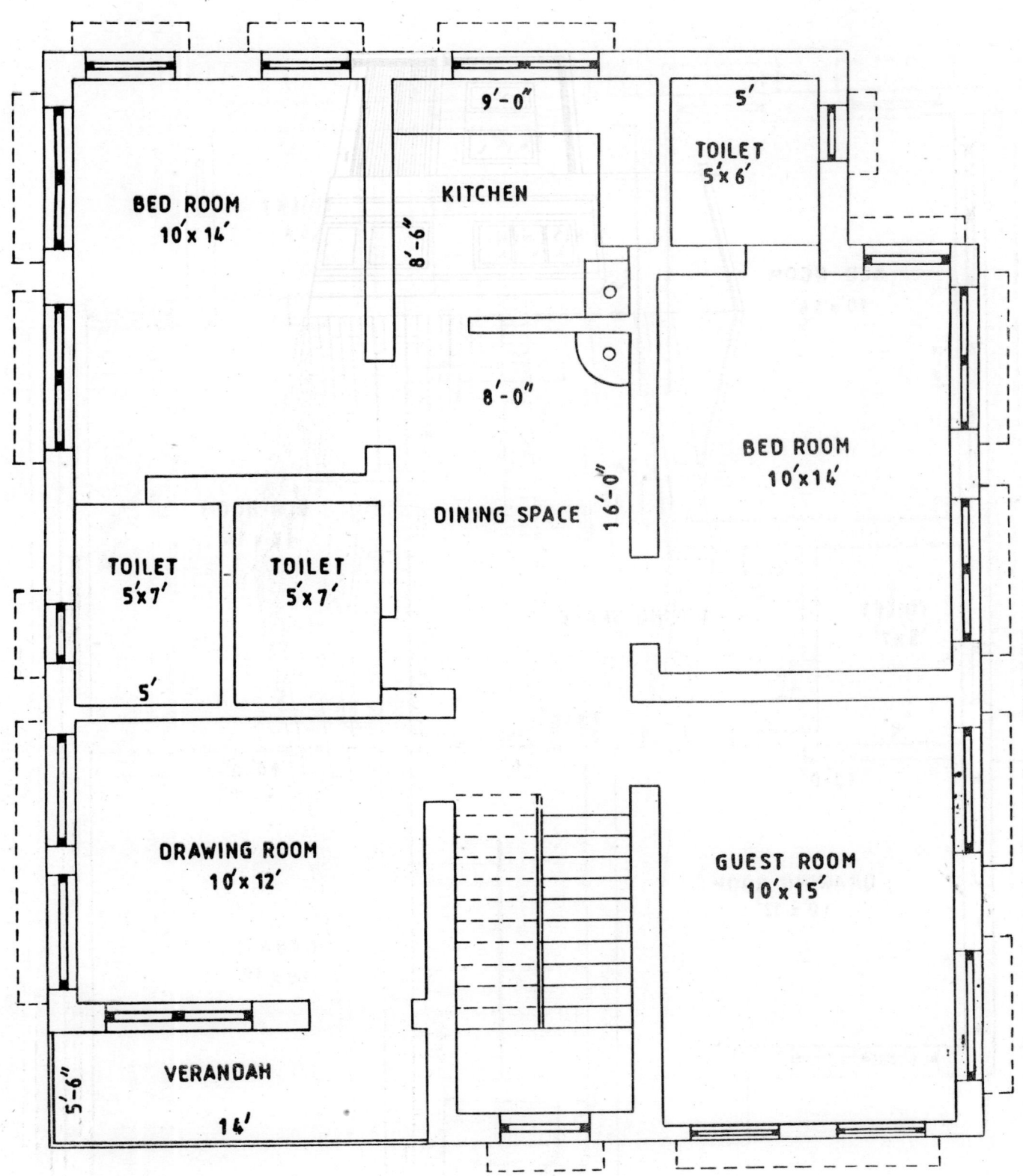

BED ROOM
10' x 14'

9'-0"

KITCHEN

5'

TOILET
5' x 6'

8'-6"

8'-0"

DINING SPACE

16'-0"

BED ROOM
10' x 14'

TOILET
5' x 7'

TOILET
5' x 7'

5'

DRAWING ROOM
10' x 12'

GUEST ROOM
10' x 15'

5'-6"

VERANDAH

14'

FIRST FLOOR PLAN

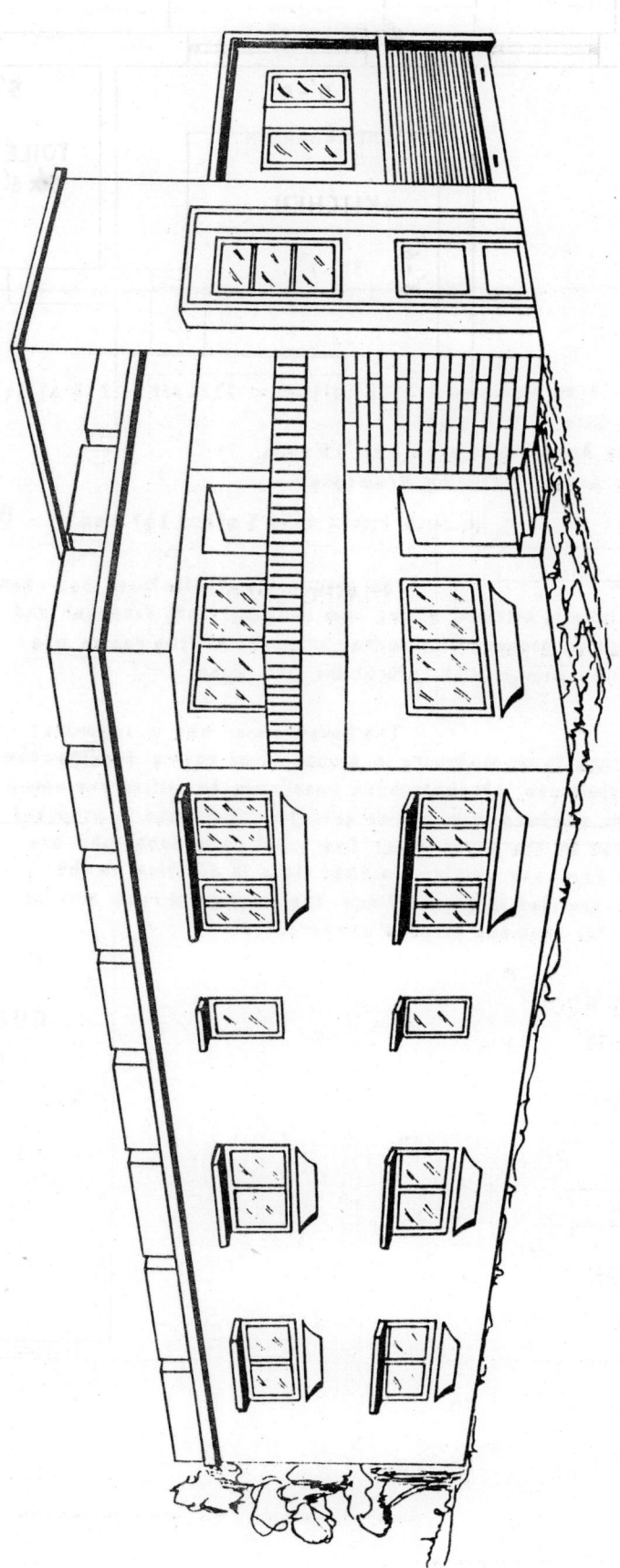

40'-0"

KITCHEN GARDEN

32'-0"

10'-0"

4'

4'

Land Area Required : 3·22 Cottahs ≈ 2320 sft ≈ 215·61 Sq. m.
Plot Size : 40'-0" x 58'-0"
Plinth Area : 1426 sft = 132·53 sq. m.
Floor Area (Including Staircase)

In Each Floor : 1158·5 sft ≈ 107·7 sq. m.

       The ground floor with three bed rooms, two toilets, kitchen, dining cum drawing room, verandah and garage is proposed for rental purpose. All the rooms are spacious enough with natural air and light.

       The house owner has a separate entrance to an enclosure in ground floor having drawing room and staircase. Attempts have been made to utilise the space to the maximum possible extent. The first floor is proposed for use by the house owner. One guest room, lobby and one more toilet are provided in first floor in addition to the units provided in ground floor. The mezzanine room may be used for business purpose or recreation.

58'-0"

45'-6"

1'-6"

12'-0"

4'

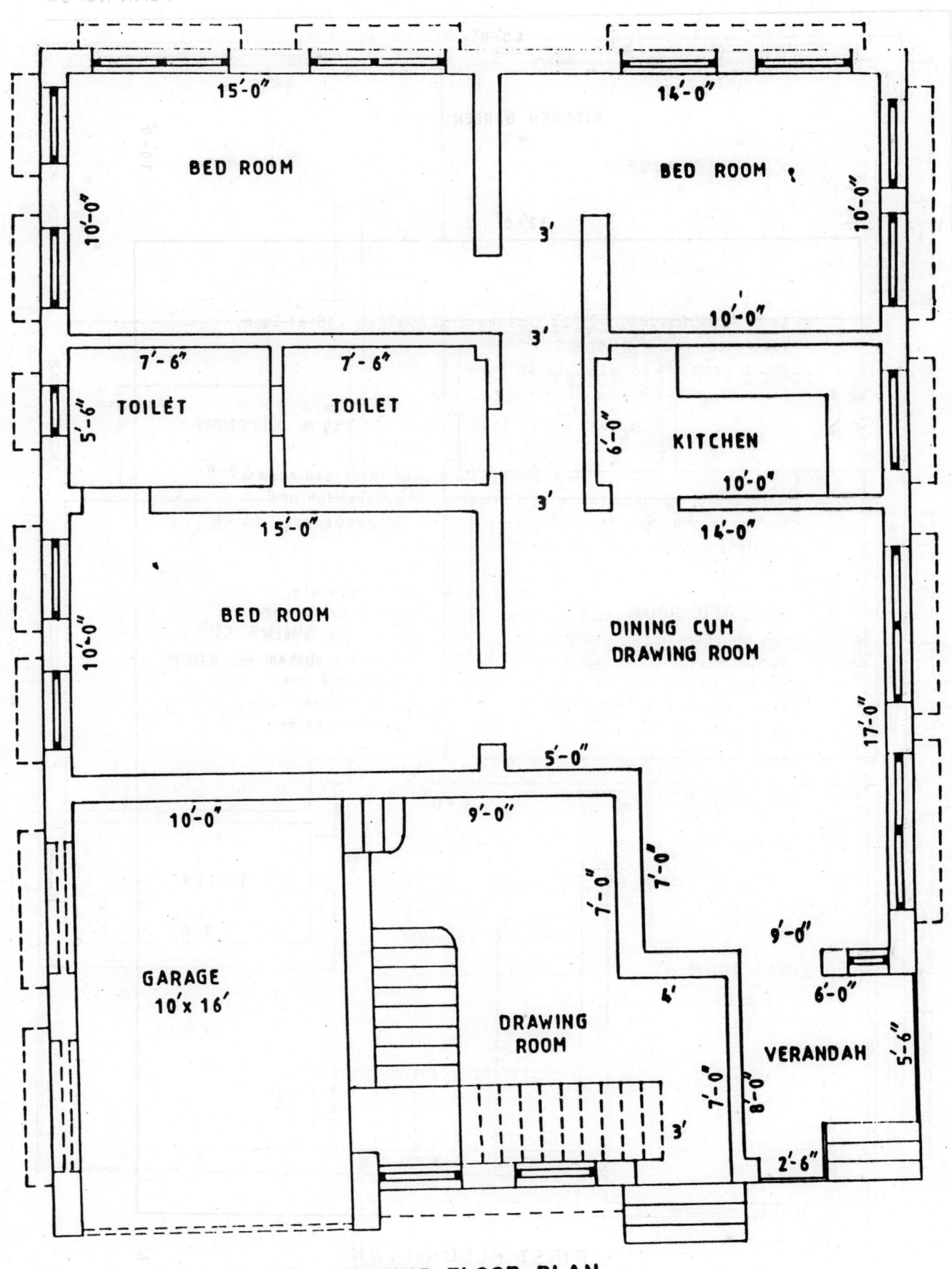

BED ROOM
15'-0"
10'-0"

BED ROOM
14'-0"
10'-0"
10'-0"

3'

3'

TOILET
7'-6"
5'-6"

TOILET
7'-6"

KITCHEN
10'-0"
14'-0"
6'-0"

3'

BED ROOM
15'-0"
10'-0"

DINING CUM DRAWING ROOM
17'-0"

5'-0"

GARAGE
10' x 16'
10'-0"

9'-0"

7'-0"
7'-0"
7'-0"

9'-0"

6'-0"

DRAWING ROOM

4'

VERANDAH
5'-6"

7'-0"
8'-0"

3'

2'-6"

GROUND FLOOR PLAN

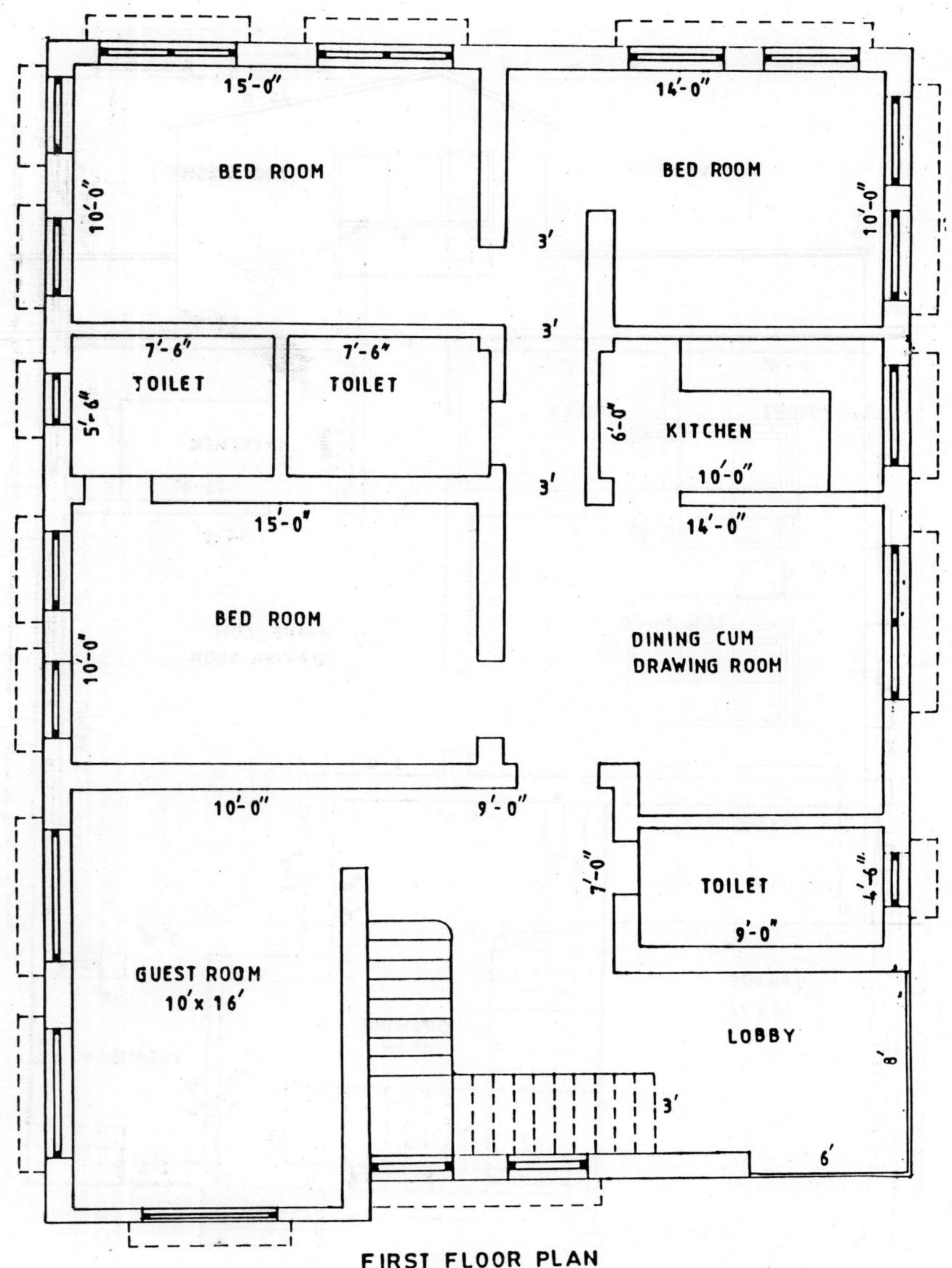

FIRST FLOOR PLAN

ELEVATION

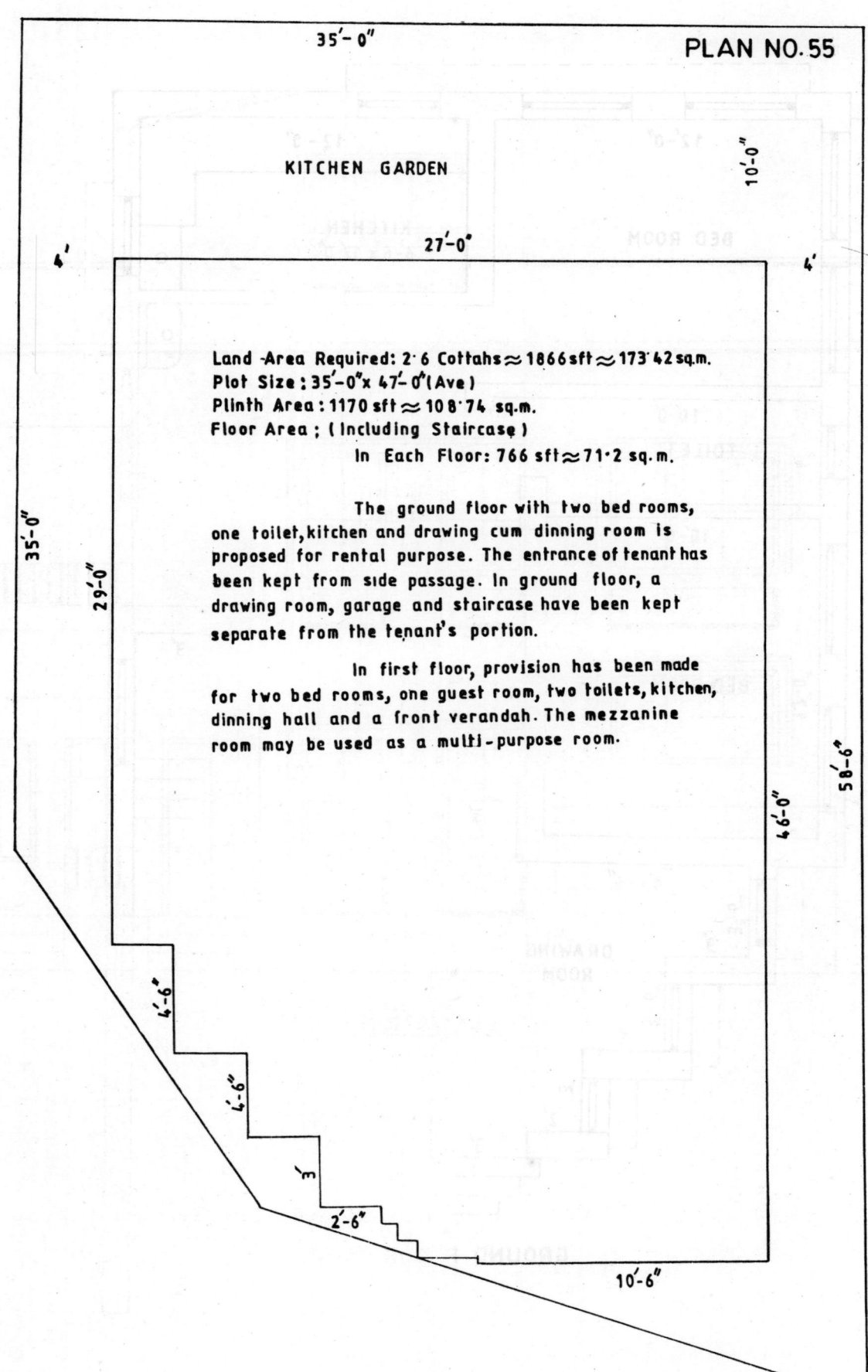

PLAN NO. 55

35'-0"

10'-0"

KITCHEN GARDEN

27'-0"

4'

4'

Land Area Required: 2˙6 Cottahs ≈ 1866 sft ≈ 173˙42 sq.m.
Plot Size : 35'-0"x 47'-0"(Ave)
Plinth Area : 1170 sft ≈ 108˙74 sq.m.
Floor Area ; (Including Staircase)

In Each Floor: 766 sft ≈ 71˙2 sq.m.

The ground floor with two bed rooms,
one toilet, kitchen and drawing cum dinning room is
proposed for rental purpose. The entrance of tenant has
been kept from side passage. In ground floor, a
drawing room, garage and staircase have been kept
separate from the tenant's portion.

In first floor, provision has been made
for two bed rooms, one guest room, two toilets, kitchen,
dinning hall and a front verandah. The mezzanine
room may be used as a multi-purpose room.

35'-0"

29'-0"

58'-6"

46'-0"

4'-6"

4'-6"

3'

2'-6"

10'-6"

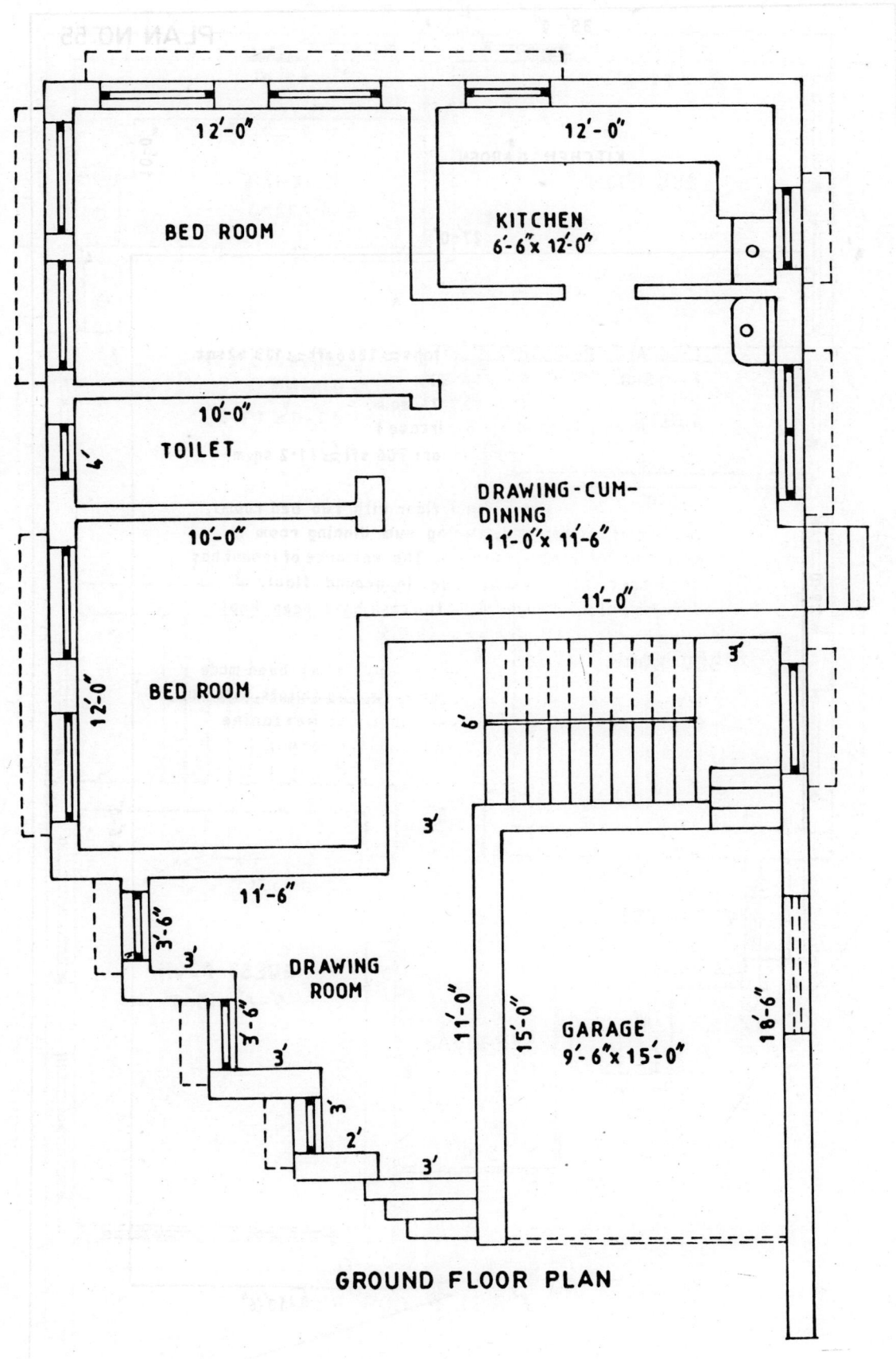

GROUND FLOOR PLAN

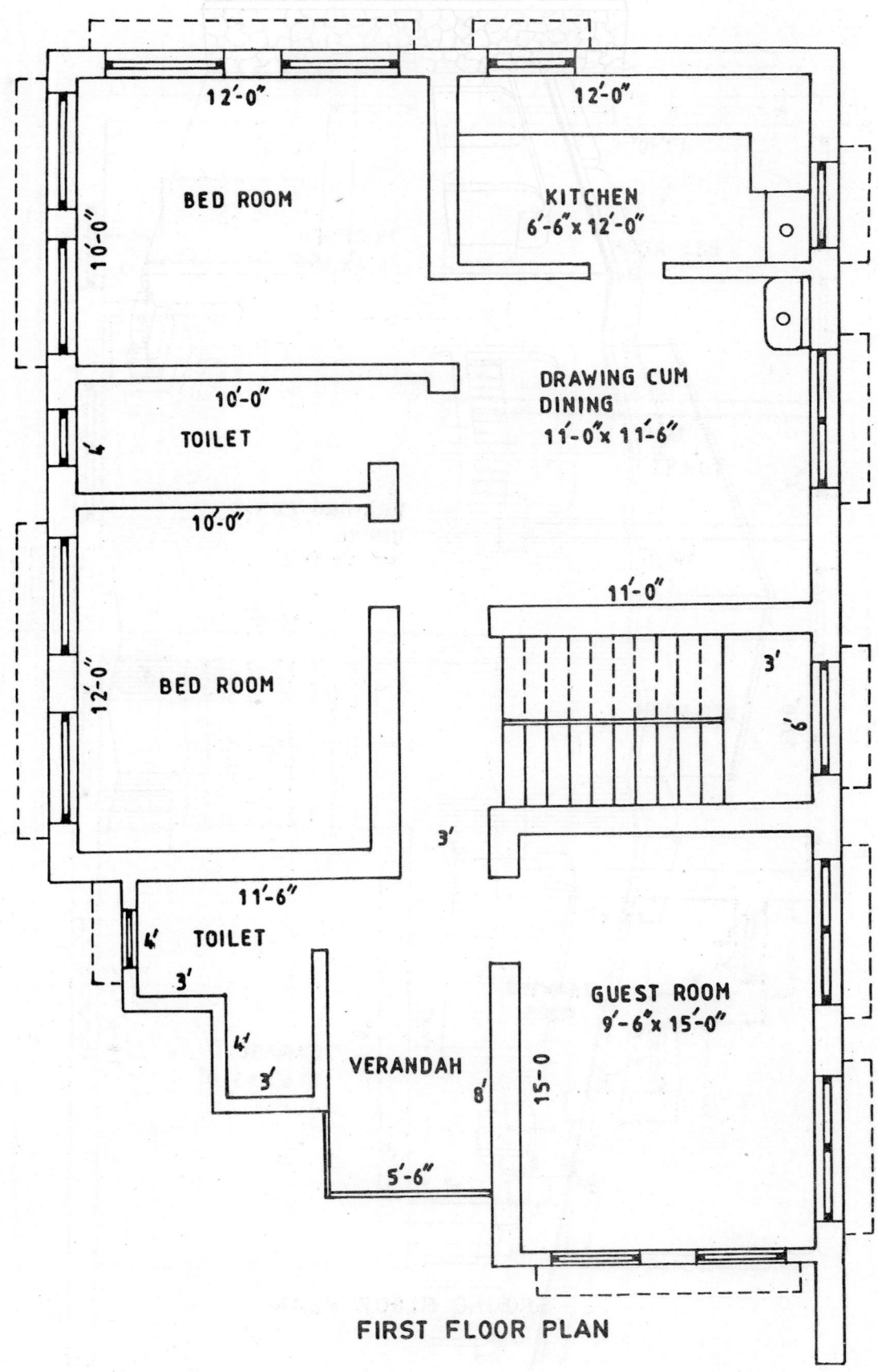

FIRST FLOOR PLAN

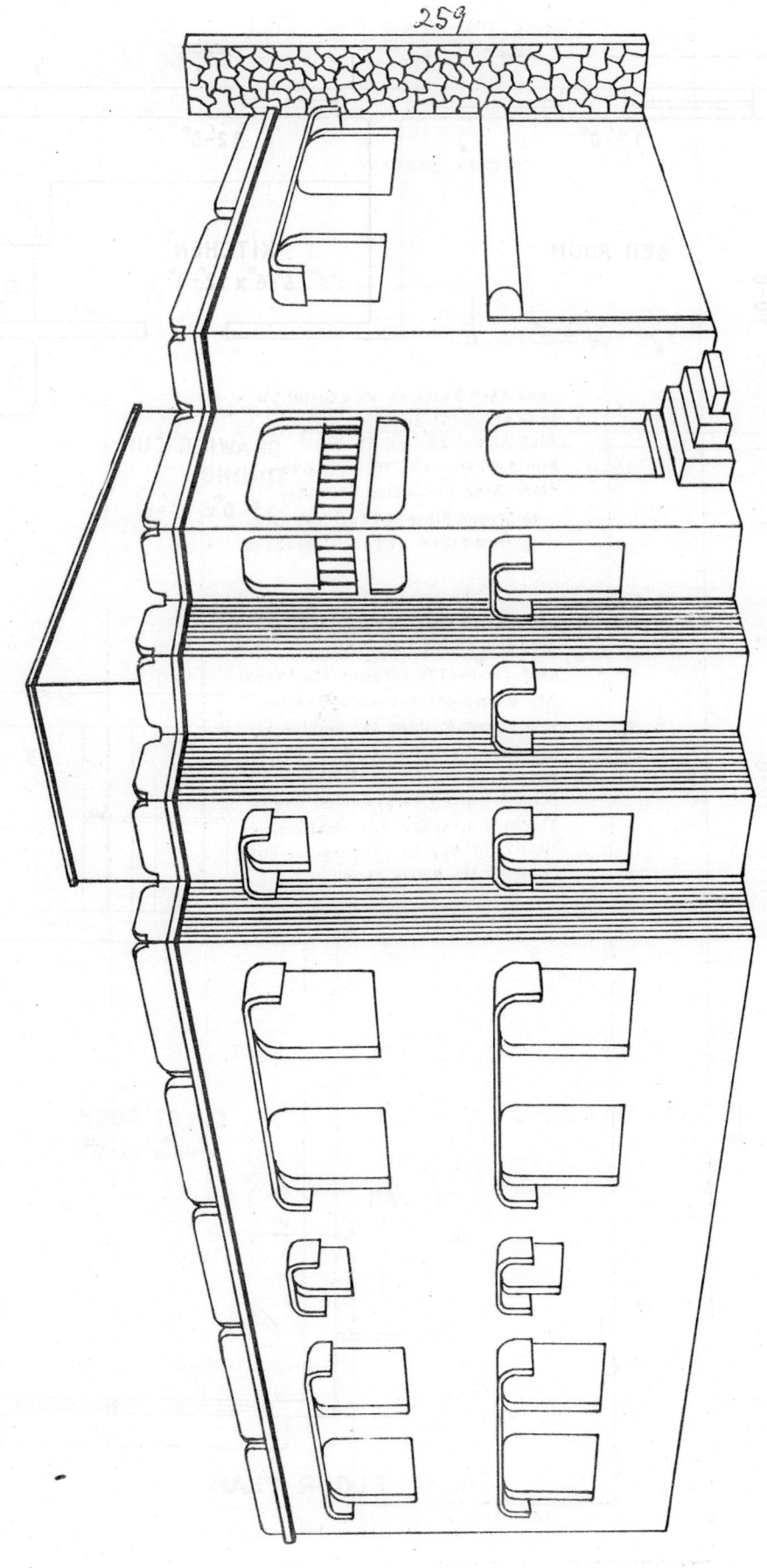

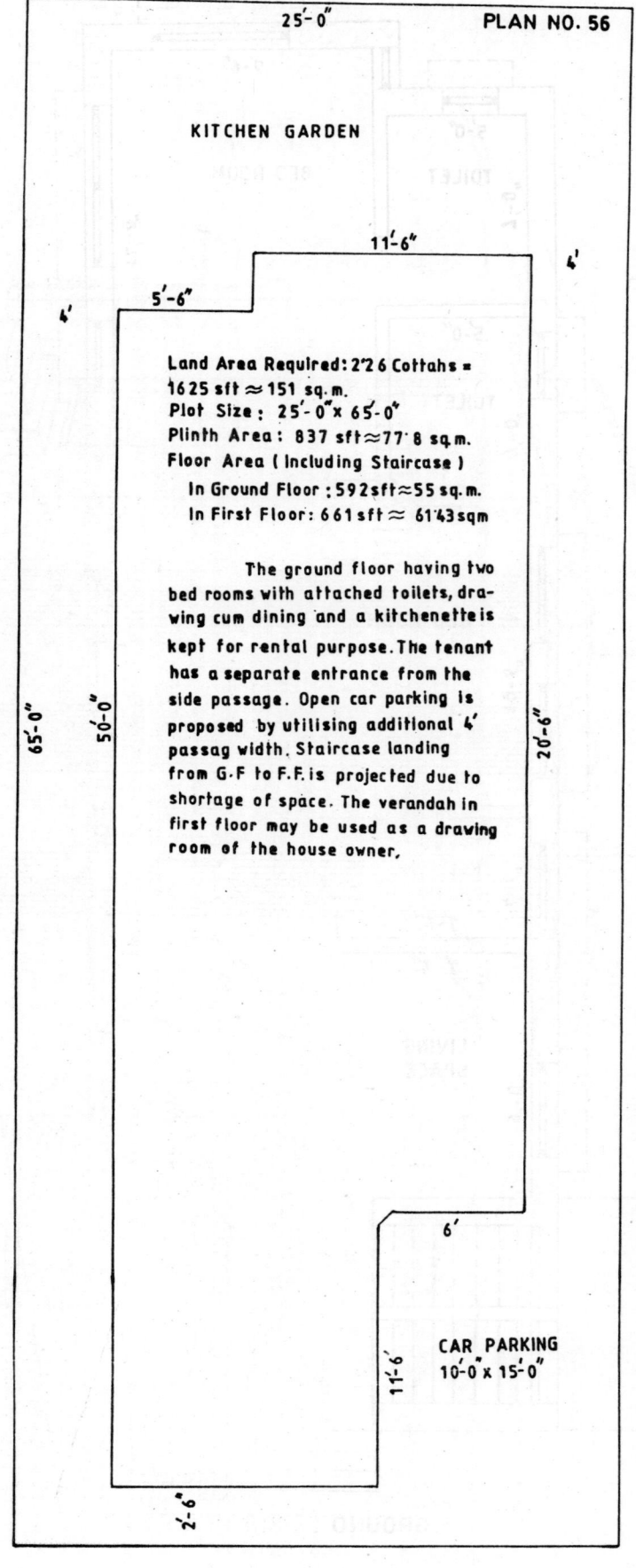

PLAN NO. 56

KITCHEN GARDEN

25'-0"

11'-6"

5'-6"

4'

Land Area Required: 2'2 6 Cottahs =
1625 sft ≈ 151 sq. m.
Plot Size : 25'-0" x 65'-0"
Plinth Area : 837 sft ≈ 77'8 sq.m.
Floor Area (Including Staircase)
  In Ground Floor : 592 sft ≈ 55 sq.m.
  In First Floor : 661 sft ≈ 61'43 sqm

       The ground floor having two
bed rooms with attached toilets, drawing cum dining and a kitchenette is
kept for rental purpose. The tenant
has a separate entrance from the
side passage. Open car parking is
proposed by utilising additional 4'
passag width. Staircase landing
from G.F to F.F. is projected due to
shortage of space. The verandah in
first floor may be used as a drawing
room of the house owner.

65'-0"

50'-0"

20'-6"

6'

11'-6"

CAR PARKING
10'-0" x 15'-0"

2'-6"

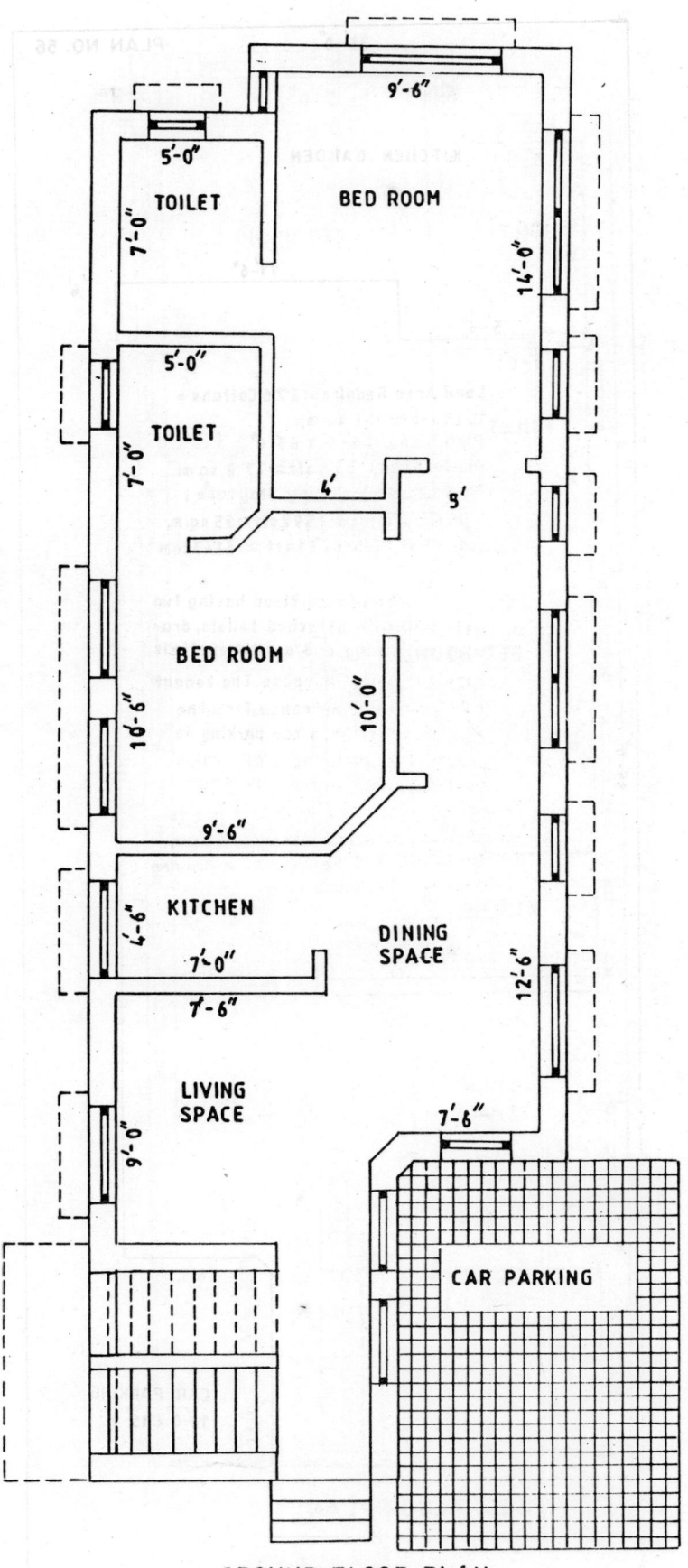

9'-6"

5'-0"
TOILET
7'-0"

BED ROOM

14'-0"

5'-0"
TOILET
7'-0"

4'

3'

BED ROOM

10'-0"

16'-6"

9'-6"

KITCHEN

4'-6"

7'-0"

7'-6"

DINING
SPACE

12'-6"

LIVING
SPACE

9'-0"

7'-6"

CAR PARKING

GROUND FLOOR PLAN

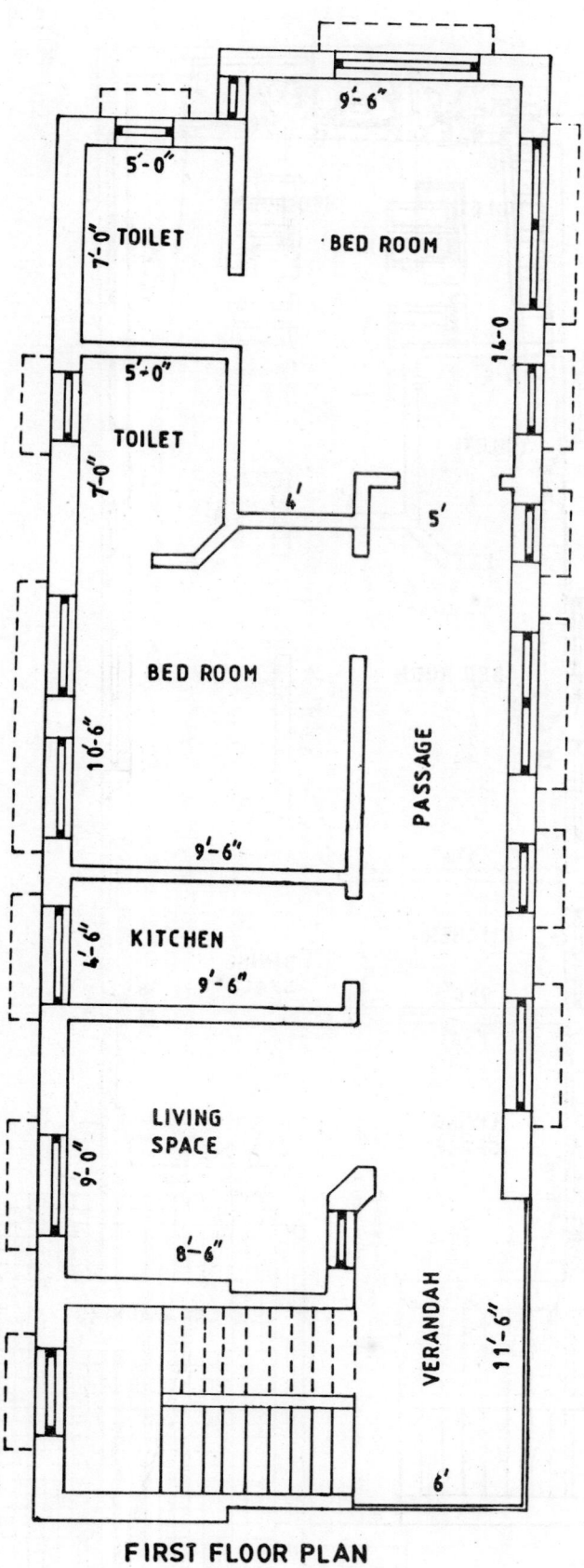

FIRST FLOOR PLAN

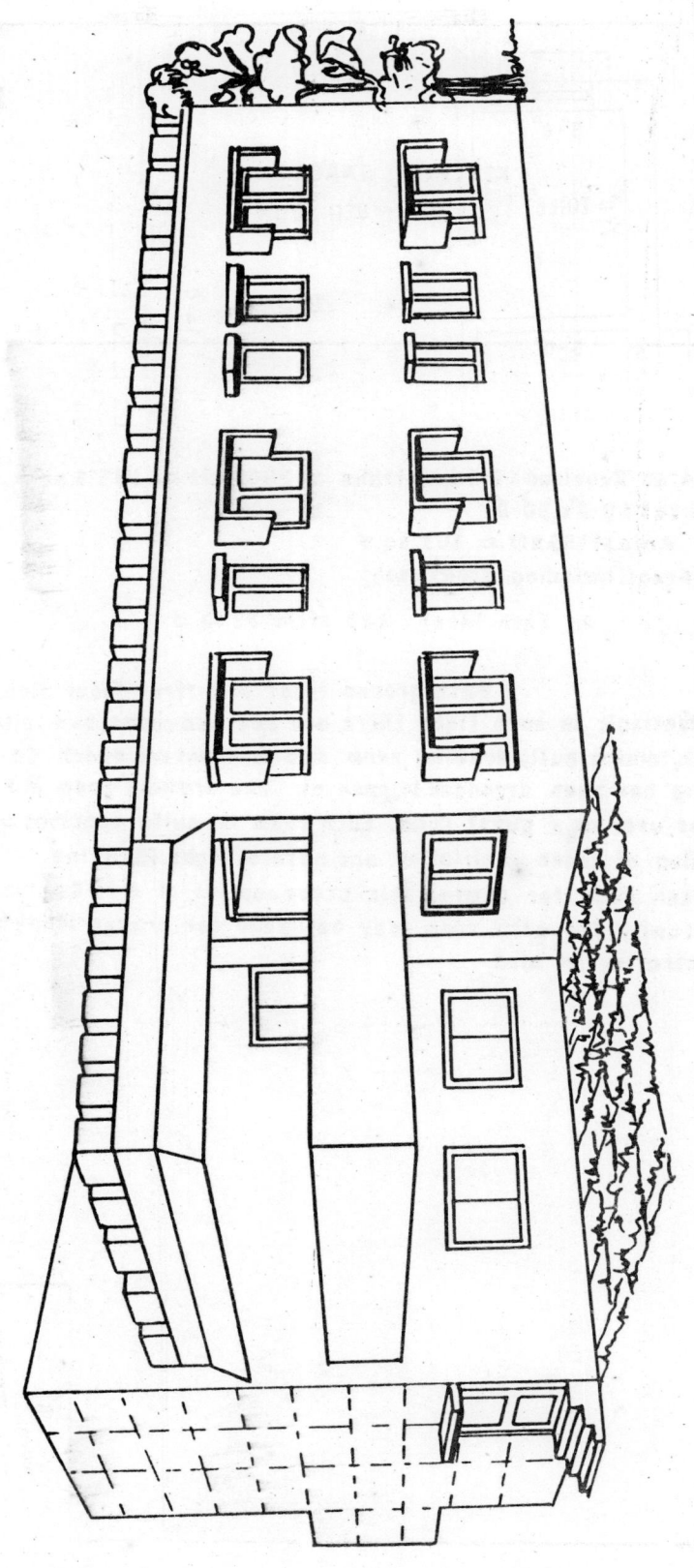

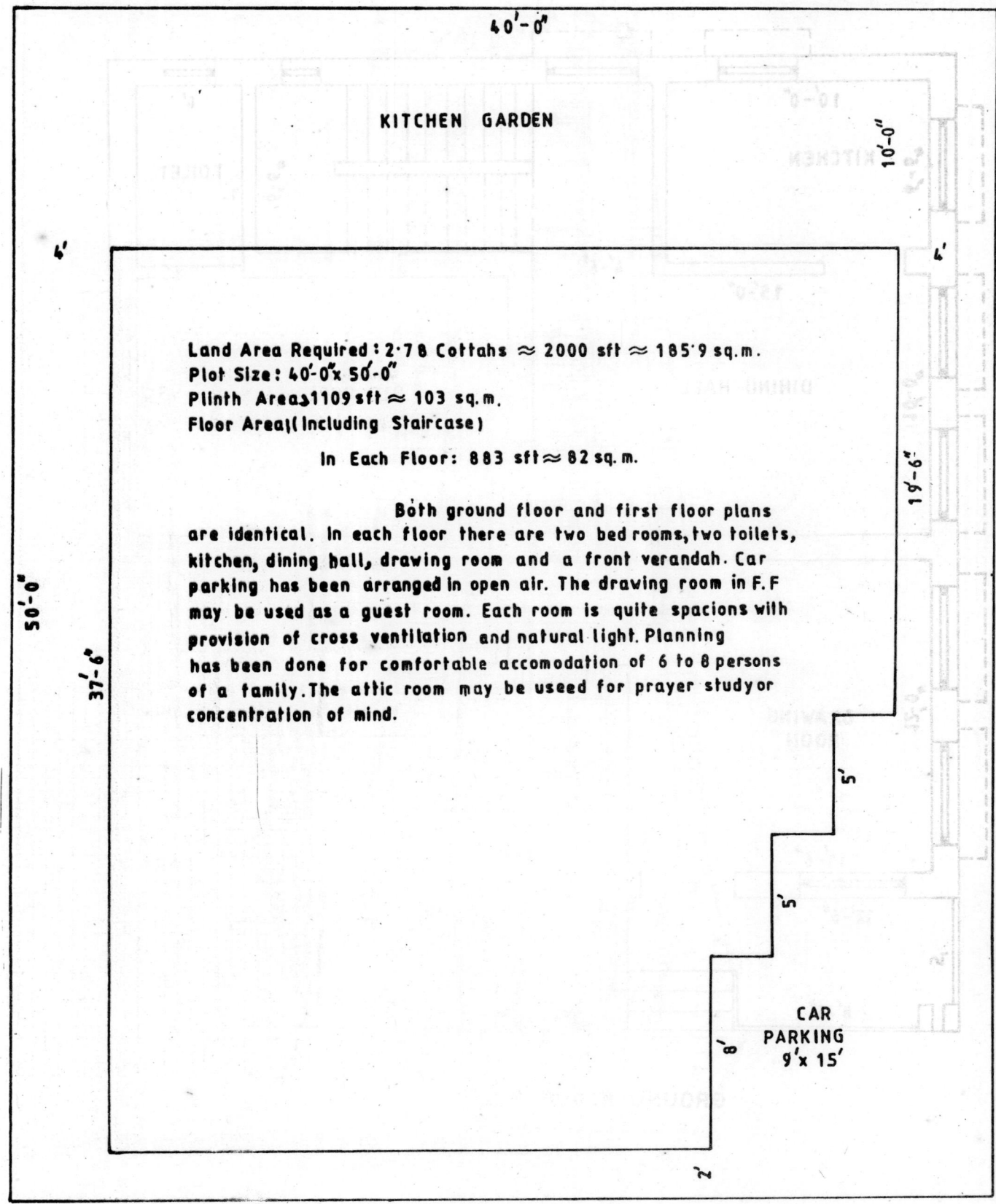

KITCHEN GARDEN

Land Area Required : 2·78 Cottahs ≈ 2000 sft ≈ 185·9 sq.m.
Plot Size : 40'-0" x 50'-0"
Plinth Area : 1109 sft ≈ 103 sq.m.
Floor Area (Including Staircase)

In Each Floor : 883 sft ≈ 82 sq.m.

Both ground floor and first floor plans
are identical. In each floor there are two bed rooms, two toilets,
kitchen, dining hall, drawing room and a front verandah. Car
parking has been arranged in open air. The drawing room in F.F
may be used as a guest room. Each room is quite spacions with
provision of cross ventilation and natural light. Planning
has been done for comfortable accomodation of 6 to 8 persons
of a family. The attic room may be useed for prayer study or
concentration of mind.

40'-0"

50'-0"

37'-6"

10'-0"

15'-0"

19'-6"

15'-0"

5'

5'

CAR
PARKING
9' x 15'

8'

2'

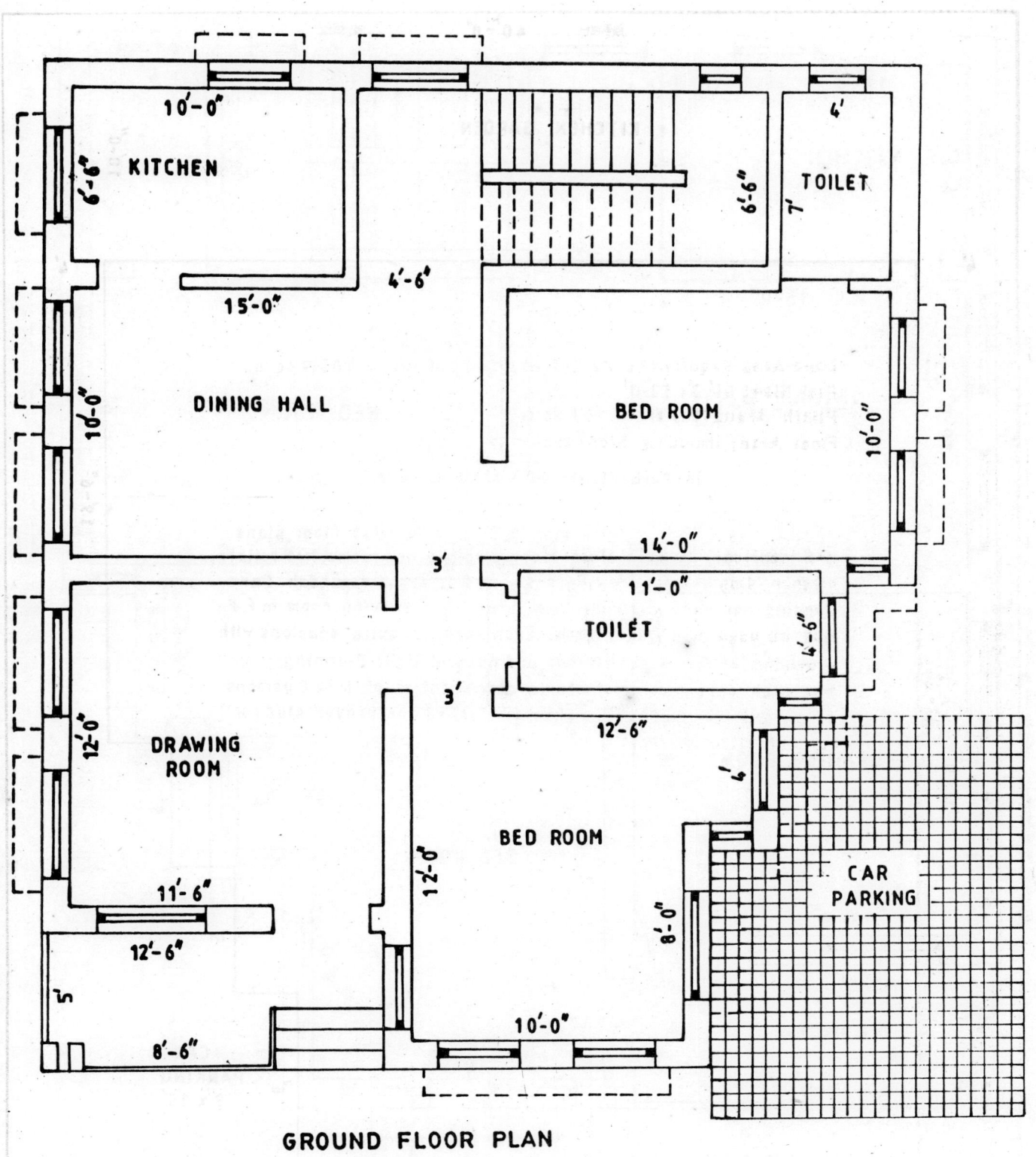

GROUND FLOOR PLAN

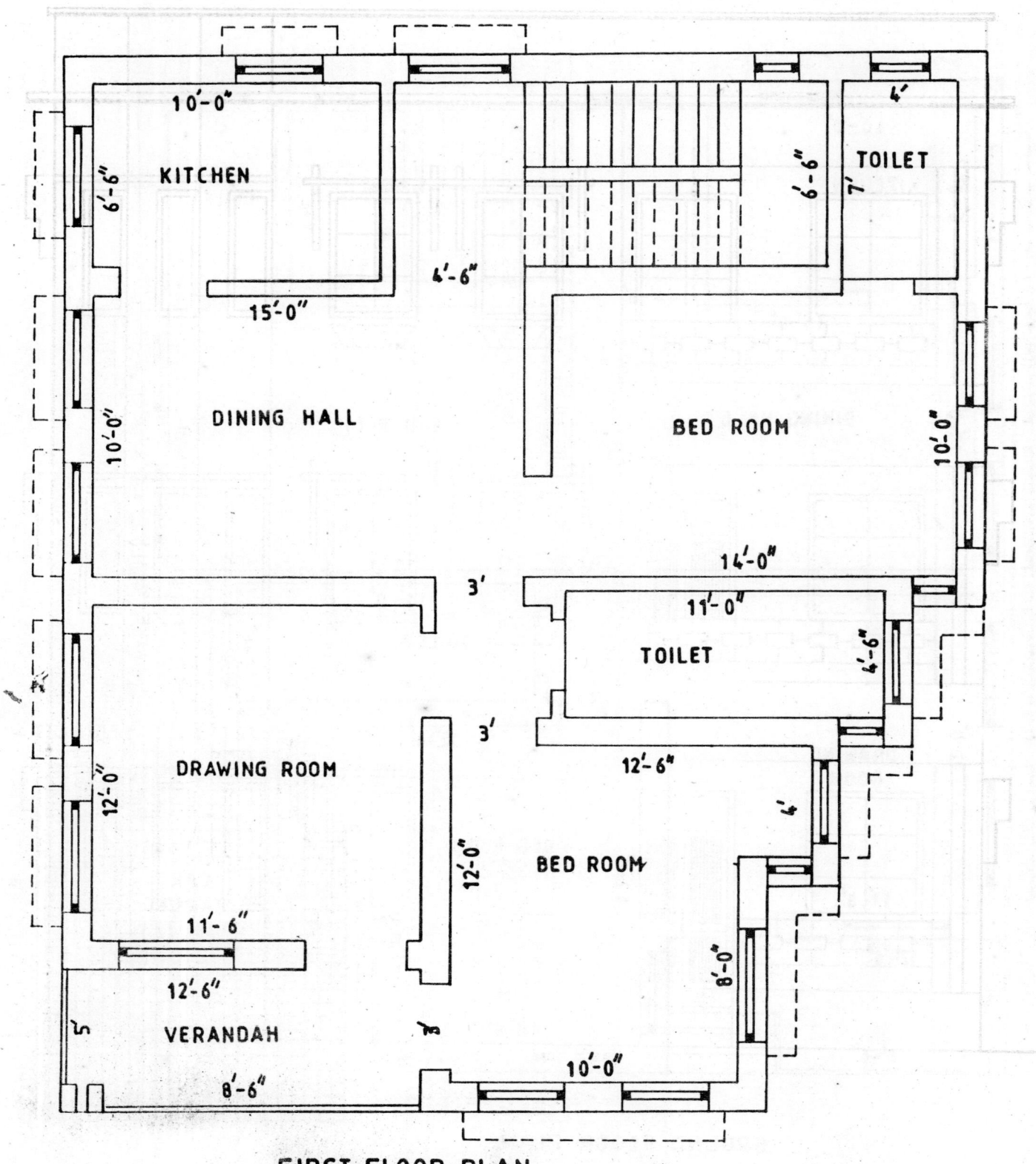

FIRST FLOOR PLAN

ELEVATION

PLAN NO 58

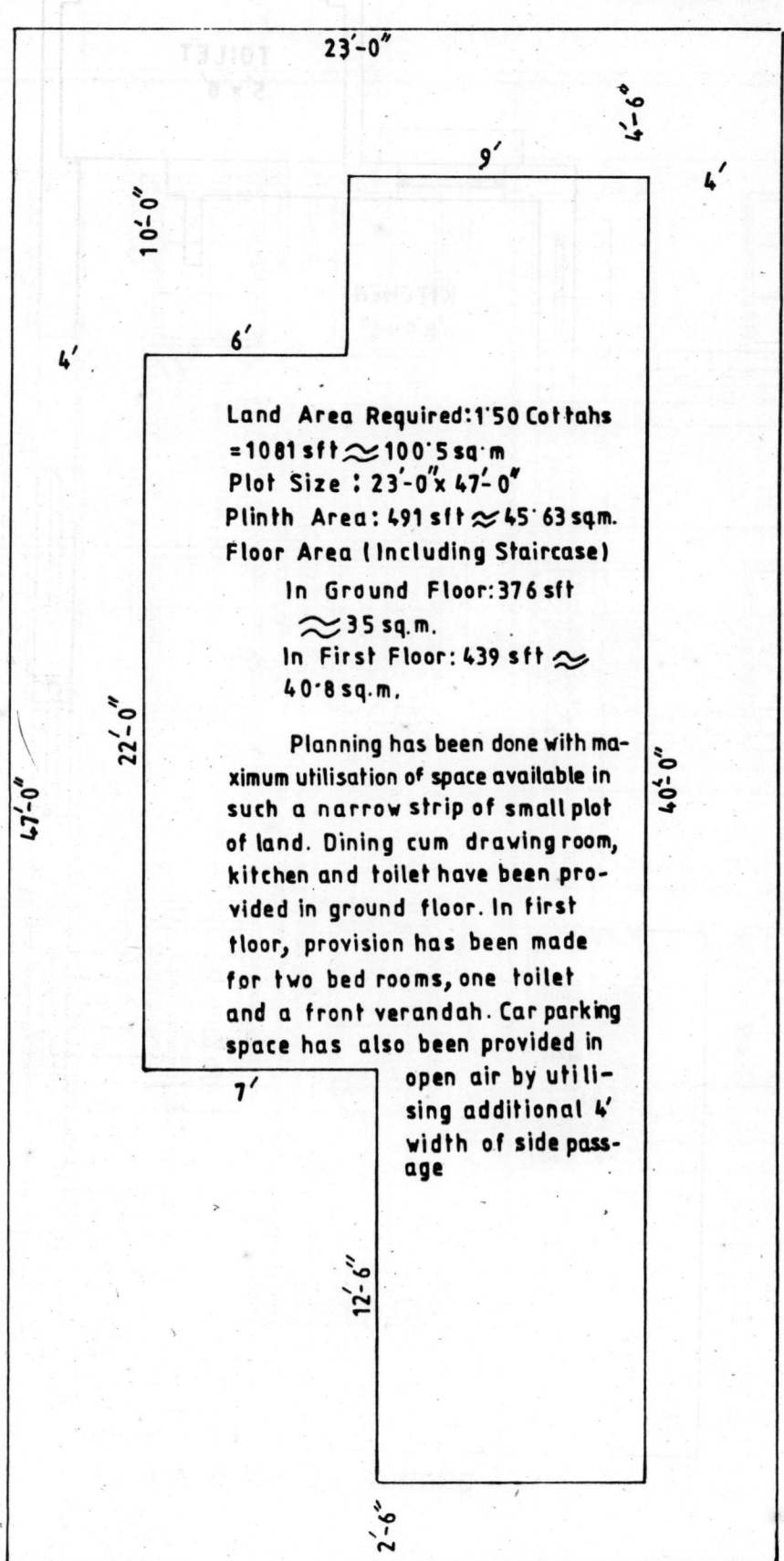

23'-0"

4'-6"

9'

4'

10'-0"

6'

4'

Land Area Required: 1'50 Cottahs
= 1081 sft ≈ 100'5 sq m
Plot Size : 23'-0" x 47'-0"
Plinth Area: 491 sft ≈ 45'63 sq.m.
Floor Area (Including Staircase)
   In Ground Floor: 376 sft
   ≈ 35 sq.m.
   In First Floor: 439 sft ≈
   40'8 sq.m.

   Planning has been done with maximum utilisation of space available in
such a narrow strip of small plot
of land. Dining cum drawing room,
kitchen and toilet have been provided in ground floor. In first
floor, provision has been made
for two bed rooms, one toilet
and a front verandah. Car parking
space has also been provided in
open air by utili-
sing additional 4'
width of side pass-
age

47'-0"

22'-0"

40'-0"

7'

12'-6"

2'-6"

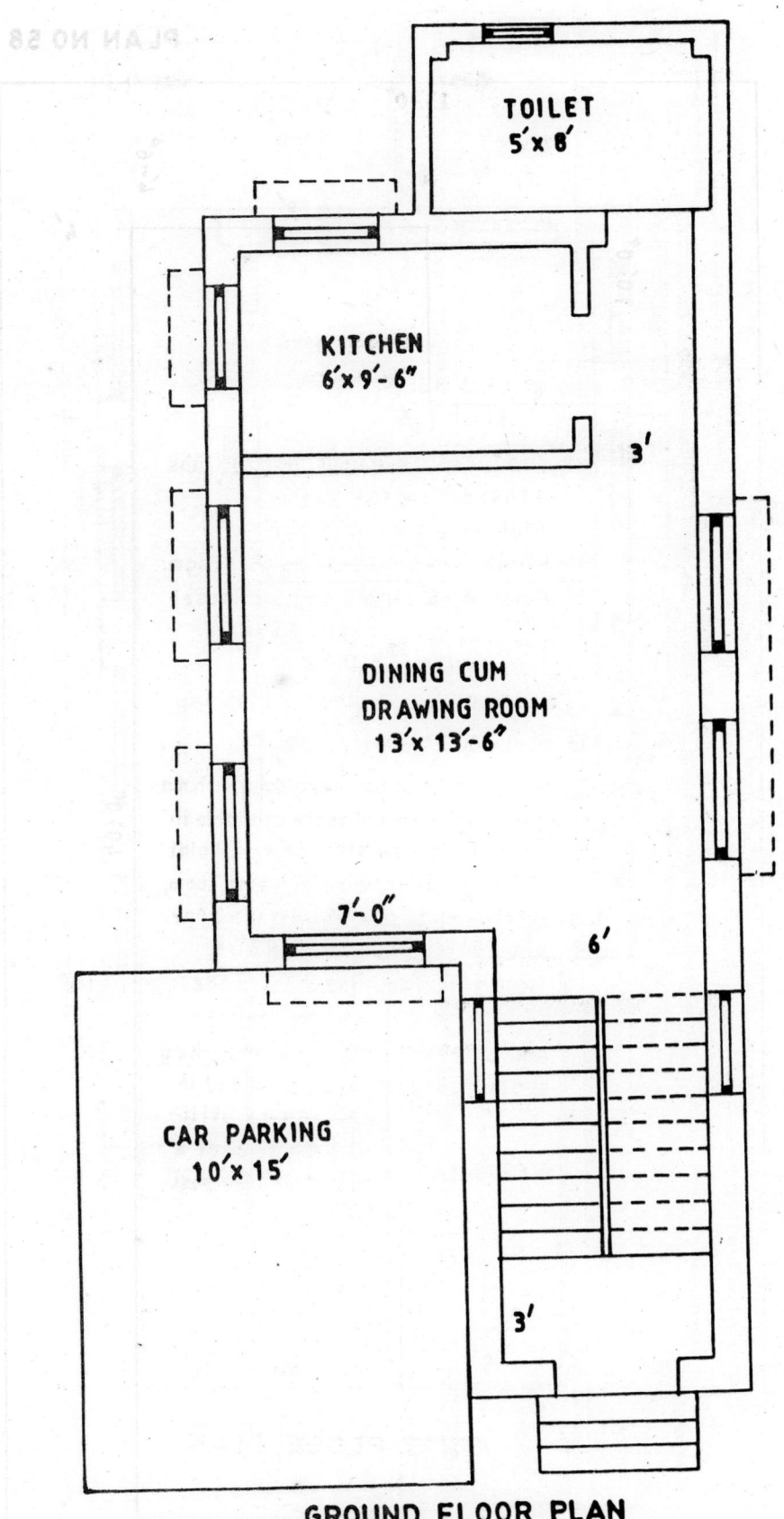

GROUND FLOOR PLAN

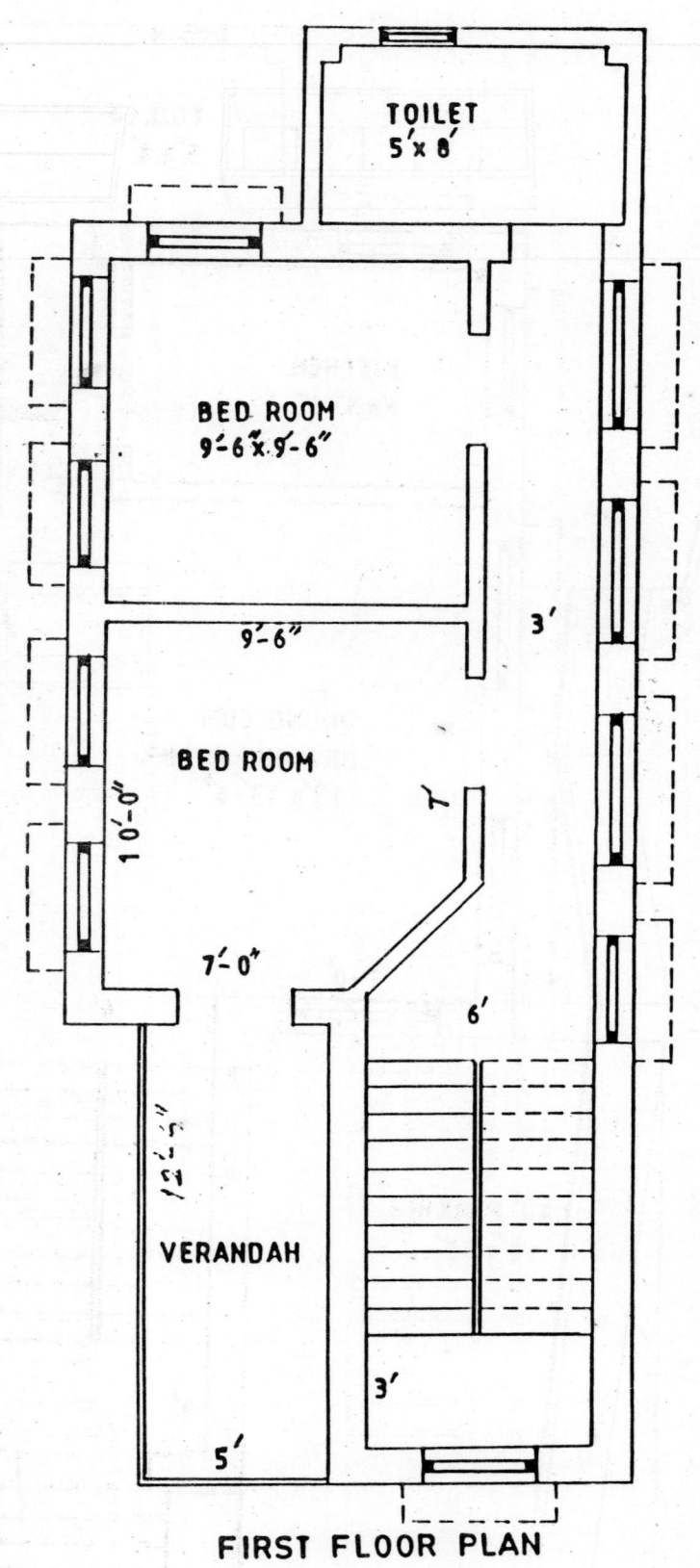

FIRST FLOOR PLAN

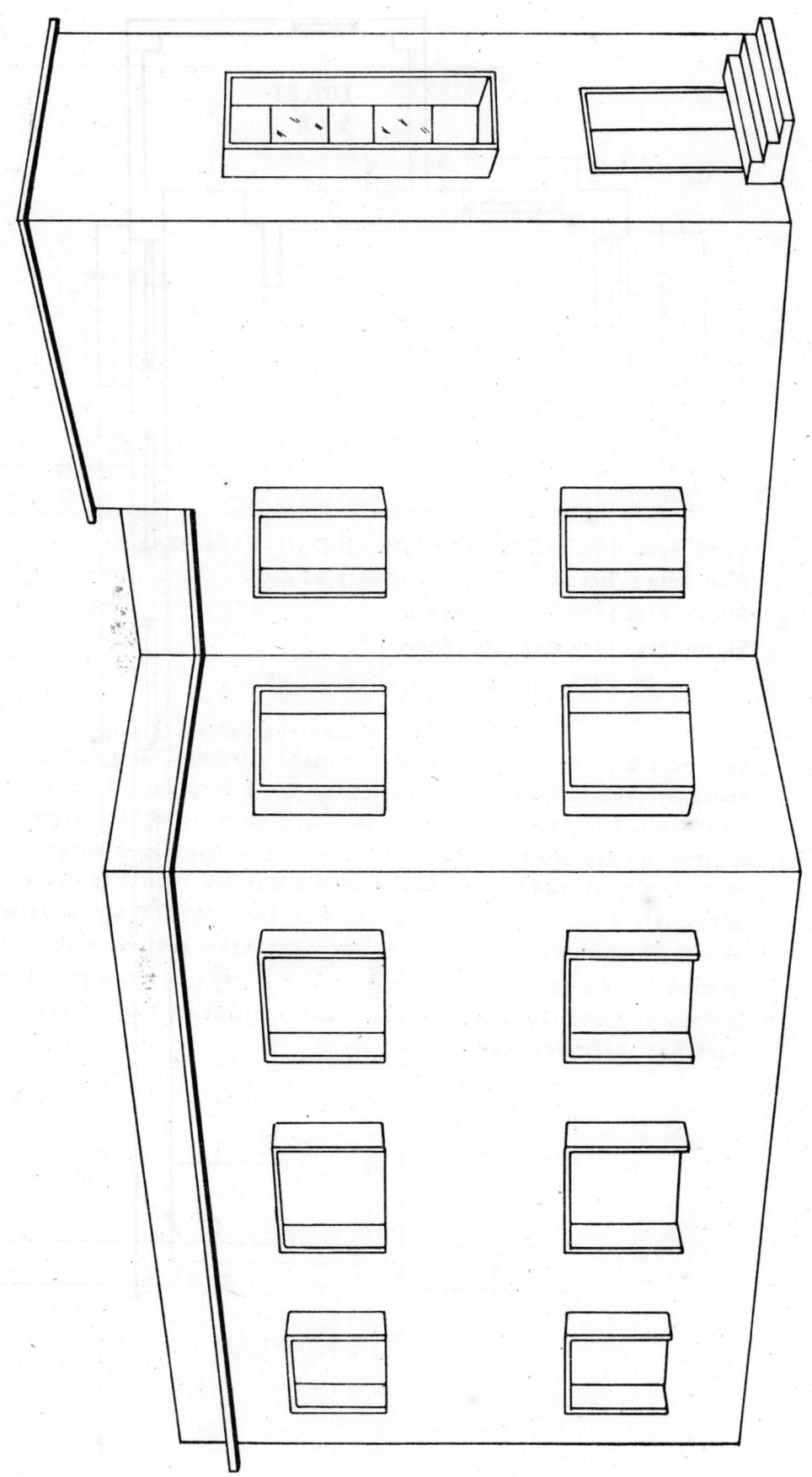

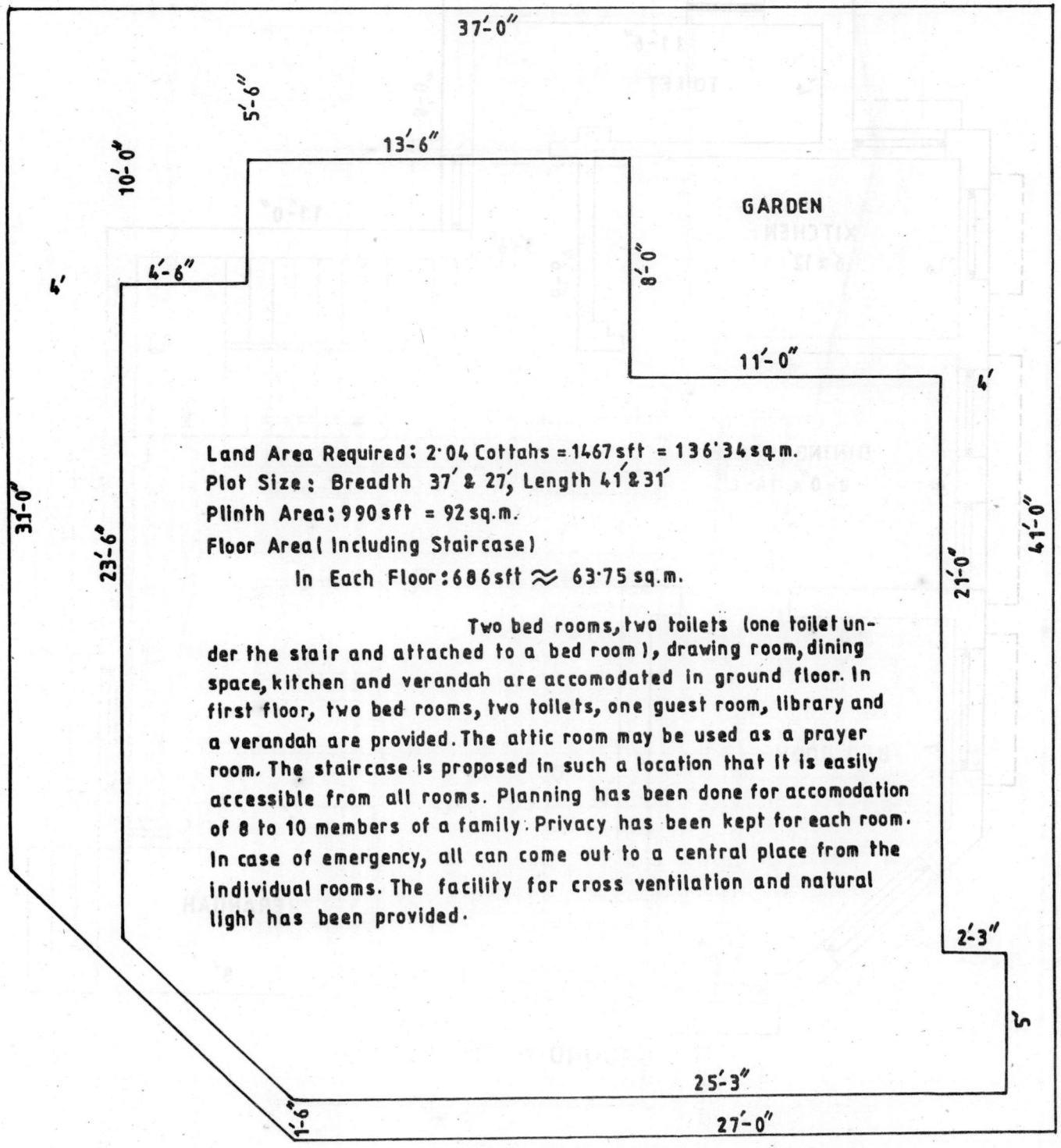

37'-0"

5'-6"

10'-0"

13'-6"

4'

4'-6"

GARDEN

8'-0"

11'-0"

4'

31'-0"

23'-6"

21'-0"

41'-0"

Land Area Required: 2·04 Cottahs = 1467 sft = 136·34 sq.m.
Plot Size: Breadth 37' & 27', Length 41' & 31'
Plinth Area: 990 sft = 92 sq.m.
Floor Area (Including Staircase)
   In Each Floor: 686 sft ≈ 63·75 sq.m.

        Two bed rooms, two toilets (one toilet under the stair and attached to a bed room), drawing room, dining space, kitchen and verandah are accomodated in ground floor. In first floor, two bed rooms, two toilets, one guest room, library and a verandah are provided. The attic room may be used as a prayer room. The staircase is proposed in such a location that it is easily accessible from all rooms. Planning has been done for accomodation of 8 to 10 members of a family. Privacy has been kept for each room. In case of emergency, all can come out to a central place from the individual rooms. The facility for cross ventilation and natural light has been provided.

2'-3"

5'

25'-3"

1'-6"

27'-0"

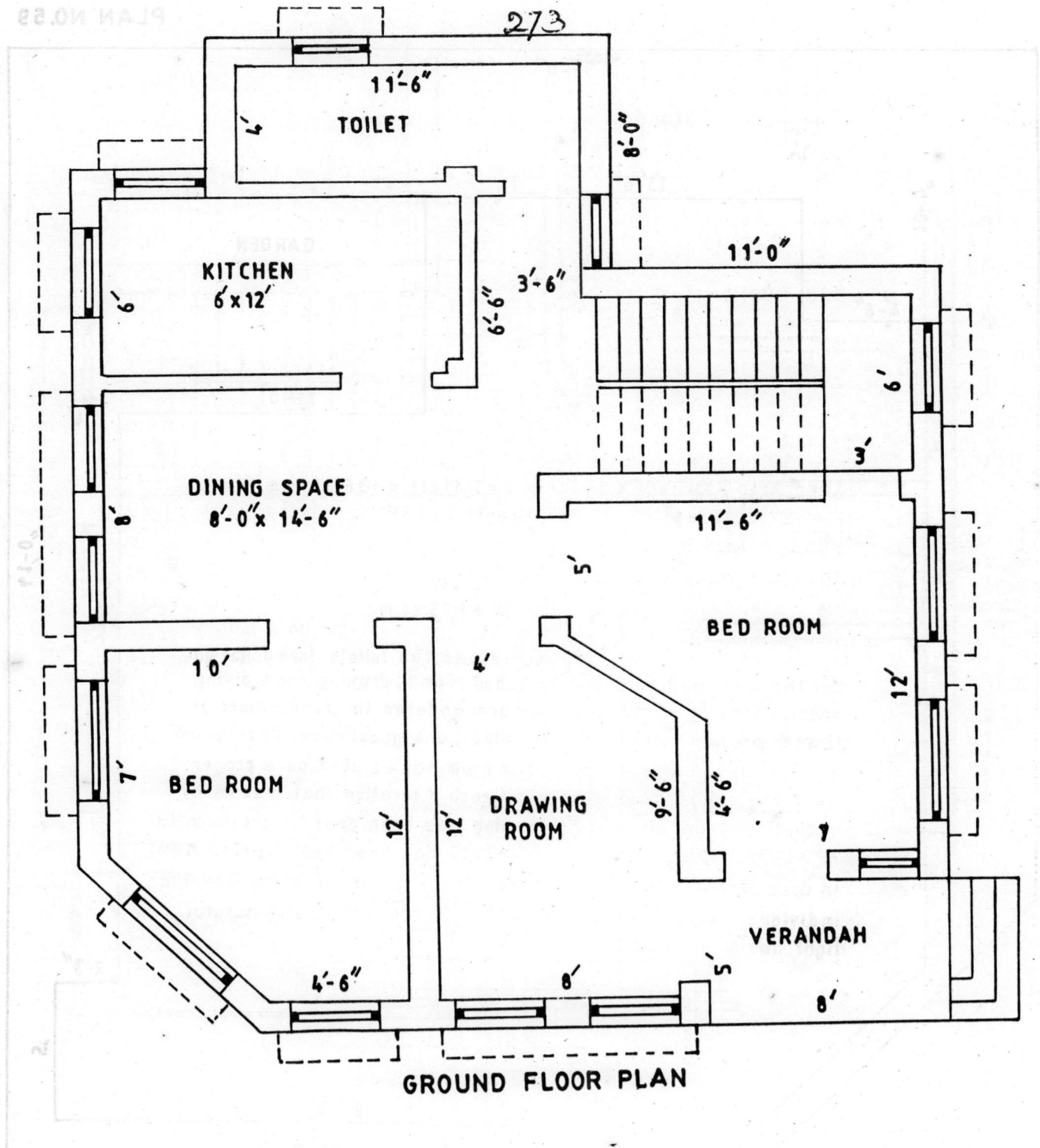

GROUND FLOOR PLAN

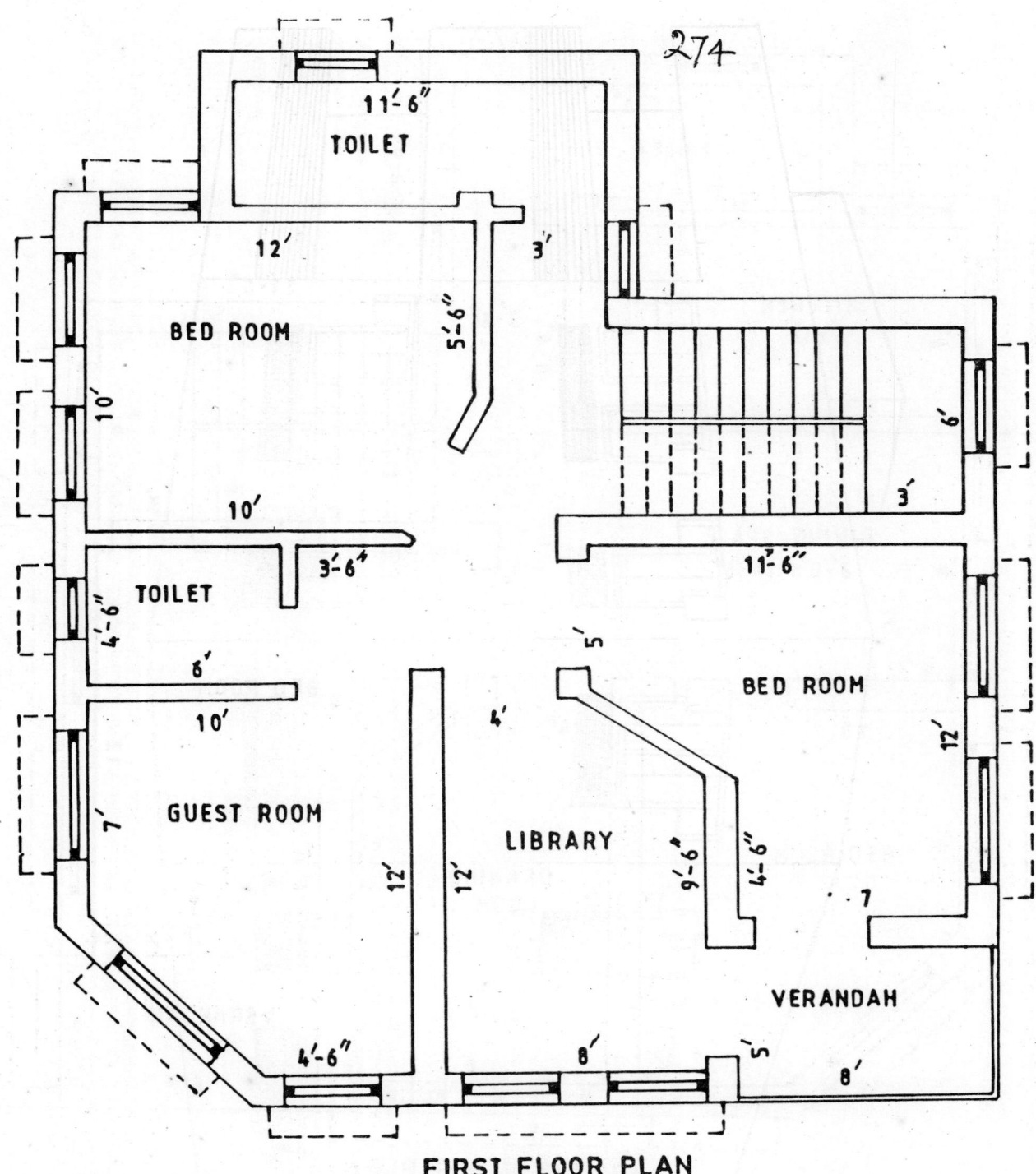

274

FIRST FLOOR PLAN

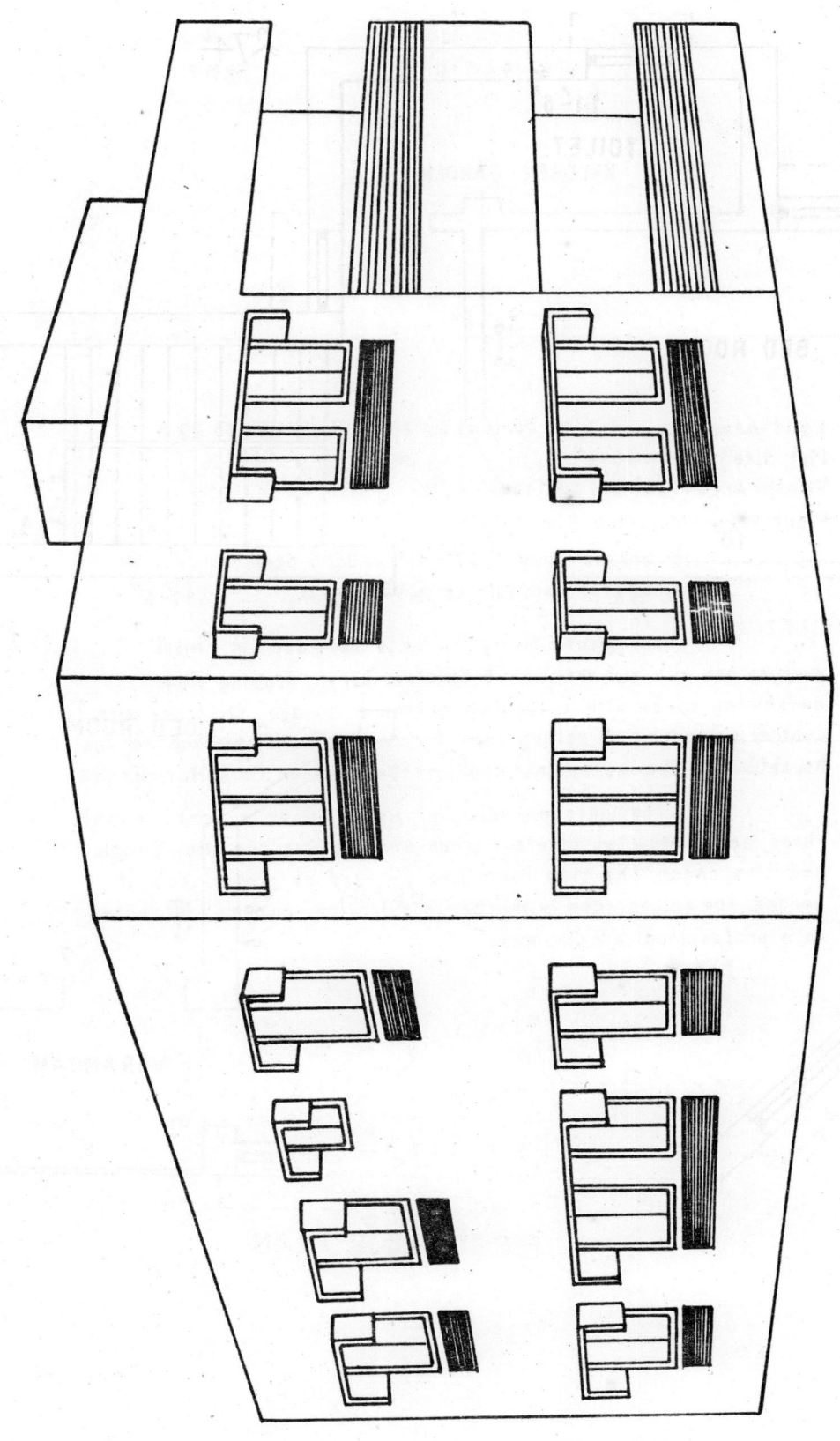

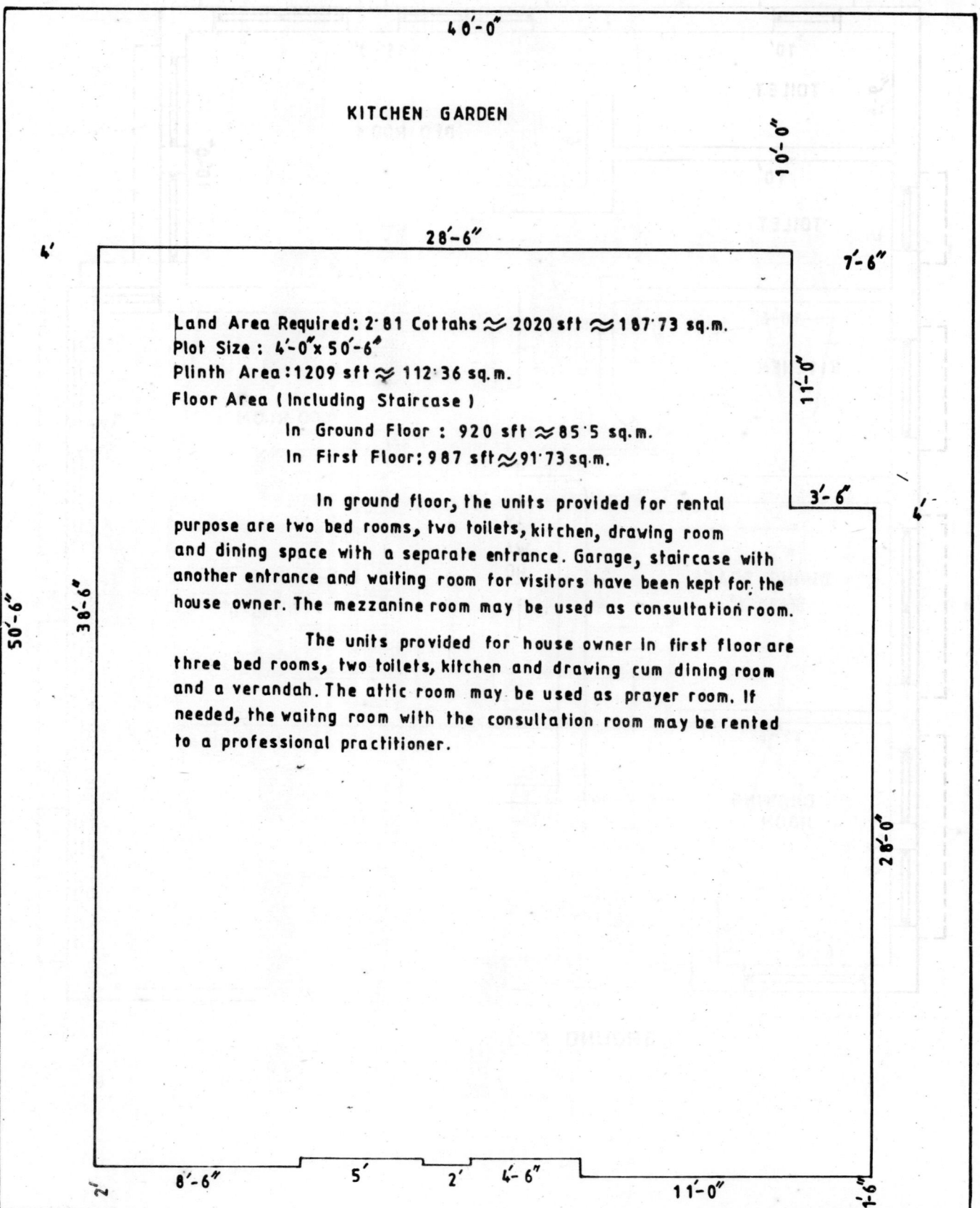

KITCHEN GARDEN

40'-0"

28'-6"

10'-0"

7'-6"

11'-0"

3'-6"

4'

Land Area Required: 2·81 Cottahs ≈ 2020 sft ≈ 187·73 sq.m.
Plot Size : 40'-0" x 50'-6"
Plinth Area : 1209 sft ≈ 112·36 sq.m.
Floor Area (Including Staircase)

In Ground Floor : 920 sft ≈ 85·5 sq.m.
In First Floor : 987 sft ≈ 91·73 sq.m.

In ground floor, the units provided for rental purpose are two bed rooms, two toilets, kitchen, drawing room and dining space with a separate entrance. Garage, staircase with another entrance and waiting room for visitors have been kept for the house owner. The mezzanine room may be used as consultation room.

The units provided for house owner in first floor are three bed rooms, two toilets, kitchen and drawing cum dining room and a verandah. The attic room may be used as prayer room. If needed, the waitng room with the consultation room may be rented to a professional practitioner.

50'-6"

38'-6"

26'-0"

2'   8'-6"   5'   2'   4'-6"   11'-0"

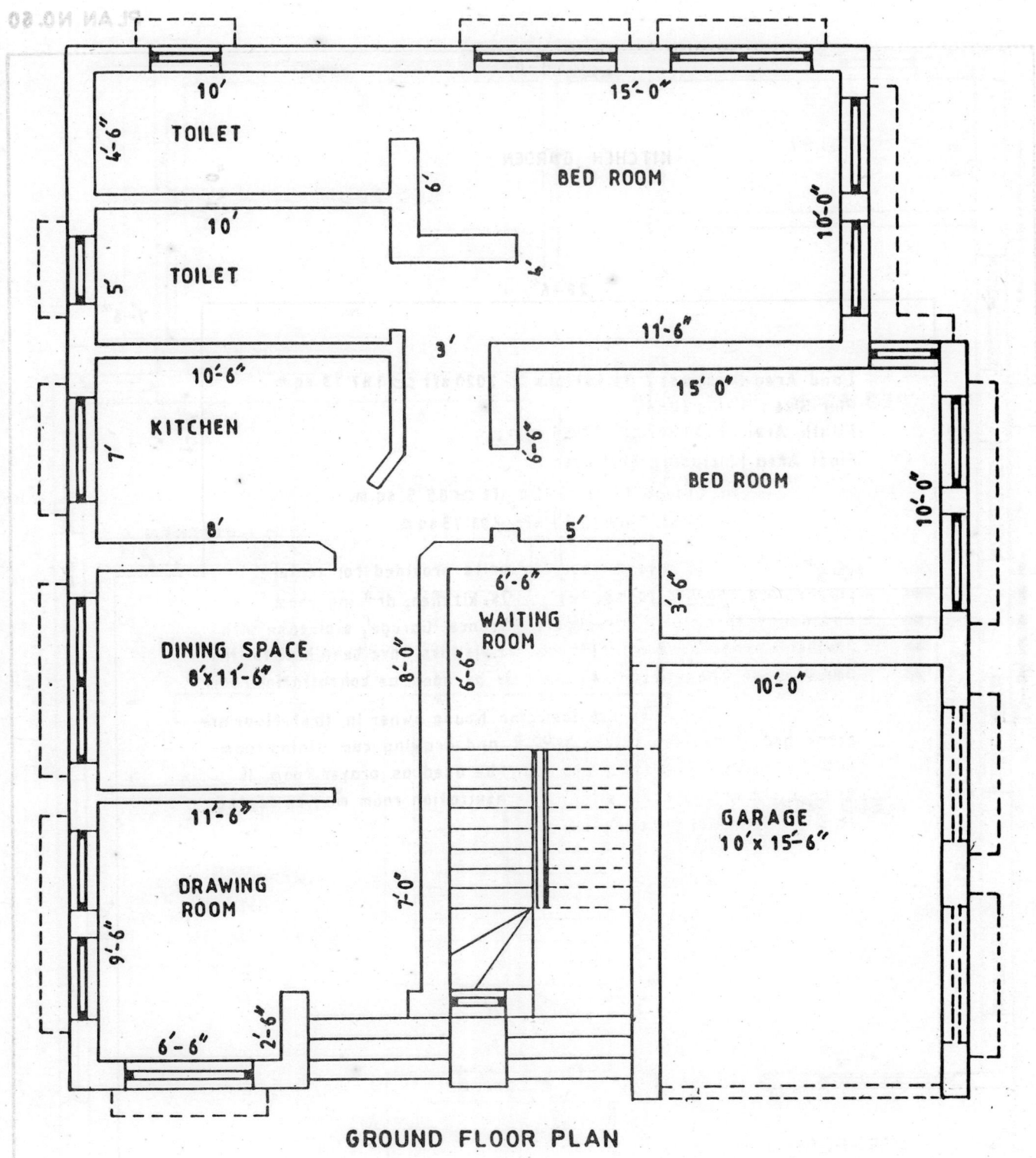

GROUND FLOOR PLAN

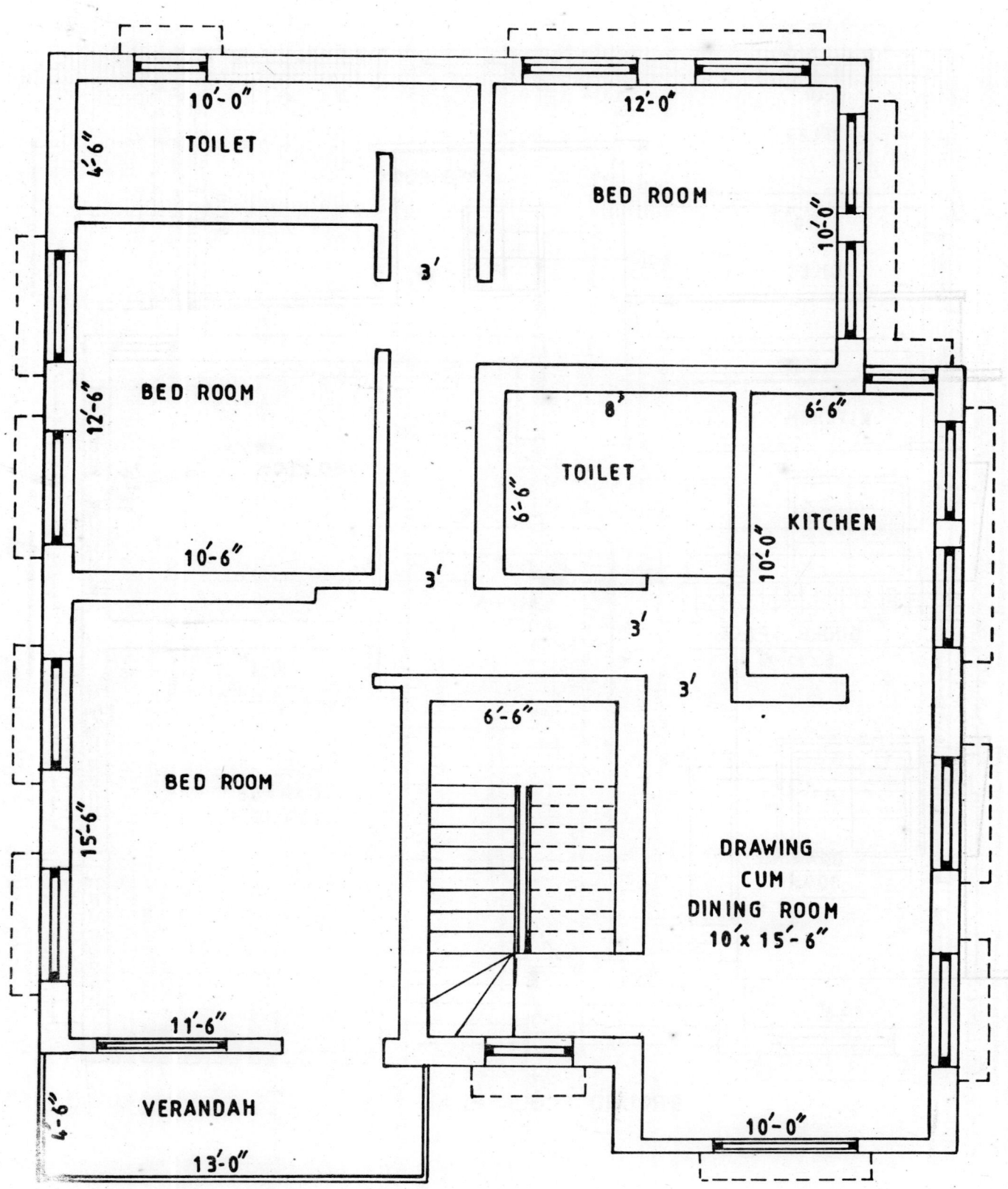

FIRST FLOOR PLAN

ELEVATION

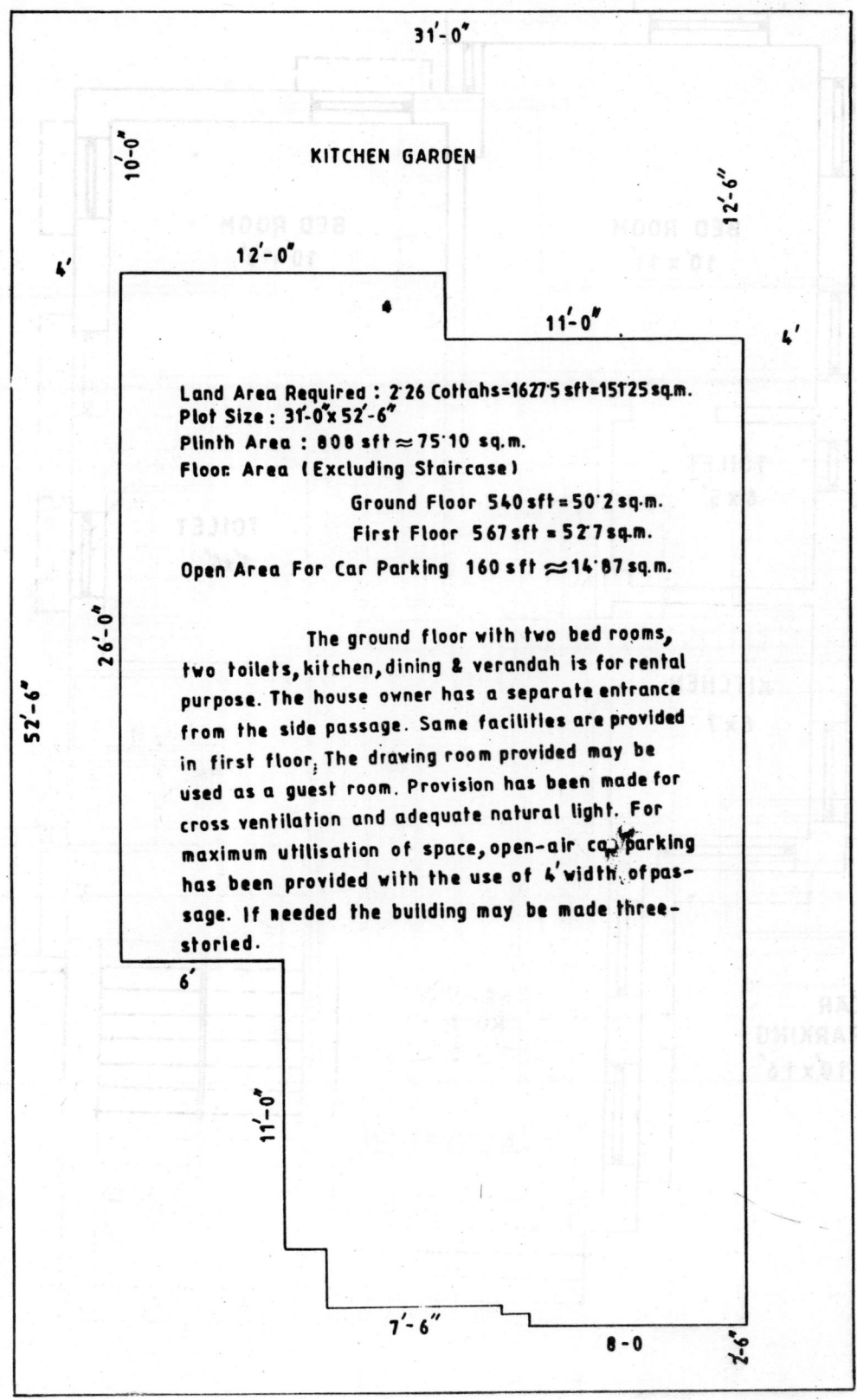

KITCHEN GARDEN

31'-0"

10'-0"

12'-6"

12'-0"

4'

11'-0"

4'

4

52'-6"

26'-0"

6'

11'-0"

7'-6"

8-0

2'-6"

Land Area Required : 2·26 Cottahs=1627·5 sft=151·25 sq.m.
Plot Size : 31'-0"x52'-6"
Plinth Area : 808 sft ≈ 75·10 sq.m.
Floor Area (Excluding Staircase)

    Ground Floor 540 sft = 50·2 sq·m.
    First Floor 567 sft = 52·7 sq.m.

Open Area For Car Parking 160 sft ≈ 14·87 sq.m.

The ground floor with two bed rooms, two toilets, kitchen, dining & verandah is for rental purpose. The house owner has a separate entrance from the side passage. Same facilities are provided in first floor. The drawing room provided may be used as a guest room. Provision has been made for cross ventilation and adequate natural light. For maximum utilisation of space, open-air car parking has been provided with the use of 4' width of passage. If needed the building may be made three-storied.

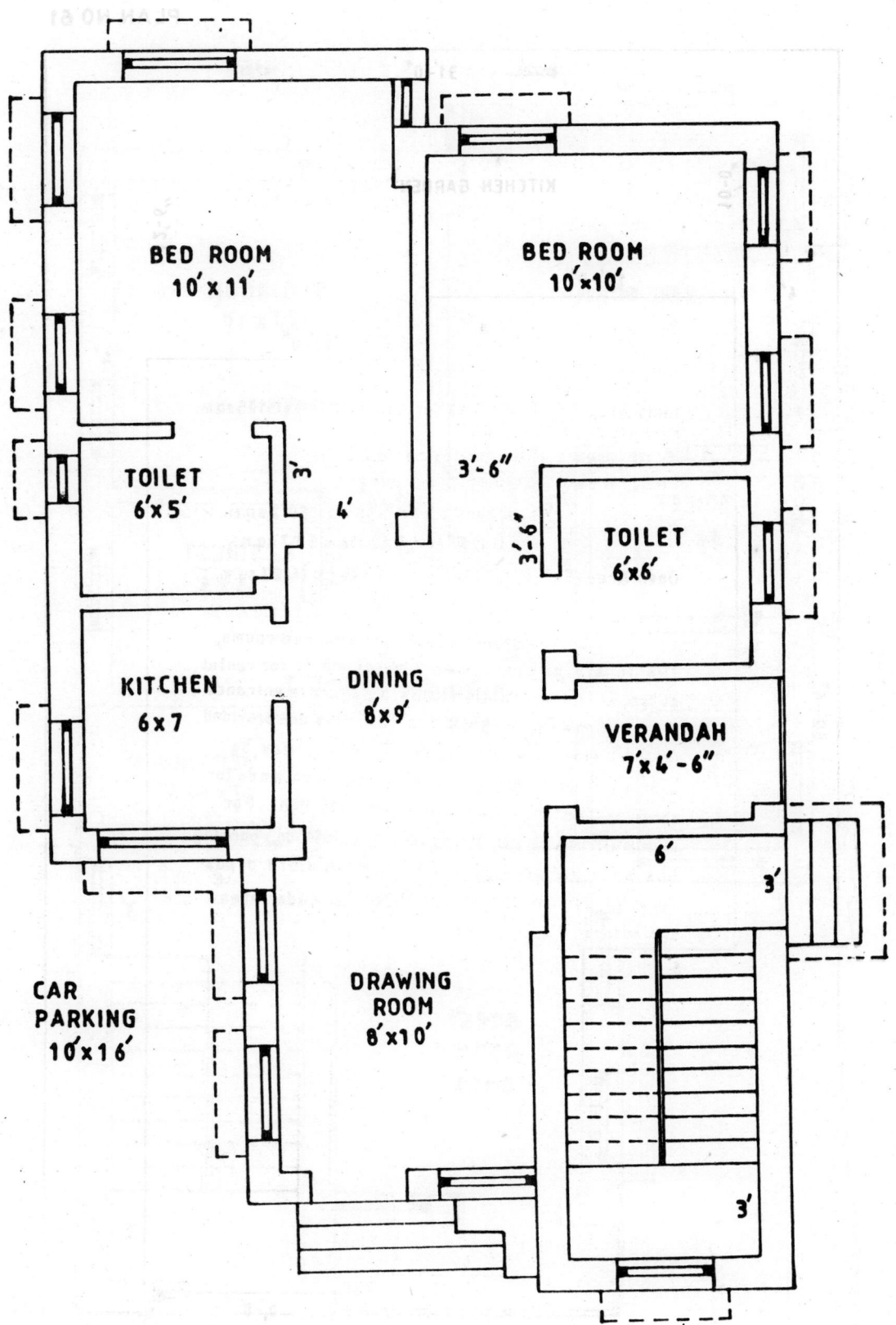

GROUND FLOOR PLAN

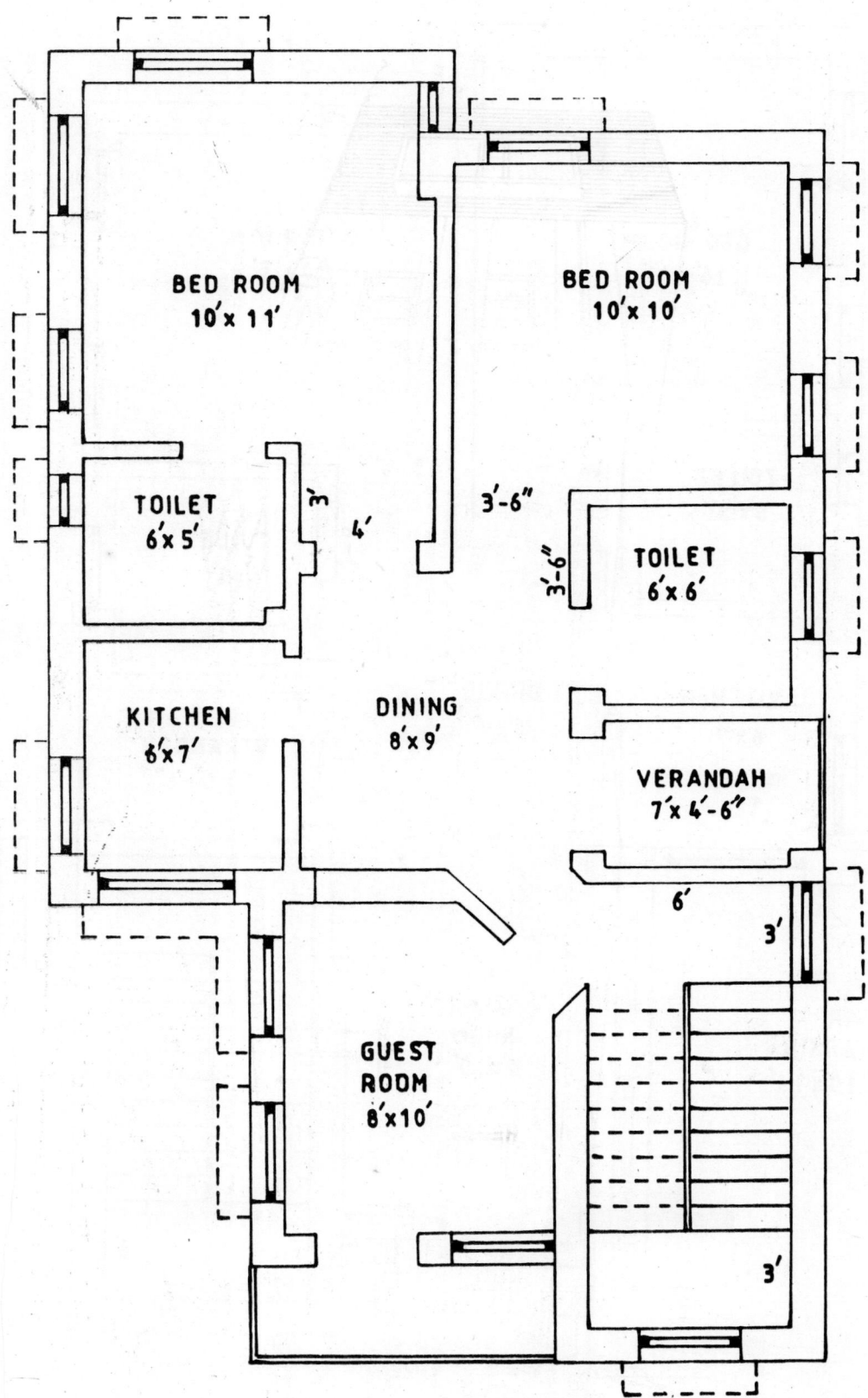

FIRST FLOOR PLAN

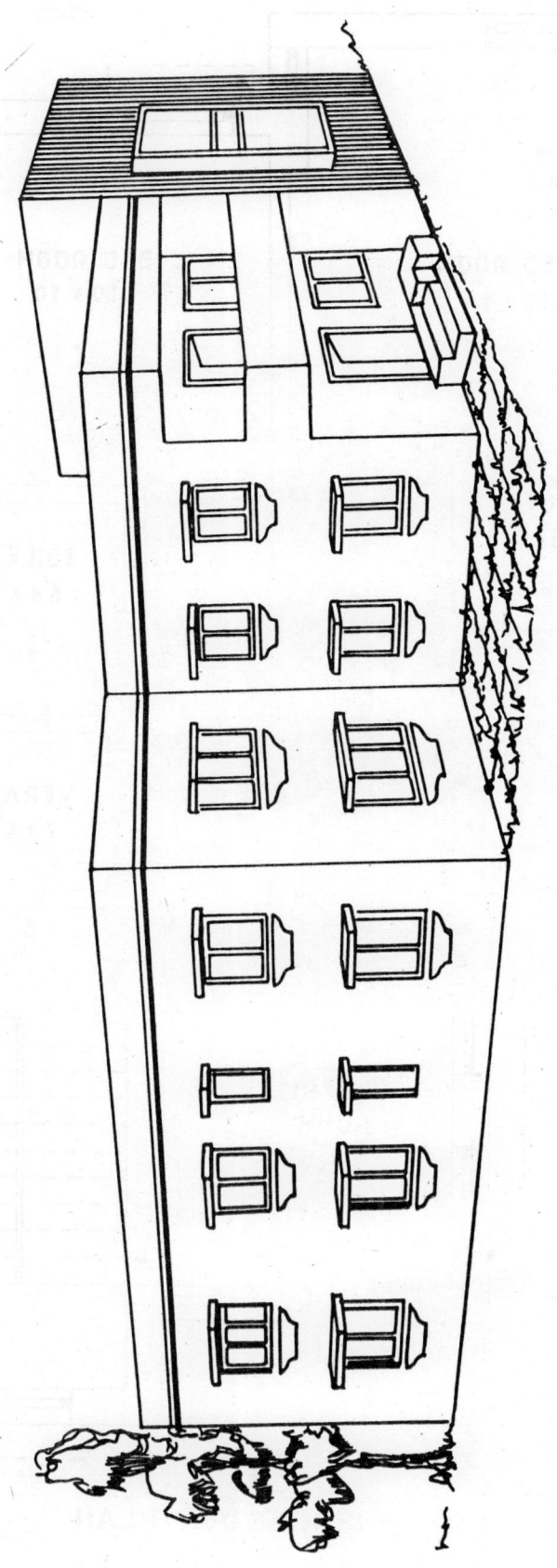